Springer Series on Atomic, Optical, and Plasma Physics

Volume 81

For further volumes:
http://www.springer.com/series/411

The Springer Series on Atomic, Optical, and Plasma Physics covers in a comprehensive manner theory and experiment in the entire field of atoms and molecules and their interaction with electromagnetic radiation. Books in the series provide a rich source of new ideas and techniques with wide applications in fields such as chemistry, materials science, astrophysics, surface science, plasma technology, advanced optics, aeronomy, and engineering. Laser physics is a particular connecting theme that has provided much of the continuing impetus for new developments in the field, such as quantum computation and Bose-Einstein condensation. The purpose of the series is to cover the gap between standard undergraduate textbooks and the research literature with emphasis on the fundamental ideas, methods, techniques, and results in the field.

Mahendra Singh Sodha

Kinetics of Complex Plasmas

Springer

Mahendra Singh Sodha
Department of Education Building
University of Lucknow
Lucknow, Uttar Pradesh
India

ISSN 1615-5653 ISSN 2197-6791 (electronic)
ISBN 978-81-322-1819-7 ISBN 978-81-322-1820-3 (eBook)
DOI 10.1007/978-81-322-1820-3
Springer New Delhi Heidelberg New York Dordrecht London

Library of Congress Control Number: 2014931383

Printed on acid-free paper

Springer is part of Springer Science+Business Media (www.springer.com)

Dedicated to
Mrs. Vijay Sodha

and

Mrs. Usha Sodha

Preface

With the rapidly increasing interest in complex plasmas many books and reviews on different aspects of this growing field have been published. However, no review or book dedicated to the kinetics of complex plasmas and associated processes is available; this book is a modest attempt to present the basic aspects of electron emission from and electron/ion accretion on the surface of dust particles and the kinetics of complex plasmas, illustrated by a few typical applications.

Over the years, the approach to the kinetics has changed from mere charge balance on the dust particles to include the number and energy balance of the constituents, size distribution of dust, quantum effects in emission from and accretion of electrons on the dust particles, statistical mechanics considerations, nonlinear interaction with electric, electromagnetic field, etc. Effort has been made in the book to introduce the readers to the contemporary concepts.

In a book like this some omissions of significant work are inevitable, for which sincere apologies are in order. As far as possible the presentation is based on charge balance on the particles and number/energy balance of the constituents.

The book should be of use to researchers, engineers, and graduate students.

Comments are welcome.

Delhi

Mahendra Singh Sodha

Acknowledgments

The author is very grateful to his associates:

Late Dr. Samiran Guha
Dr. Sanjay Mishra
Dr. Shikha Misra
Dr. Sweta Srivastava
Dr. Sujeet Agarwal
Dr. Amrit Dixit
Dr. M. P. Verma
Dr. Lalita Bhasin,

with whom he has learnt the subject; special thanks are due to Dr. S. K. Mishra for writing Chap. 6. He is also grateful to the Department of Science and Technology, Government of India for financial support toward writing the book. Thanks are also due to coauthors, other authors, and corresponding publishers, whose tables/diagrams have been reproduced in the book with their generous permission/policy. The typing and organization of the contents was done by Mr. Ram Shanker, whose efforts are appreciated. The pains taking assistance in proof correction by Dr. Rashmi Mishra is gratefully acknowledged.

Mahendra Singh Sodha

Contents

Part I Basics

About the Book

The presentation in the book is based on charge balance on dust particles, number, and energy balance of the constituents and atom–ion–electron interaction in the gaseous plasma. Size distribution of dust particles, statistical mechanics, Quantum effects in electron emission from and accretion on dust particles, and nonlinear interaction of complex plasmas with electric and electromagnetic fields have been discussed in the book. The book introduces the reader to the basic concepts and typical applications.

The book should be of use to researchers, engineers, and graduate students.

Part I
Basics

Chapter 1
Introduction

1.1 Complex Plasma

It is well-known that about 99 % of the matter in the universe is in the plasma state, comprising of electrons, ions, and neutral atoms/molecules. Very often the plasmas have a suspension of dust and in case the dust significantly affects the properties of the plasma, the dust plasma system is known as a colloidal plasma, dusty plasma, or complex plasma. Despite significant variation in the characterization, we have in this book referred to plasma with suspended dust system as complex plasma, in case the dust significantly affects the properties of the plasma.

1.2 Occurrence in Space

1.2.1 Planetary Magnetospheres

Dust plasma interactions in the planetary magnetospheres have been studied for a long time. The period of vigorous research on the role of complex plasma in the magnetospheres started with the highly significant observations by spacecraft on interesting phenomena in the magnetospheres of giant planets in the early 1980s (*For a review* see Horanyi [4]). It is instructive to take a look at some specific cases.

1.2.1.1 Saturn's Spokes

The approximately radial features (Spokes) across the dense B ring, observed intermittently by Voyager spacecraft 1 and 2 as they flew by Saturn have attracted a great amount of attention.

All theories have proposed a sporadic increase of the plasma density, and subsequent charging of fine dust to a negative potential and consequent levitation. Thus the charging of the dust turns out to be an important part of the dynamics of the spokes.

M. S. Sodha, *Kinetics of Complex Plasmas*, Springer Series on Atomic, Optical, and Plasma Physics 81, DOI: 10.1007/978-81-322-1820-3_1,

1.2.1.2 High speed dust streams from Jupiter and Saturn

The observation [3] of collimated quasi-periodic and high-speed streams of fine dust particles, emanating from Jupiter by the Ulysses spacecraft highlighted the role of the electrodynamic force in the dynamics of fine dust particles; the charge on the particles is an important parameter in the process. Subsequently, such streams emanating from Saturn were also discovered by the Cassini spacecraft [5].

1.2.1.3 Differential Collection of Charged Dust by Planetary Satellites

Mendis and Axford [6] have proposed that the two-tone appearance of the satellites could be explained in terms of the differential collection of charged dust by the leading and trailing faces of the satellite. Recent Cassini observations support the model of Mendis and Axford [6]. As in other cases, the charge on the dust is a critical parameter.

1.2.2 Cometary Magnetosphere

The solar radiation and solar wind cause a variety of phenomena, associated with dust plasma interaction in the magnetosphere of a comet, approaching the sun. These include electrostatic levitation of dust from the bare cometary nucleus at large distance from the sun and electrostatic disruption and electrodynamic transport in the induced cometary magnetosphere as the distance from the sun gets decreased.

1.2.3 Interplanetary Dust

Particulates of size between a few centimeters to few nanometers in space with appreciable solar wind in the solar system are referred to as interplanetary dust. The main sources of interplanetary dust in the inner region (within a distance of 5 A.U. from the sun) of the solar system are the asteroidal debris and ejecta from comets.

1.2.4 Interstellar Dust

The interstellar medium (ISM) is highly nonuniform and it contains a suspension of fine (mm. to nm.) dust in regions of very low (H I) and very high (H II) ionization. Dust plays an important role in the dynamics and thermodynamics of ISM and in the secondary star formation. The dust tends to be negatively charged by the impact of low energy cosmic rays while the ultraviolet radiation causes photoelectric emission from the dust particles, making these positively charged. Dust exists in the H I region and localized regions in H II.

1.2.5 *Polar Mesospheric Clouds*

Noctilucent clouds (NLC) and polar mesospheric summer echoes (PMSE, discussed in Chap. 9) are of considerable interest on account of their relevance to global warming. The low temperatures (110–130 K) at the low ionospheric altitudes in the polar region cause condensation of water vapor as ice particles; the resulting system of suspended ice dust with size from 3 nm to 0.1 μm are known as polar mesospheric clouds (PMC); this term includes both the NLC (dust size from 20/30 nm to 0.1 μm) and PMSE regions (dust size from 3 to 20/30 nm). These dust particles get charged on account of the accretion of ionospheric plasma electrons/ions on the surface of the particles and photoelectric emission from dirty ice particles, if present. PMCs are a well-known manifestation of dust plasma interaction in the terrestrial atmosphere.

1.3 Flames

The early experiments by Sugden and Thrush [9] and Shuler and Weber [7] showed that the electron density in rich hydrocarbon flames is far in excess of what can be explained by the application of Saha's equation to gaseous species. Further, they have shown that the observed electron density can be explained on the basis of the thermionic emission of electrons from the hot solid carbon particles, present in the flames.

Complex plasmas are also encountered in the exhausts of the rockets that use solid propellants [1, 2, 8]. It has been known for long that the flight of some rockets, using solid propellants is accompanied by electromagnetic effects that interfere with communications and guidance by radio and radar. One of the commonest types of interference is due to free electrons in the neighborhood of the rocket. The electrons causing these effects originate from thermal ionization of atoms and molecules, and probably also from the thermionic emission from hot solid particles; particles of alumina and aluminum are present in the combustion of aluminized propellants and should contribute significantly to the electron density.

1.4 Complex Plasma in Laboratory and Industry

Dust occurs in d.c. and more so in r.f. discharges. The role of dust in plasma processing reactors, commonly employed in microelectronics/thin film industry is significant and in many cases crucial. The role of significant amount of dust, mainly produced by sputtering on the walls in fusion machines is important; the radioactive nature of the dust is another concern.

1.5 Kinetics

While there are a number of books and reviews on different aspects of complex plasma, there is no review or book on the kinetics which is essential for understanding any aspect of the physics of complex plasmas or for a rational design of an application. This book is an attempt to provide a text on the kinetics of complex plasmas, which will serve as an introduction to the basic processes and illustrative applications to specific situations.

The initial approach to the kinetics comprised of determining the charge on a particle from the charge balance on an isolated particle, which implied that the plasma parameters were not affected by the presence of dust.

The next improvement in the approach was the incorporation of the number balance of the constituents; the results indicated significant departure from the results of the isolated particle theory.

It was realized that the rates of different processes in the kinetics were dependent on the temperature of different species, and hence energy balance of different constituents had also to be taken into account. In addition, the maintenance of the plasma (without dust) has also to be accounted for; in particular ionization of atoms, electron–ion recombination electron attachment, etc. should also be incorporated.

Thus a satisfactory formulation of the kinetics of complex plasmas should include:

(1) Maintenance of gaseous plasma
(2) Number balance of constituents
(3) Energy balance of constituents and
(4) Charge balance on the dust particles.

In this book, as far as possible all the four aspects have been kept in mind.

1.6 Organization of the Book

Chapter 2 is a presentation of the theory of thermionic, photoelectric, electric field, light induced field, and secondary electron emission from charged spherical and cylindrical surfaces of solids (particles); quantum effects have been also discussed. Chapter 3 is a discussion of electron and ion accretion on charged spherical and cylindrical particles; quantum effects and effect of ion trapping have been included. Chapter 4 is a discussion of charge distribution on particles in an electron dust cloud; both thermal equilibrium and steady state nonequilibrium situation, corresponding to photoelectric emission from the dust particles have been considered. The kinetics of a complex plasma with uniform size dust has been analyzed in Chap. 5 under thermal equilibrium and steady state illumination (causing photoelectric emission from the dust) situations; the analysis incorporates charge balance on the particles, maintenance of the plasma and number/energy balance of the constituents. The kinetics of a complex plasma with a flow velocity larger than ion speed and much less than electron speed has been analyzed in Chap. 6. Chapter 7 is discussion of the kinetics of a complex plasma, with a size distribution of dust, which is based on the uniform potential theory; the case of illuminated complex plasma with small size dust particles (when the uniform potential theory is not valid) has also been highlighted. The nonlinear theory of transport phenomena in the presence of a high electrical field has been given in Chap. 8. The results of Chap. 8 have been used in Chap. 9 to explore linear and nonlinear electromagnetic propagation in a dark complex plasma, nonlinear effects in PMSE and self focusing of an e.m. beam in a complex plasma. The theory of fluctuation of charge on dust in a complex plasma is given in Chap. 10.

The rest of the chapters describe the application of the basics (Chaps. 2–10) to some interesting situations. The kinetics of polar mesospheric clouds, cometary plasma, plasma in Saturn E ring, interplanetary medium, and interstellar medium, as presented in Chap. 11 are illustrative applications of the basics to space environment. Use of Ar-BaO complex plasma in MHD generators has been explored in Chap. 12. Chapter 13 explores complex plasma effects in rocket exhaust plasma and its interaction with the ionosphere. Chapter 14 investigates the reduction of electron density in a plasma by a water spray; the loss of negative ions from the surface of the droplets by evaporation is a unique phenomenon. Chapter 15 discusses a simple situation of growth of particles in a plasma and is illustrative of the application of complex plasma kinetics to plasma reactors, used in microelectronics and thin film industry. Chapter 16 discusses the working of an electrostatic precipitator via interaction of corona ion plasma with dust.

References

1. R. Friedman, L.W. Fagg, T.K Miller, W.D. Charles, M.C. Hughes, in *Progress in Astronautics and Aeronantics*, vol. 12, ed. by K.E. Shuler, T.B. Fenn (Academic Press, New York, 1963)
2. R.M. Fristrom, F.A. Oxhus, G.H. Albrecht, Am. Rocket Soc. J. **32**, 1729 (1962)
3. E. GrÜn et al., Nature **362**, 428 (1992)
4. M. Horanyi et al., Rev. Geophys. **42**, RG 4002 (2004)
5. S. Kempf, R. Srama, M. Horanyi, M.E. Burton, S. Elfort, G. Maragos-Klostermeyer, M. Roy, E. GrÜn, Nature **433**, 289 (2005)
6. D.A. Mendis, W.A. Axford, Rev. Earth Planetar. Sci. **2**, 419 (1974)
7. K.E. Shuler, J. Weber, J. Chem. Phys. 22, 491 (1954)
8. F.T. Smith, C.R. Gatz, *Progress in Astronautics and Aeronautics*, vol. 12 (Academic Press, New York, 1963)
9. T.M. Sugden, B.A. Thrush, Nature **168**, 703 (1951)

Chapter 2
Electron Emission from Dust

2.1 Free Electron Model

2.1.1 Basic Model

The free electron model, which is applicable to metals, is based on the fact that the valence electrons in a metal get detached from atoms and are free to move around in the metal and that their motion is not affected by the ions and other electrons. The confinement of the free electrons in the metal is ensured by a potential energy barrier of height W_a at the surface, corresponding to energy, much larger than the mean energy of the free electrons in the metal. The potential energy of the electrons within the metal is assumed to be uniform; it is often, without loss of generality taken as zero or $-W_a$.

2.1.2 Density of Electronic States

Wave mechanics provides an effective basis for the understanding of processes on the atomic scale. The wave mechanical approach is necessary to appreciate a host of phenomena, related to the behavior of electrons in a solid.

It is well known that electrons with a momentum $\boldsymbol{p}$ display wave phenomena with an associated wave vector $\boldsymbol{k}$, given by

$$\boldsymbol{k} = (2\pi/h)\boldsymbol{p}, \tag{2.1}$$

where h is *Planck's* constant.

One may associate a parameter ψ with the wave motion so that the probability of occurrence of an electron in the volume element $\mathrm{d}x\mathrm{d}y\mathrm{d}z$ is $\psi\psi^*\mathrm{d}x\mathrm{d}y\mathrm{d}z$; the function ψ satisfies *Schrödinger's* equation

$$\nabla^2\psi + (8\pi^2 m_e/h^2)[E - V(x, y, z)]\psi = 0, \tag{2.2}$$

M. S. Sodha, *Kinetics of Complex Plasmas*, Springer Series on Atomic, Optical, and Plasma Physics 81, DOI: 10.1007/978-81-322-1820-3_2,

where E and V denote the total and potential energy of the electron and m_e is the mass of the electron.

Schrödinger's equation has stood the test of time and led to a host of experimentally verified results.

In a metallic crystal (with dimensions of many atomic spacings) having a periodicity L_x, L_y, and L_z in the x, y, and z directions, ψ has a periodic solution of the form (for $V = 0$)

$$\psi(x, y, z) = A\ \exp[i(k_x x + k_y y + k_z z)], \tag{2.3a}$$

such that

$$\psi[(x + L_x),\ (y + L_y),\ (z + L_z)] = \psi(x, y, z). \tag{2.3b}$$

Equations (2.3a) and (2.3b) lead to $k_x L_x = 2n_x\pi, k_y L_y = 2n_y\pi$ and $k_z L_z = 2n_z\pi$. Substituting for $\boldsymbol{k}$, from (2.1) in the above relations one gets

$$p_x = n_x h/L_x, \quad p_y = n_y h/L_y \quad \text{and} \quad p_z = n_z h/L_z, \tag{2.4}$$

where n_x, n_y, and n_z are integers.

In the volume element $L_x L_y L_z$ an electronic state is characterized by a set of integral values of n_x, n_y, and n_z. In this volume consider an element $\Delta n_x \Delta n_y \Delta n_z$ such that n_x lies between n_x and $n_x + \Delta n_x$, n_y lies between n_y and $n_y + \Delta n_y$ and n_z lies between n_z and $n_z + \Delta n_z$; within this element n_x can have Δn_x values, n_y can have Δn_y values and n_z can have Δn_z values. Hence by *Pauli's* exclusion principle (which stipulates a unique set of four quantum numbers for an electron) the number of electronic states (characterized by a definite combination of n_x, n_y, n_z and the spin quantum number) in this element is $2\Delta n_x \Delta n_y \Delta n_z$, because the electron spin quantum number has two values. Therefore, the number of electronic states per unit volume, characterized by n_x, n_y, and n_z lying between n_x and $n_x + \Delta n_x$, n_y and $n_y + \Delta n_y$ and n_z and $n_z + \Delta n_z$, respectively is

$$2\Delta n_x \Delta n_y \Delta n_z / L_x L_y L_z$$

Substituting for (n_x/L_x), (n_y/L_y), and (n_z/L_z) from (2.4) in the above expression the number of electronic energy states $\mathrm{d}^3 n_s$ per unit volume corresponding to electron momenta between $\boldsymbol{p}$ and $\boldsymbol{p}+\mathrm{d}\boldsymbol{p}$ is given by

$$\mathrm{d}^3 n_s = (2/h^3)\mathrm{d}p_x \mathrm{d}p_y \mathrm{d}p_z. \tag{2.5}$$

To evaluate the number of states with momenta between p and $p + \mathrm{d}p$, the above equation has to be integrated such that the volume element $\mathrm{d}p_x \mathrm{d}p_y \mathrm{d}p_z$ lies between spheres of radius p and $p + \mathrm{d}p$ because

$$p_x^2 + p_y^2 + p_z^2 = p^2$$

represents a sphere in the p_x, p_y, p_z space. Since the volume of the p space between spheres of radius $p$and $p + \mathrm{d}p$ is $4\pi p^2\mathrm{d}p$, the number of electronic energy states with momenta between $p$and $p + \mathrm{d}p$ per unit volume is

$$\mathrm{d}n_s = (2/h^3)4\pi p^2\mathrm{d}p = (8\pi p^2/h^3)\mathrm{d}p, \tag{2.6}$$

An alternate derivation of (2.5) and (2.6), based on the uncertainty relation has been given by Seitz [37].

2.1.3 Distribution Function

The probability $P(E)$ of the occupation of a state of energy $E = p^2/2m_e$ by electrons is given by the *Fermi-Dirac* distribution, viz.

$$P(E) = F_D[(E - E_F)/k_BT], \tag{2.7}$$

where

$F_D(\Omega) = [1 + \exp\,\Omega]^{-1}$,
k_B is *Boltzmann's* constant,
T is the temperature of the electrons and
E_F is the *Fermi* energy.

Therefore from (2.5), (2.6), and (2.7) the number of electrons per unit volume with momenta between $\boldsymbol{p}$ and $\boldsymbol{p}+\mathrm{d}\boldsymbol{p}$ is given by

$$\mathrm{d}^3n_e = (2/h^3)F_D[(p^2/2m_ek_BT) - (E_F/k_BT)]\mathrm{d}p_x\mathrm{d}p_y\mathrm{d}p_z. \tag{2.8a}$$

and correspondingly

$$\mathrm{d}n_e = (8\pi p^2/h^3)F_D[(p^2/2m_ek_BT) - (E_F/k_BT)]\mathrm{d}p, \tag{2.8b}$$

is the number of electrons per unit volume with momentum between $\boldsymbol{p}$ and $\boldsymbol{p}+\mathrm{d}\boldsymbol{p}$.

The number of electrons d^3n_1, having momenta between $\boldsymbol{p}$ and $\boldsymbol{p}+\mathrm{d}\boldsymbol{p}$, which are incident on the surface (from inside) per unit area per unit time is given by

$$\mathrm{d}^3n_1 = (p_x/m_e)\mathrm{d}^3n_e, \tag{2.8c}$$

where $\boldsymbol{x}$ is normal to the surface.

If p_t is the magnitude of the transverse component (normal to $\boldsymbol{x}$) of the electron momentum,

$$p_t^2 = p_y^2 + p_z^2$$

and $dp_y dp_z$ may be integrated as $2\pi p_t dp_t$, which is the area between two circles of radii p_t and $p_t + dp_t$, respectively. Hence from (2.8c) and (2.8a), one obtains

$$d^2 n_1 = (4\pi/h^3) F_D[(p^2/2m_e k_B T) - (E_F/k_B T)](p_x/m_e) p_t dp_x dp_t.$$

Putting $\varepsilon_x = (p_x^2/2m_e k_B T$, $\varepsilon_t = (p_t^2/2m_e k_B T)$ and $\varepsilon_F = (E_F/k_B T)$, the above equation may be expressed as:

$$d^2 n_1 = (4\pi m_e k_B^2 T^2/h^3) F_D(\varepsilon_x + \varepsilon_t - \varepsilon_F) d\varepsilon_x d\varepsilon_t = (A_0 T^2/e) F_D(\varepsilon_x + \varepsilon_t - \varepsilon_F) d\varepsilon_x d\varepsilon_t \tag{2.9}$$

where $A_0 = (4\pi e m_e k_B^2/h^3) = 120\ \mathrm{A/cm^2K^2}$ and $-e$ is the electronic charge.

In the above equation, ε_x and ε_t represent the normalized normal and transverse kinetic energies on account of the normal (p_x) and transverse (p_t) components of the electron momentum in the metal. The parameter $\varepsilon_x + \varepsilon_t$ represents the total normalized electron energy, while ε_F is the normalized Fermi energy. The parameters ε_x, ε_t, and ε_F are all positive.

2.1.4 Fermi Energy

Putting $E = (p^2/2m_e)$ in (2.8b) one obtains

$$dn_e = (8\pi/h^3)(2m_e)^{1/2} F_D[(E - E_F)/k_B T] E^{1/2} dE$$

or

$$n_e = (8\pi/h^3)(2m_e)^{1/2} \int_0^\infty E^{1/2} F_D[(E - E_F)/k_B T] dE, \tag{2.10}$$

where n_e is the number of electrons per unit volume, i.e., the electron density.

The above equation can be used to evaluate the *Fermi energy* E_F, corresponding to a given electron density n_e and the temperature T.

A simple expression for Fermi energy E_{F0} can be obtained at the temperature 0 K, corresponding to

$$F_D = 0 \quad \text{for}\, E > E_{F0}$$

and

$$F_D = 1 \quad \text{for}\, E < E_{F0}.$$

Thus, the electron density

$$n_e = \int_0^{E_{F0}} (8\pi/h^3)(2m_e)^{1/2}E^{1/2}\mathrm{d}E,$$

which leads to

$$E_{F0} = (h^2/8m_e)(3n_e/8\pi)^{2/3}. \tag{2.11}$$

Using (2.10), one may obtain [37] an approximate expression for the Fermi energy E_F corresponding to an electron density n_e and finite temperature T, as

$$\begin{aligned} E_F &\approx E_{F0}[1 - (\pi^2/8)(k_BT/E_{F0})]^{2/3} \\ &\approx E_{F0}[1 - (\pi^2/12)(k_BT/E_{F0})]. \end{aligned} \tag{2.12}$$

where E_{F0} is given by (2.11) and is much larger than k_BT.

2.2 Basic Concepts of Electron Emission

2.2.1 Potential Energy of an Electron Near the Plane Surface of a Metal

The attractive force on an electron, with charge $-e$ at a distance x from a plane metallic surface is the same as that exerted by a particle with an equal and opposite charge at the same distance from the surface on the other side; it may be noted that the two charges and their positions ensure zero electric potential on the surface. Hence the force, termed as image force on the electron $F_i(x)$ is given by

$$F_i(x) = -e^2/(x+x)^2 = -e^2/4x^2.$$

and the associated potential energy $V_i(x)$ is given by

$$V_i(x) = -\int_\infty^x F_i(x)\mathrm{d}x = -e^2/4x.$$

According to the free electron theory of metals, the potential energy of an electron inside a metal is $-W_a$; to ensure a uniform electric potential $(-V_0/e)$ on the surface (or a potential energy $V_0 - W_a$ in the metal) and to take into account the applied electric field $F$and the image force, the potential energy $V(x)$ of an electron $(x > 0)$ is given by [37]

$$V(x) = V_0 - eFx - e^2/[(e^2/W_a) + 4x]. \tag{2.13a}$$

Neglecting the term (e^2/W_a) in (2.13a)

$$V(x) = V_0 - eFx - e^2/4x. \tag{2.13b}$$

The position x_m and magnitude V_m of the maximum potential energy is given by putting $dV/dx = 0$ and using (2.13b); the condition $\mathrm{d}^2V/\mathrm{d}x^2 < 0$ at $x = x_m$ (for a maximum) can be easily verified. Thus,

$$x_m = (e/F)^{1/2}/2$$

and

$$V_m = V_0 - (e^3F)^{1/2}. \tag{2.14}$$

It is also seen that $(e^2/W_a) << x_m$ for usual values of the parameters and hence (2.13b) is valid around $x = x_m$. Thus, the effective height of the potential energy barrier W_a' is less than W_a by $(e^3F)^{1/2}$ in the presence of an electric field F. Thus

$$W_a' = V_m - V(x = 0) = W_a - (e^3F)^{1/2} \tag{2.15}$$

The additional electron emission due to the reduction in the potential energy surface barrier is known as Schottky Emission [36]. The electron potential energy model, represented by (2.13a) is too cumbersome, to be used conveniently for the study of electron emission. Hence, one may adopt a simpler model represented by Fig. 2.1f, g, which incorporates the essential feature of (2.13a) viz. the reduction of the surface energy barrier. The potential energy models, adopted for the evaluation of the transmission coefficient of an electron through the surface in different papers are illustrated in Fig. 2.1a–g and represented analytically by (2.16a) to (2.16g) in Table 2.1. From classical considerations, the transmission coefficient is unity when the normal electron energy due to the velocity component normal to the surface exceeds W_a and is zero otherwise. However, from wave mechanical considerations, discussed later the results are substantially different. The figures correspond to different regions, (characterized in Table 2.1) and negatively charged surface except Fig. 2.1, which corresponds to a positively charged surface.

Here W_a is the height of the surface potential energy barrier, F is the electric field outside the metallic surface, x is the distance normal to the surface, $(-e^2/4x)$ is the potential energy of the electron due to the image force and $W_a'\left[= W_a - (e^3F)^{1/2}\right]$ is the reduced height of the surface potential energy barrier on account of the *Schottky* effect (2.15). The three region model incorporates a field free region, which is physically realistic. If Region-III consists of the anode $V(x) = -W_a'$ in region III.

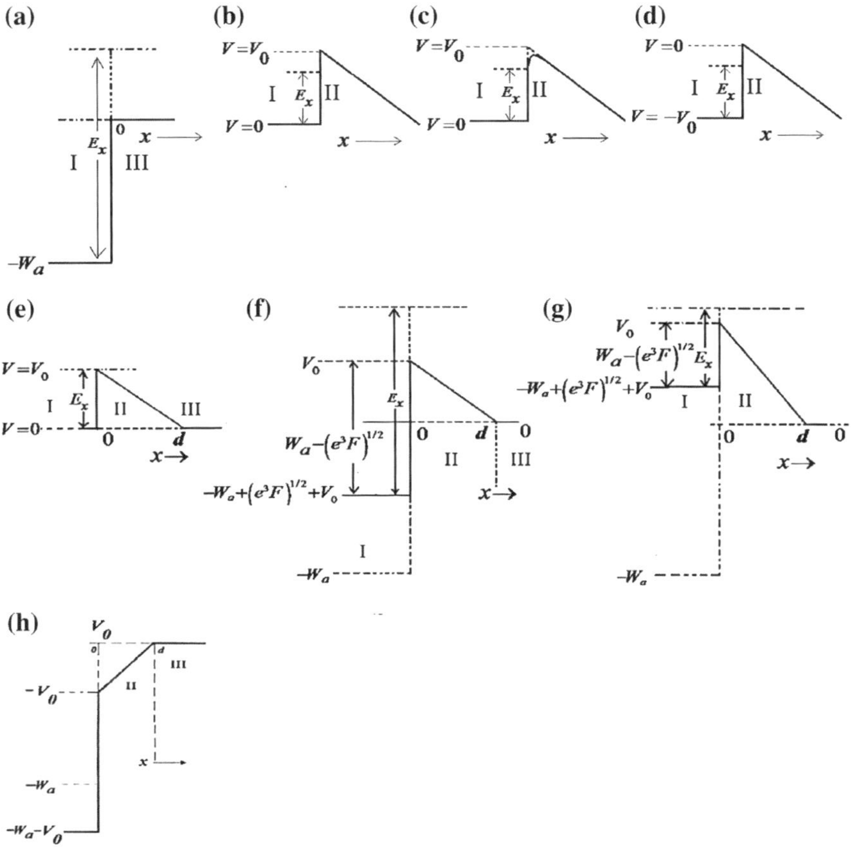

Fig. 2.1 Potential energy of an electron near the surface of a metallic plate. The labels are as follows: **a** Uncharged surface, **b** two region Fowler's model [15] (negatively charged), **c** two region. Schottky/Nordheim [32] model (negatively charged), **d** two region Forbes and Deane [12] model (negatively charged) **e** three region model by Sodha et al. [40] (negatively charged), **f** negatively charged surface with $V_0 < W_a$ and **g** negatively charged surface with $V_0 > W_a$; **f** and **g** correspond to the present three region model and **h** positively charged surface ($V_0 < 0$) (after Agarwal et al. [1]; curtsey authors and publishers NRC Press)

2.2.2 Transmission Coefficient Across Metallic Plane Surfaces: Uncharged Surface

The transmission coefficient of an electron through the surface can be evaluated from matching the solution (ψ and $\mathrm{d}\psi/\mathrm{d}x$) of *Schrödinger's* equation at the interface of the two regions.

The derivation of an expression for the transmission coefficient $D(E_x)$ corresponding to an uncharged surface with electron potential energy, described by

Table 2.1 Models of potential energy of an electron near a negatively charged plane metallic surface ($x = 0$)

Fig. 2.1a	Region-I	$V(x) = -W_a$	$x < 0$	
Simple model (uncharged)	Region-II	$V(x) = 0$	$x > 0$	(2.16a)
Fig. 2.1b	Region-I	$V(x) = 0$	$x < 0$	
Fowler and Nordheim [15]	Region-II	$V(x) = V_0 - eF(x)$	$x > 0$	(2.16b)
Fig. 2.1c	Region-I	$V(x) = -W_a$	$x < 0$	
Forbes and Deane [12]	Region-II	$V(x) = -eF(x)$	$x > 0$	(2.16c)
Fig. 2.1d	Region-I	$V(x) = 0$	$x < 0$	
Schottky/ Nordheim [32]	Region-II	$V(x) = V_0 - eF(x) - (e^2/4x)$	$x > 0$	(2.16d)
Fig. 2.1e	Region-I	$V(x) = 0$	$x < 0$	
Sodha and Dixit [40]	Region-II	$V(x) = V_0 - eF(x)$	$0 < x < d(= V_0/eF)$	
	Region-III	$V(x) = 0$	$x > d$	(2.16e)
Fig. 2.1f ($V_0 > W_a$)	Region-I	$V(x) = V_0 - W_a'$	$x < 0$	
Fig. 2.1g ($V_0 < W_a$)	Region-II	$V(x) = V_0 - eF(x)$	$0 < x < d(= V_0/eF)$	
(Three region model)	Region-III	$V(x) = 0$	$x > d$	(2.16f, g)
Fig. 2.1h ($V_0 < 0$)	Region-I	$V(x) = V_0 - W_a$	$x < 0$	(2.16h)
	Region-II	$V(x) = V_0 - V_0(x/d)$	$x > 0$	
	Region-III	$V(x) = 0$	$x > d$	

After Agarwal et al. [1]; curtsey authors and publishers NRC Press

(2.16a) is given in many books on Quantum Mechanics (e.g., [17]. Thus, the transmission coefficient $D_0(E_x)$ is given by

$$\begin{aligned} D_0(E_x) &= 4E_x^{1/2}(E_x - W_a)^{1/2} \Big/ [E_x^{1/2} + (E_x - W_a)^{1/2}]^2 \\ &= 4\varepsilon_x^{1/2}(\varepsilon_x - w_a)^{1/2} \Big/ [\varepsilon_x^{1/2} + (\varepsilon_x - w_a)^{1/2}]^2, \end{aligned} \tag{2.17}$$

where

$$\varepsilon_x = E_x/k_BT,$$

$$w_a = W_a/k_BT$$

and E_x is the normal component of kinetic energy of the electron in the metal.

To a first approximation (2.17) is valid for negatively charged surfaces, when W_a is replaced by $W_a–e^{3/2}F^{1/2}$ (2.15) and E_x is significantly larger than W_a. For a positively charged surface ($V_0 < 0$) (2.17) is an approximation when W_a is replaced by $W_a–V_0$; the surface is then charged to an electric potential ($–V_0/e$).

2.2.2.1 Negatively Charged Surface (After Agarwal et al. [1])

Referring to Fig. 2.1f, g, one notices that from wave mechanical considerations the transmission coefficient, $D(E_x)$ is also finite for $W_a–V_0–e^{3/2}F^{1/2} < E_x < W_a–e^{3/2}F^{1/2}$, when $W_a > V_0$ and $0 < E_x < W_a–e^{3/2}F^{1/2}$ for $W_a < V_0$ on account of tunneling, where V_0 is the potential energy of electrons at the surface; this accounts for the electric field emission. A critique of the theories of electric field emission has been given by [12, 40]. Little attention has however been given to $D(E_x)$ (for a negatively charged surface) for electrons with $E_x > W_a–e^{3/2}F^{1/2}$, which are responsible for thermionic and photoelectric emission. This may be due to the electric field emission studies being mainly limited to low temperatures.

The frequently used [15] model, which is the basis of the famous *Fowler–Nordheim equation* for the electric field emission, is illustrated in Fig. 2.1d. This model ignores the fact that $V(x) = (-W_a + V_0 + e^{3/2}F^{1/2})$ in Region-I and also the existence of Region-III where $V = 0$. The vast amount of work on two region models, corresponding to Fig. 2.1d has been critically reviewed by Forbes and Deane [12]. Some models (Fig. 2.1e) have taken into account (e.g., [40] Region-III but have ignored the fact that in Region-I, $V(x) = (-W_a + V_0 + e^{3/2}F^{1/2})$. Nordheim [32] has used a model similar to that in Fig. 2.1b except that in Region-II, the potential energy takes into account the image force (Fig. 2.1d). Because of ignoring the surface energy barrier in these models, the corresponding evaluated transmission coefficients and the electron emission currents, lack a sound foundation.

In what follows an expression for $D(E_x)$ corresponding to a negatively charged surface has been derived and the dependence of $D(E_x)$on electron energy, electric field, and height of the surface energy barrier has been graphically illustrated; it is seen that $D(E_x)$ is an increasing function of E_x and hence the thermionic and photoelectric currents should also increase with increasing field F. The dependence of thermionic and photoelectric current density on the electric field and the height of the surface energy barrier has also been investigated and the results have been graphically illustrated in later sections. For the sake of completeness, the currents on account of the electric field emission and light-induced field emission as a function of W_a and F have also been evaluated; the earlier studies pertained to low temperatures.

It is seen that the electric field significantly enhances the transmission coefficient of an electron across the surface and hence for the evaluation of the electron emission from a metal an appropriate expression for $D(E_x)$, (derived herein), should be used for negatively charged surfaces. The electron emission, corresponding to a positively charged surface is also briefly discussed later for the sake of completeness.

2.2.2.2 Expression for $D(E_x)$

The potential energy of an electron inside and outside a negatively charged plane metallic surface may be modeled by (2.16f, g).

The time independent Schrödinger equation is

$$\nabla^2\psi(x,y,z)+\frac{8\pi^2 m_e}{h^2}\,[E-V(x)]\,\psi(x,y,z)=0. \tag{2.18}$$

Using the method of separation of variables by substituting $\psi = \psi_x(x)\psi_y(y)\psi_z(z)$, the above equation reduces to

$$\begin{aligned}&(\psi_x)^{-1}(\mathrm{d}^2\psi_x/\mathrm{d}x^2)+(\psi_y)^{-1}(\mathrm{d}^2\psi_y/\mathrm{d}y^2)\\&\quad+(\psi_z)^{-1}(\mathrm{d}^2\psi_z/\mathrm{d}z^2)+(8\pi^2 m_e/h^2)[E-V(x)]=0\end{aligned} \tag{2.19}$$

The first and fourth terms are functions of x only, while the second and third terms are respectively functions of only y and z. Hence, one can write

$$(\psi_y)^{-1}(\mathrm{d}^2\psi_y/\mathrm{d}y^2)=-k_y^2 \quad \text{and} \quad (\psi_z)^{-1}(\mathrm{d}^2\psi_z/\mathrm{d}z^2)=-k_z^2, \tag{2.20}$$

where k_y and k_z are recognizable as the components of the wave vector $\boldsymbol{k}$, associated with the wave function ψ. Use of the relation $\boldsymbol{k}=(2\pi/h)\boldsymbol{p}$, (2.3a, 2.3b) and (2.4) leads to

$$\mathrm{d}^2\psi_x/\mathrm{d}x^2+(8\pi^2 m_e/h^2)[E_x'-V(x)]\psi_x=0, \tag{2.21}$$

where $\boldsymbol{p}$ is the electron momentum and $E_x' = E-(p_y^2/2m_e)-(p_z^2/2m_e)$ represents the difference between the total energy and the kinetic energy due to the transverse components p_y and p_z of the momentum $\boldsymbol{p}$.

If $E_x = (p_x^2/2\,m_e)$ denotes the normal kinetic energy of the electron inside $[V(x) = -W_a + (e^3F)^{1/2} + V_0]$ the metal on account of the normal (i.e., x) component of the momentum, one can write

$$E_x' = E_x - W_a + (e^3F)^{1/2} + V_0. \tag{2.22}$$

Thus, (2.21) can be expressed as

$$\mathrm{d}^2\psi_x/\mathrm{d}x^2+(8\pi^2 m_e/h^2)[E_x-W_a+(e^3F)^{1/2}+V_0-V(x)]\psi_x=0, \tag{2.23}$$

From (2.16f, g) and (2.23), one obtains the following set of dimensionless equations for the three-region model

$$\frac{\mathrm{d}^2\psi_x}{\mathrm{d}\xi^2}+\varepsilon_x\psi_x=0, \quad \xi<0 \quad \text{Region-I} \tag{2.24a}$$

$$\frac{\mathrm{d}^2\psi_x}{\mathrm{d}\xi^2}+[\varepsilon_x-w_a+\alpha f^{1/2}+f\xi]\psi_x=0, \quad 0<\xi<\xi_0 \quad \text{Region-II} \tag{2.24b}$$

and

$$\frac{\mathrm{d}^2\psi_x}{\mathrm{d}\xi^2}+[\varepsilon_x+\upsilon_0-w_a+\alpha f^{1/2}]\psi_x=0, \quad \xi>\xi_0 \quad \text{Region-III} \tag{2.24c}$$

where $\varepsilon_x = E_x/k_BT, \upsilon_0 = V_0/k_BT = eFd/k_BT, w_a = W_a/k_BT$, $f = heF/(\pi k_BT\sqrt{8m_ek_BT})$ is the dimensionless electric field strength in Region-II,

$$\alpha = (8\pi^2 m_e/h^2 k_B T)^{1/4} e,$$

$$\xi = (\pi\sqrt{8m_ek_BT}/h)x,$$

$\xi_0 = (\pi\sqrt{8m_ek_BT}/h)d, (V_0/eF) = \upsilon_0/f$ and T is the temperature of the metal.

The solution of (2.24a), (2.24b), and (2.24c) may be written as

$$\psi_x(\xi) = A_1 \exp(ik_1\xi) + A_2 \exp(-ik_1\xi), \quad \text{Region-I} \tag{2.25a}$$

$$\begin{aligned}\psi_x(\xi) =& B_1 Ai[-(\varepsilon_x - w_a + \alpha f^{1/2} + f\xi)/f^{2/3}] \\ &+ B_2 Bi[-(\varepsilon_x - w_a + \alpha f^{1/2} + f\xi)/f^{2/3}] \quad \text{Region-II}\end{aligned} \tag{2.25b}$$

and

$$\psi_x(\xi) = C_1 \exp(ik_3\xi), \quad \text{Region-III} \tag{2.25c}$$

where Ai and Bi are Airy functions,

$$k_1^2 = \varepsilon_x \quad \text{and} \quad k_3^2 = [\varepsilon_x + \upsilon_0 - w_a + \alpha f^{1/2}].$$

The constants a_1, a_2, b_1, b_2, and c_1 are related by the fact that the wave function and its derivative is continuous at the boundaries. viz., at $\xi = 0$ and $\xi = \xi_0$. Thus at $\xi = 0$

$$a_1 + a_2 = b_1 l_2 + b_2 m_2, \tag{2.26a}$$

and

$$ik_1 a_1 - ik_1 a_2 = -b_1 l_2' f^{1/3} - b_2 l_2' f^{1/3}. \tag{2.26b}$$

and at $\xi = \xi_0$

$$b_1 l_1 + b_2 m_1 = C_1 \exp(ik_3\xi_0) \tag{2.26c}$$

and

$$-b_1 l_1' f^{1/3} - b_2 m_1' f^{1/3} = ik_3 c_1 \exp(ik_3\xi_0), \tag{2.26d}$$

where

$$
\begin{aligned}
l_1 &= Ai[-(\varepsilon_x + \upsilon_0 - w_a + \alpha f^{1/2})/f^{2/3}],\ l_1' = Ai'[-(\varepsilon_x + \upsilon_0 - w_a + \alpha f^{1/2})/f^{2/3}] \\
l_2 &= Ai[-(\varepsilon_x - w_a + \alpha f^{1/2})/f^{2/3}],\ l_2' = Ai'[-(\varepsilon_x - w_a + \alpha f^{1/2})/f^{2/3}], \\
m_1 &= Bi[-(\varepsilon_x + \upsilon_0 - w_a + \alpha f^{1/2})/f^{2/3}],\ m_1' = Bi'[-(\varepsilon_x + \upsilon_0 - w_a + \alpha f^{1/2})/f^{2/3}], \\
m_1' &= Bi'[-(\varepsilon_x + \upsilon_0 - w_a + \alpha f^{1/2})/f^{2/3}] \\
m_2 &= Bi[-(\varepsilon_x - w_a + \alpha f^{1/2})/f^{2/3}] \quad \text{and} \\
m_2' &= Bi'[-(\varepsilon_x - w_a + \alpha f^{1/2})/f^{2/3}].
\end{aligned}
$$

From (2.26a), (2.26b), (2.26c), and (2.26d), one obtains

$$
\frac{c_1}{a_1} = \frac{2ik_1\delta_5 f^{1/3}\exp(-ik_3\xi_0)}{\{k_1k_3\delta_1 + \delta_2 f^{2/3}\} + i(k_1\delta_4 + k_3\delta_3)f^{1/3}} \tag{2.27}
$$

where $(m_1l_2 - l_1m_2) = \delta_1$, $(m_1'l_2' - m_2'l_1') = \delta_2$, $(m_1l_2' - m_2'l_1) = \delta_3$, $(m_2l_1' - m_1'l_2) = \delta_4$ and $(m_1l_1' - m_1'l_1) = \delta_5$.

From (2.27), one obtains

$$
\frac{c_1c_1^*}{a_1a_1^*} = \frac{4k_1^2\delta_5^2 f^{2/3}}{\{k_1k_3\delta_1 + \delta_2 f^{2/3}\}^2 + (k_1\delta_4 + k_3\delta_3)^2 f^{2/3}} \tag{2.28}
$$

Further, the probability current density normal to the surface is given by

$$
j_x = \frac{ih}{4\pi m_e}\left(\psi_x \frac{\partial}{\partial\xi}\psi_x^* - \psi_x^* \frac{\partial}{\partial\xi}\psi_x\right) \tag{2.29}
$$

Hence, the probability current density in Region-I is given by

$$
(j_x)_I = \frac{k_1h}{2\pi m_e}[a_1^*a_1 - a_2^*a_2]
$$

In the above expression, the first term on the right-hand side represents the incident current while the second term represents the reflected current. Thus, the incident current density in Region-I is given by

$$
(j_x)_i = \frac{k_1h}{2\pi m_e}a_1^*a_1 \tag{2.30}
$$

Similarly, the transmitted current density in Region-III is given by

$$
(j_x)_t = \frac{k_3h}{2\pi m_e}c_1c_1^* \tag{2.31}
$$

Hence, the transmission coefficient D across the potential energy barrier at the surface is given by

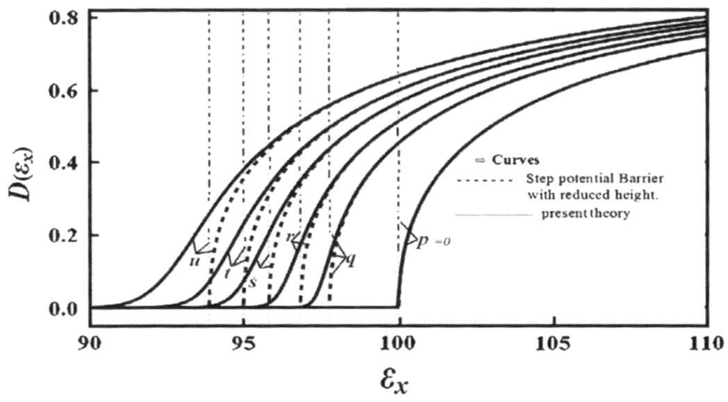

Fig. 2.2 Dependence of Transmission-coefficient $D(\varepsilon_x)$ on ε_x for the parameters $T = 1000$ K, $w_a = 100$ and $\upsilon_0 = 10^4$. The different values of field strength has been shown on the graph by letters; p, q, r, s, t and u correspond to $f = 0$, 0.2, 0.4, 0.7, 1.0 and 1.5 respectively. The portion of the *curves* corresponding to *left* and *right-hand side* of the *vertical lines* refer to field emission and thermionic emission respectively (after Agarwal et al. [1]; curtsey authors and publishers NRC Press)

$$D = \frac{(j_x)_t}{(j_x)_i} = \frac{k_3}{k_1}\frac{c_1c_1^*}{a_1a_1^*} \tag{2.32}$$

which using (2.28) can be reduced to

$$D(\varepsilon_x) = \frac{4k_1k_3\delta_5^2f^{2/3}}{\{k_1k_3\delta_1 + \delta_2f^{2/3}\}^2 + (k_1\delta_4 + k_3\delta_3)^2f^{2/3}}, \tag{2.32a}$$

According to Fowler [13], a fraction of electrons get their normal energy enhanced by an amount $h\nu$, when the surface is irradiated by light of frequency ν. Hence the transmission coefficient $D_{\text{ph}}(\varepsilon_x)$of such electrons can be obtained by substituting $(\varepsilon_x + \varepsilon_\nu)$for ε_x in (2.32a), where $\varepsilon_\nu = (h\nu/k_BT)$. Thus

$$D_{\text{ph}}(\varepsilon_x) = D(\varepsilon_x + \varepsilon_\nu). \tag{2.32b}$$

Equation (2.32a) is also valid in the case of positively charged surfaces when one substitutes $(-\upsilon_0)$ for υ_0 and $-f$ for f in the inherent coefficients in the final expression for $D(\varepsilon_x)$.

The set of Fig. 2.2 and 2.3 illustrate the dependence of transmission coefficient $[D(\varepsilon_x)]$ on the normal energy of electrons ε_x, as a function of dimensionless field strength (f) (Fig. 2.2) and the potential energy barrier height (w_a) (Fig. 2.3). It is seen that $D(\varepsilon_x)$ increases monotonically with increasing ε_x and f.

The straight vertical dashed lines are indicative of the effective surface energy barrier $(w_a - \alpha f^{1/2})$, so that the points to the left of these lines contribute to the electric field emission while those to the right contribute to thermionic emission; the broken curves correspond to transmission coefficient for step potential barrier with reduced barrier height $D_0(\varepsilon_x)$. It is interesting to notice that contribution of the

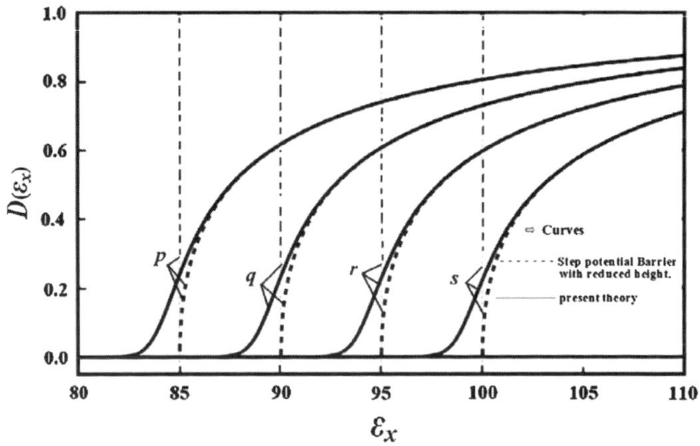

Fig. 2.3 Dependence of Transmission-coefficient $D(\varepsilon_x)$ as a function of ε_x for the parameters $T = 1000$ K, $f = 1$ and $\upsilon_0 = 10^4$. The different values of w_a parameter have been shown on the graph by letters. The letters p, q, r and s refer to $w_a = 90$, 95, 100 and 105 respectively (after Agarwal et al. [1]; curtsey authors and publishers NRC Press)

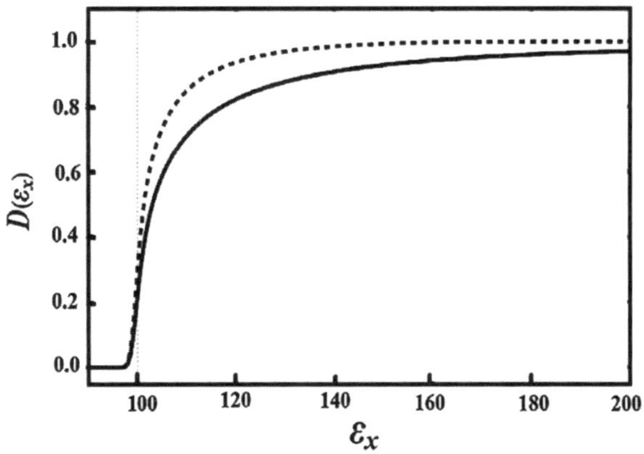

Fig. 2.4 Dependence of $D(\varepsilon_x)$ on ε_x for the Case-I as stated in the text, for the parameters $T = 1000$ K, $f = 1$, $w_a = 120$ and $\upsilon_0 = 1.0 \times 10^4$; here the role of image force is ignored. The *solid curve* corresponds to the present three region model while the *broken curve* refers to Forbes and Deane's approach (after Agarwal et al. [1]; curtsey authors and publishers NRC Press)

field emission and departure of $D(\varepsilon_x)$ from $D_0(\varepsilon_x)$ increase with increasing f. The figure also reflects the fact that $D_0(\varepsilon_x)$ is only meaningful for $\varepsilon_x > (w_a - \alpha f^{1/2})$. Figure 2.3 indicates that $D(\varepsilon_x)$ decreases with increasing potential energy barrier height (w_a). It is seen that the surface potential energy υ_0 has no significant effect on the energy dependence of $D(\varepsilon_x)$. It is seen from Fig. 2.4 that the expression due to Forbes and Dean [12], based on the two region model overestimates the transmission coefficient except for very high and low values of ε_x compared to w_a.

2.2.3 Thermionic and Electric Field Emission of Electrons from a Plane Surface

2.2.3.1 Negatively Charged Surface

From classical considerations electrons hitting the surface with normal energy $\varepsilon_x > (w_a - \alpha f^{1/2})$ have a unit probability of crossing the surface or of emission; such an emission is known as thermionic emission. However, wave mechanical considerations, outlined before lead to a probability of emission $D(\varepsilon_x)$, given by (2.32a). The number of electrons hitting the surface per unit area per unit time and having normal energy between ε_x and $\varepsilon_x + \mathrm{d}\varepsilon_x$ and transverse energy between ε_t and $\varepsilon_t + \mathrm{d}\varepsilon_t$ is

$$\mathrm{d}^2 n_1 = (A_0/e)T^2 F_D(\varepsilon_x + \varepsilon_t - \varepsilon_F)\mathrm{d}\varepsilon_x \mathrm{d}\varepsilon_t, \tag{2.9}$$

Hence, the number n_{th} of electrons emitted from the surface per unit area per unit time and the corresponding electric current J_{th} is given by

$$\begin{aligned} J_{\mathrm{th}} &= -\,en_{\mathrm{th}} = -e \int\limits_{\varepsilon_x=(w_a-\alpha f^{1/2})}^{\infty} \int\limits_{\varepsilon_t=0}^{\infty} D(\varepsilon_x)\mathrm{d}^2 n_1 \\ &= -\,A_0 T^2 \int\limits_{(w_a-\alpha f^{1/2})}^{\infty} \int\limits_{0}^{\infty} D(\varepsilon_x)\cdot F_D(\varepsilon_x + \varepsilon_t - \varepsilon_F)\mathrm{d}\varepsilon_x \mathrm{d}\varepsilon_t. \end{aligned} \tag{2.33}$$

For electrons of interest to thermionic emission, $\varepsilon_x > (w_a - \alpha f^{1/2})$,

$$\eta = (\varepsilon_x + \varepsilon_t - \varepsilon_F) > (w_a - \varepsilon_F + \varepsilon_t) - \alpha f^{1/2}$$

$[\approx \varphi + \varepsilon_t - \alpha f^{1/2}]$ is much larger than unity, where $\varphi = (w_a - \varepsilon_F) = (W_a - E_F)/k_B T = \Phi/k_B T$, and Φ is known as the work function of the metal. Hence $F_D(\eta) \to \exp(-\eta)$ and (2.9) simplifies to

$$d^2 n_1 = (A_0/e)T^2 \exp(-\varepsilon_x - \varepsilon_t + \varepsilon_F)\mathrm{d}\varepsilon_x \mathrm{d}\varepsilon_t, \tag{2.9a}$$

Thus (2.33) reduces to

$$\begin{aligned} J_{\mathrm{th}} &= -\,en_{\mathrm{th}} = -e(A_0 T^2/e) \int\limits_{(w_a-\alpha f^{1/2})}^{\infty} \int\limits_{0}^{\infty} D(\varepsilon_x)\cdot\exp(-\varepsilon_x - \varepsilon_t + \varepsilon_F)\mathrm{d}\varepsilon_x \mathrm{d}\varepsilon_t \\ &= -\,A_0 T^2 \int\limits_{(w_a-\alpha f^{1/2})}^{\infty} D(\varepsilon_x)\cdot\exp(-\varepsilon_x + \varepsilon_F)\mathrm{d}\varepsilon_x \end{aligned} \tag{2.34a}$$

In most of investigations on thermionic emission, the transmission coefficient is assumed to be unity; as evident from Figs. 2.2, 2.3 and similar calculations for other parameters this is indeed a poor approximation.

However, putting $D(\varepsilon_x) = 1$ (2.34a) reduces to

$$J'_{\text{th0}} = -A_0T^2\exp(-w_a + \alpha f^{1/2} + \varepsilon_F) = -A_0T^2\exp(-\varphi + \alpha f^{1/2}) \tag{2.34b}$$

and

$$J_{\text{th}}/J'_{\text{th0}} = \exp(\varphi - \alpha f^{1/2}) \int\limits_{(w_a - \alpha f^{1/2})}^{\infty} D(\varepsilon_x)\exp(-\varepsilon_x + \varepsilon_F)\mathrm{d}\varepsilon_x \tag{2.34c}$$

For an uncharged surface $f = 0$ and if $D(\varepsilon_x) = 1$ (2.34b) reduces to

$$J_{\text{th0}} = -A_0T^2\exp(-\varphi) = -A_0T^2\exp(-\Phi/k_BT). \tag{2.35a}$$

The above equation is known as *Richardson Dushman* equation.

For metals, as well as semiconductors the experimental data conforms to the relation

$$J_{\text{th0}} = -AT^2\exp(-\Phi/k_BT), \tag{2.35b}$$

where the constant A is known as *Richardson's Constant* and is in general different from A_0.

For an uncharged surface and $D(\varepsilon_x) = 1$, the mean energy ε_{th0} of the emitted electrons at the surface is given by

$$\varepsilon_{\text{th0}} = \int\limits_{w_a}^{\infty}\int\limits_{0}^{\infty} (\varepsilon_x + \varepsilon_t - w_a)d^2n_1 \Bigg/ \int\limits_{w_a}^{\infty}\int\limits_{0}^{\infty} d^2n_1$$

Using (2.9) and putting $\varepsilon'_x = \varepsilon_x - w_a$ and $\varepsilon' = \varepsilon'_x + \varepsilon_t$

$$\begin{aligned}\varepsilon_{\text{th0}} &= \int\limits_{0}^{\infty}\int\limits_{0}^{\infty} (\varepsilon'_x + \varepsilon_t)\exp(-\varepsilon'_x - \varepsilon'_t)\mathrm{d}\varepsilon'_x\mathrm{d}\varepsilon_t \Bigg/ \int\limits_{0}^{\infty}\int\limits_{0}^{\infty} \exp(-\varepsilon'_x - \varepsilon_t)\mathrm{d}\varepsilon'_x\mathrm{d}\varepsilon_t \\ &= \int\limits_{0}^{\infty} (\varepsilon'^2)\exp(-\varepsilon')\mathrm{d}\varepsilon' \Bigg/ \int\limits_{0}^{\infty} \varepsilon'\exp(-\varepsilon')\mathrm{d}\varepsilon' = 2 \quad \text{(just outside the surface)}\end{aligned} \tag{2.36a}$$

In writing the above equation, the following identity has been used

$$\int\limits_{0}^{\infty}\int\limits_{0}^{\infty} f(x_1 + x_2)\mathrm{d}x_1\mathrm{d}x_2 = \int\limits_{x_0}^{\infty} xf(x)\mathrm{d}x \tag{2.37}$$

$$x_1 + x_2 > x_0$$

For a negatively charged surface at a potential $(-V_0/e)$, the mean electron energy far away from the surface is

$$\varepsilon_{th} = 2 + \upsilon_0. \tag{2.36b}$$

The values of *Richardson's Constant* for different metals are different from one another as well as from A_0. The reasons for departure of A from A_0, as given by Seitz [37] are as follows:

I. The concept of effective mass of an electron is phenomenological; putting effective mass equal to free electronic mass is another approximation.
II. The interaction of electrons (mutual and with ions) in the metals has been neglected.
III. The relation $E = (p_x^2 + p_y^2 + p_z^2)/2\ m_e$ is at best an approximation, within the metal.
IV. The assumption of a perfectly plane surface is idealistic.
V. The coefficient of transmission from the surface for the electrons with $(p_x^2/2m) > W_a - (e^3F)^{1/2}$ has been assumed to be unity $[D(\varepsilon_x) = 1]$.
VI. The temperature dependence of Φ has been neglected. If one assumes $\Phi = \Phi_0 + \Phi_1\ T$, substitution for Φ as above in (2.3b) leads to (2.3c) with $A = A_0 \exp(-\Phi_1/k_B)$. This explains A being larger or smaller than A_0, depending on the sign of Φ_1.

The work function Φ and other parameters for some materials has been listed in Table 2.2.

2.2.3.2 Electric Field Emission

As discussed before electrons with $\varepsilon_x < (w_a - \alpha f^{1/2})$ have a finite probability $D(\varepsilon_x)$ of tunneling through the modified surface energy barrier and thus contributing to electron emission, known as electric field emission. Hence proceeding as in the derivation of (2.34a) the field emission current density is given by

$$J_{fe} = -e \int_0^{w_a - \alpha f^{1/2}} \int_0^{\infty} D(\varepsilon_x) d^2 n_1 = -A_0 T^2 \int_0^{(w_a - \alpha f^{1/2})} D(\varepsilon_x) \ln[1 + \exp(-\varepsilon_x + \varepsilon_F)] d\varepsilon_x \quad \text{for } \upsilon_0 > w_a \tag{2.38a}$$

and

$$J_{fe} = -A_0 T^2 \int_{w_a - \alpha f^{1/2} - \upsilon_0}^{(w_a - \alpha f^{1/2})} D(\varepsilon_x) \ln[1 + \exp(-\varepsilon_x + \varepsilon_F)] d\varepsilon_x \quad \text{for } \upsilon_0 < w_a \tag{2.38b}$$

Table 2.2 Electron emission data

S. No.	Materials	E_F (eV)	W_a (eV)	Φ (eV)	A (Acm^{-2}K^{-2})
	A. Metals				
1	Li	4.74	7.67	2.93	
2	Na	3.24	5.60	2.36	
3	K	2.12	4.41	2.29	
4	Rb	1.85	4.11	2.26	
5	Cs	1.59	3.54	1.95	160
6	Cu	7.00	12.1–11.53	5.10–4.48	
7	Ag	5.49	10.23–10.01	4.74–4.52	
8	Au	5.53	11.0–10.84	5.47–5.31	
9	Be	14.3	19.28	4.98	
10	Mg	7.08	10.74	3.66	
11	Ca	4.69	7.56	2.87	
12	Sr	3.93	6.42	2.59	
13	Ba	3.64	6.16	2.52	60
14	Nb	5.32	9.27- 9.84	3.95–4.87	
15	Fe	11.1	15.77- 5.91	4.67–4.81	
16	Mn	10.9	15.0	4.1	
17	Zn	9.47	13.10–14.37	3.63–4.9	
18	Cd	7.47	11.55	4.08	
19	Hg	7.13	11.60	4.47	
20	Al	11.7	15.76–15.96	4.06–4.26	
21	Ga	10.4	14.72	4.32	
22	In	8.63	12.72	4.09	
23	Tl	8.15	11.99	3.84	
24	Sn	10.2	14.62	4.42	
25	Pb	9.47	13.72	4.25	
26	Bi	9.9	14.24	4.34	
27	Sb	10.9	15.45–15.6	4.55–4.7	
28	Mo			4.36–4.95	55
29	Ni			5.04–5.35	60
30	Ta			4.0–4.8	60
31	W			4.45–5.22	80
32	Pt			5.22–5.93	170
33	Th			3.4	70
34	Zr			4.05	330
35	Hf			3.9	14.5
36	Pd			5.22–5.6	60
37	Ca			2.86	2.6
38	Sr			2.67	0.14
39	Ba			2.45	16
	B. Borides				
40	La			2.66	29
41	Ce			2.58	3.6
42	Th			2.92	0.5

(continued)

Table 2.2 (continued)

S. No.	Materials	E_F (eV)	W_a (eV)	Φ (eV)	A (Acm^{-2}K^{-2})
	C. Carbides				
43	Ta			3.14	0.3
44	Ti			3.35	25
45	Zr			2.18	0.3
46	Th			3.5	550
47	Ur			3.3	40
48	Th on W			2.63	3.0
49	Th on Mo			2.58	1.5
50	Zr on W			3.14	5.0
51	La, Ce on W			2.71	8.0
52	Cs on W			1.36	3.2
53	O on W			9.2	5×10^{11}
54	Cs on O on W			0.72	0.003
	D.Mono molecular films				
55	Ba on W			1.56	1.5
56	Ba on O on W			1.34	0.18
57	Ice			8.7	

Note The values for E_F, $W_a$and Φ for materials 1–36have been taken from *Lide* [25]
The values of A for materials 5, 13, 28–36 and 48–56are from Fowler [14]
The values of Φ for materials 48–56are from Fowler [14]
The values of A and Φ for materials 37–47 are from Jenkins and Trodder [22]
The value of Φ for ice is an indirect reference from Klumov et al. [24]

The limits can be appreciated by looking at Fig. (2.1f, g).

2.2.3.3 Numerical Results and Discussion

For an appreciation of the results computations were made, corresponding to the parameters $T = 1000$ K when

$$w_a = 38.64(W_a \,\text{in eV}), f = 4.7 \times 10^{-6}(F \,\text{in V/cm}) \,\text{and}\, \upsilon_0 = 38.64(\eta_0 \text{in eV});$$

η_0 is surface potential energy of an electron. The dependence of the field emission ($J_{\text{fe}}/J_{\text{th0}}$) and thermionic ($J_{\text{th}}/J_{\text{th0}}$) emission currents on the dimensionless field strength (f) for different values of the parameter w_a has been illustrated in Fig. 2.5. It is seen that the field emission current (J_{fe}) strongly depends on the applied field and significantly contributes to the total current for large fields. The thermionic current (J_{th}) also increases monotonically with increasing f, with a slower rate than J_{fe}. The dependence of these currents on parameter w_a can easily be understood in terms of transmission coefficient dependence on potential energy barrier height. It may be noticed that the total emission current ($J_t = J_{\text{th}} + J_{\text{fe}}$) gets enhanced by a factor of about eight for $f = 2$ from its initial value (for $f = 0$).

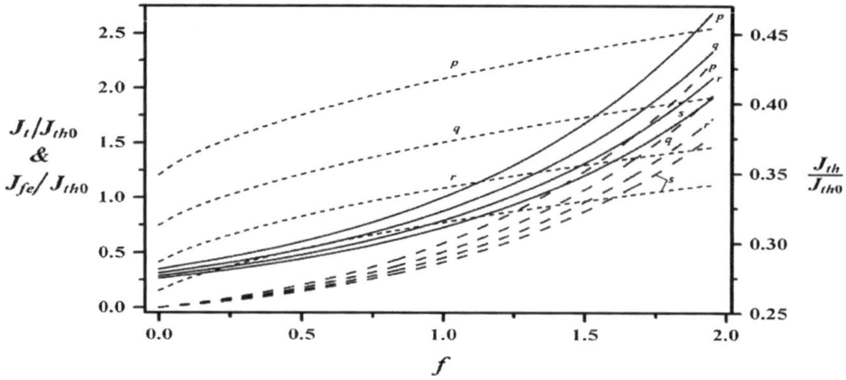

Fig. 2.5 Dependence of $(J_{\mathrm{th}}/J_{\mathrm{th0}}), (J_{\mathrm{fe}}/J_{\mathrm{th0}}) and (J_t/J_{\mathrm{th0}})$ with dimensionless electric field (*f*) for the parameters $T = 1000$ K and $\upsilon_0 = 1.0 \times 10^4$. The labels on the curves *p*, *q*, *r* and *s* correspond to $w_a = 60$, 80, 100 and 120 respectively. Dashed (*left hand* scale), *short dashed* (*right hand* scale) and *solid curves* (*right hand* scale) correspond to the field $(J_{\mathrm{fe}}/J_{\mathrm{th0}})$, thermionic $(J_{\mathrm{th}}/J_{\mathrm{th0}})$ and total $(J_{\mathrm{th}}/J_{\mathrm{th0}})$ emission current density respectively (after Agarwal et al. [1]; curtsey authors and publishers NRC Press)

2.2.3.4 Positively Charged Surface ($V_0 < 0$)

An electron can be emitted from a surface, charged to positive electric potential $(-V_0/e)$only when its normal energy exceeds $W_a - V_0$.

Hence the thermionic emission current J_{th} and the mean energy of the emitted electrons far away from the surface is in the approximation $D(\varepsilon_x) = 1$ given by

$$\begin{aligned} J_{\mathrm{th}} &= -A_0 T^2 \exp[-(\varphi - \upsilon_0)] = -A_0 T^2 \exp[-(\Phi - V_0)/k_B T] \\ &= J_{\mathrm{th0}} \exp(V_0/k_B T) \end{aligned} \tag{2.35c}$$

and

$$(\varepsilon_{\mathrm{th}})_{\mathrm{far\ away}} = 2 \tag{2.36c}$$

In contrast to (2.36b), $\varepsilon_{\mathrm{th}}$ does not depend on υ_0. This is because the net effect of the positive potential is to enhance the work function and $\varepsilon_{\mathrm{th0}}$ is independent of the work function. In fact in the evaluation of J_{th} and $\varepsilon_{\mathrm{th}}$ only electrons with energy in excess of $W_a - V_0$ are considered.

2.2.4 Photoelectric and Light Induced Field Emission from a Plane Surface

A quantitative theory for the rate of photoelectric emission from a plane metallic surface was first formulated by *Fowler* [13, 14] on the basis of the following explicit and implicit assumptions:

I. The free electron theory is applicable.
II. Free electrons incident on a unit area of the surface (from inside) have a probability of absorbing a photon, which is proportional to the number of incident photons per unit area per unit time (the absorption occurs on the surface of the metal and does not alter the number of incident electrons on the surface from inside).
III. The normal energy of an electron (due to the velocity component, normal to the surface) gets enhanced by $h\upsilon$ after absorption of a photon of frequency v at the surface, where h is Planck's constant.
IV. Electrons with a normal energy, exceeding the potential energy barrier at the surface get emitted.

A little later than Fowler [13], DuBridge [11] presented a similar theory for photoelectric emission; however instead of assumptions (ii) and (iii) DuBridge [11] assumed that

(iia) The absorption of a photon by the electron occurs inside the metal and the number of such absorptions per unit volume is proportional to the number density of free electrons; the implicit assumption that this number is also proportional to the number of incident photons per unit area per unit time was not mentioned by him. Since he did not consider collisions of the electrons after absorption of photon, it is implied that the absorption was assumed to occur close to the surface, within a thickness much less than the mean free path of the electrons and
(iiia) The total (not the normal) energy of an electron gets enhanced by $h\upsilon$ after the absorption of the photon.

In his derivation of the energy distribution of emitted photoelectrons and the emitted photoelectric current, DuBridge [11] made two further simplifications without justification viz.

(i) The energy of the electron after absorption of a photon is much larger than $h\upsilon$ (footnote on p. 735 of his paper) and
(ii) The normal energy of electrons after absorption of photons is just equal to the potential energy barrier at the surface [paragraph after (25) of his paper].

Usually the environment in which the photoelectric emission is of interest is at a temperature of 300 K, which corresponds to a thermal energy of electrons of the order of 0.04 eV, which is negligible as compared to the work function of the metal (which is the minimum value of $h\upsilon$ of interest). Hence in general the ratio of

Table 2.3 Characteristics of four models of photoelectric emission

S.No.	Model	Absorption of photon	Enhancement of energy
Ia.	Fowler [13] plus transmission coefficient	Surface	Normal
IIa.	Modified *DuBridge* [11] (without simplifications)	Inside	Total
Ib.	A	Inside	Normal
IIb.	B	Surface	Total

the energy of the electron after absorption of the photon to $h\upsilon$ is of the order of unity and certainly not much larger than one, as assumed by DuBridge [11]. The second assumption implies the neglect of electrons of normal energy, significantly higher than that corresponding to the potential energy barrier at the surface. This also severely limits the applicability of the theory.

Fowler [13] has analyzed the effect of the dependence of the boundary transmission coefficient and the probability of absorption of a photon by an electron on the energy of the emitted electrons. It was seen that the observed dependence of the photoemission current on the temperature and frequency was in slightly better agreement with experiments, when these parameters were assumed to be independent of the electron energy than the case when the energy dependence was taken into account. A critique of the later work which justifies the adequacy of the assumptions of Fowler [13] and DuBridge [11] has been given by Dewdney [9].

In view of the arbitrariness of the assumptions, made by Fowler [13] and DuBridge [11] it is desirable, to also investigate the problem, based on the following alternate (but equally arbitrary) assumptions:

A. Electrons absorb a photon inside the metal and the normal (not total) energy gets enhanced by $h\upsilon$.
B. Electrons absorb a photon at the surface and the total energy (not normal energy) gets enhanced by $h\upsilon$.

The characteristics of four models for photoelectric emission are outlined in Table 2.3.

Detailed computations highlight the fact that the important aspect is the mode of absorption of the photons by the free electrons (enhancement of normal or total energy) and not the site of absorption viz. inside the metal or on the surface. Hence, only cases Ia and IIa viz. *Fowler's* and *Dubridge's* models have been considered in this chapter.

A number of papers have been published on different aspects of photoelectric emission. Most of these are concerned with basic derivation of $\beta(E, \upsilon)$, the probability of absorption of a photon by an electron of energy E, based on different models and mathematical techniques. Some are concerned with surface states, multiphoton absorption, discussion of specific emitters and the band structure of the material, Auger process and elastic as well as inelastic collisions of electrons, within the emitting material. These theories are material-specific and do not have the general applicability of the models in Table 2.3. Moreover, Fowler [13] presented results in a user friendly mode; these have been used frequently by the investigators. Most of these investigations assumed that the probability of emission of an electron having a normal energy greater than the surface potential energy barrier is unity; this assumption is valid from classical considerations but is untenable in wave mechanics.

2.3 Fowler's Theory (Case: Ia) (After Fowler et al. [13])

2.3.1 Uncharged Surface

In view of *Fowler's* assumption (ii), the electrons, hitting the surface per unit time per unit area d^2n_1 have a probability $\beta(v)\Lambda(v)$ of absorbing a photon of frequency v, where $\Lambda(v)$ is the number of incident photons per unit area per unit time and $\beta(v)$ is indicative of the efficiency of the absorption of photons by electrons, incident on the surface; the energy of the absorbed photon enhances the normal energy of the electrons. Hence, using (2.9) the number of photoelectrons (electrons, which have absorbed a photon), hitting the surface per unit area per unit time is

$$\mathrm{d}^2n_{\mathrm{ph}} = (A_0/e)T^2\beta(v)\Lambda(v)F_D(\varepsilon_x + \varepsilon_t - \varepsilon_F)\mathrm{d}\varepsilon_x\mathrm{d}\varepsilon_t.$$

Putting $\varepsilon'_x = \varepsilon_x + \varepsilon_v$ (where $\varepsilon_v = h\upsilon/k_BT$ where ε'_x is the normal energy of an electron with energy ε_x after absorption of a photon) in the above equation one obtains

$$\begin{aligned}\mathrm{d}^2n_{\mathrm{ph}} &= (A_0/e)T^2\beta(v)\Lambda(v)F_D(\varepsilon'_x + \varepsilon_t - \varepsilon_v - \varepsilon_F)\mathrm{d}\varepsilon'_x\mathrm{d}\varepsilon_t.\\ &= (A_0/e)T^2\beta(v)\Lambda(v)F_D(\varepsilon''_x + \varepsilon_t + w_a - \varepsilon_v - \varepsilon_F)\mathrm{d}\varepsilon''_x\mathrm{d}\varepsilon_t,\end{aligned} \tag{2.39}$$

where $\varepsilon''_x = \varepsilon'_x - w_a$ is the energy of a photoelectron after crossing the surface energy barrier w_a.

Integrating $\mathrm{d}^2n_{\mathrm{ph}}$ over $0 < \varepsilon_t < \infty$ one obtains

$$\mathrm{d}n_{\mathrm{ph}} = (A_0/e)T^2\beta(v)\Lambda(v)\ln[1 + \exp(\varepsilon_v + \varepsilon_F - \varepsilon'_x)]\mathrm{d}\varepsilon'_x. \tag{2.40}$$

Hence the number of photoelectrons emitted due to the photoelectric effect is given by

$$\begin{aligned}n_{\mathrm{ph}} = (-J_{\mathrm{ph}}/e) &= (A_0/e)T^2\beta(v)\Lambda(v)\int_{w_a}^{\infty} D(\varepsilon'_x)\ln[1 + \exp(\varepsilon_v + \varepsilon_F - \varepsilon'_x)]\mathrm{d}\varepsilon'_x,\\ &= (A_0/e)T^2\beta(v)\Lambda(v)\int_{0}^{\infty} D(\varepsilon''_x + w_a)\ln[1 + \exp(\xi - \varepsilon''_x)]\mathrm{d}\varepsilon''_x,\end{aligned} \tag{2.41a}$$

where $\varepsilon_x'' = \varepsilon_x' - w_a$ is the energy of a photoelectron after emission,

$$\xi = \varepsilon_v - [w_a - \varepsilon_F] = \varepsilon_v - \varphi = (hv - \Phi)/k_B T$$

and $\Phi = W_a - E_F$ is the work function of the material.

Putting $D(\varepsilon_x'' + w_a) = 1$ as is usually the practice and $\exp(\xi - \varepsilon_x'') = \zeta$ one obtains the current density due to photoelectron emission

$$J_{\rm ph} = -en_{\rm ph} = -A_0 T^2 \beta(v) \Lambda(v) \Phi_0(\xi), \tag{2.41b}$$

where

$$\Phi_0(\xi) = \int_0^{\exp \xi} \frac{\ln[1 + \zeta]}{\zeta} d\zeta. \tag{2.41c}$$

As discussed earlier, in general $D(\varepsilon_x')$ is not unity but given by (2.17) for an electrically neutral surface.

However in case $D(\varepsilon_x'' + w_a) = 1$, as is usually assumed the mean energy of the emitted photoelectrons, just outside the surface is

$$\begin{aligned}\langle \varepsilon'' \rangle &= \int_0^\infty \int_0^\infty (\varepsilon_x'' + \varepsilon_t) \mathrm{d}^2 n_{\rm ph} \Bigg/ \int_0^\infty \int_0^\infty \mathrm{d}^2 n_{\rm ph} \\ &= \int_0^\infty \int_0^\infty (\varepsilon_x'' + \varepsilon_t) F_D(\varepsilon_x'' + \varepsilon_t - \xi) \mathrm{d}\varepsilon_x'' \mathrm{d}\varepsilon_t \Bigg/ \int_0^\infty \int_0^\infty F_D(\varepsilon_x'' + \varepsilon_t - \xi) \mathrm{d}\varepsilon_x'' \mathrm{d}\varepsilon_t.\end{aligned}$$

Using identity (2.37) the above equation reduces to

$$\begin{aligned}\langle \varepsilon'' \rangle &= \int_0^\infty \varepsilon''^2 F_D(\varepsilon'' - \xi) \mathrm{d}\varepsilon \Bigg/ \int_0^\infty \varepsilon'' F_D(\varepsilon'' - \xi) \mathrm{d}\varepsilon'' \\ &= \frac{1}{\Phi_0(\xi)} \int_0^\infty 2\varepsilon'' \ln[1 + \exp(\xi - \varepsilon'')] \mathrm{d}\varepsilon''.\end{aligned} \tag{2.42a}$$

It is useful to define photoelectric efficiency $\chi(v)$ as the number of photoelectrons emitted per incident photon. Thus

$$\chi(v) = n_{ph}/\Lambda(v).$$

The best fit of experimental data for dependence of $\chi(v)$ on v is as follows

$\chi(v)/\chi_m = (729/16)(v_0/v)^4(1 - v_0/v)^2$ (Spitzer [64]; Sodha et al. [49] and $\chi(v)/\chi_m = (1 - v_0/v)^2$ [7] where v_0 is threshold frequency and χ_m is the maximum value of χ.

It is seen that for $\xi > 5$, $\langle \varepsilon'' \rangle \approx 0.472 + 0.657\xi$ to an excellent approximation.

2.3.1.1 Negatively Charged Surface ($\upsilon_0 > 0$)

Equations (2.41a), (2.41b) and (2.42a, 2.42b) for the current density and mean energy of electrons are valid for a negatively charged surface when w_a is replaced by $w_a - \alpha f^{1/2}$ and appropriate data for $D(\varepsilon'_x)$ is used. Further the mean energy away from the surface is $\varepsilon''_{\text{faraway}} = \langle\varepsilon''\rangle + \upsilon_0$.

2.3.2 *Positively Charged Surface ($\upsilon_0 < 0$)*

As in the case of thermionic emission the expressions for J_{ph} and $\langle\varepsilon''\rangle$ (far away from the surface) is obtained by putting $w_a - \upsilon_0$ for w_a. Thus

$$J_{\text{ph}} = J_{\text{ph}} = -en_{\text{ph}} = -A_0 T^2 \beta(v) \Lambda(v) \Phi_0(\xi + \upsilon_0) \tag{2.41d}$$

and

$$\langle\varepsilon''\rangle_{\text{far away}} = \frac{1}{\Phi_0(\xi + \upsilon_0)} \int_0^\infty 2\varepsilon'' \ln[1 + \exp(\xi + \upsilon_0 - \varepsilon'')] \mathrm{d}\varepsilon'' \tag{2.42b}$$

2.3.3 *Light Induced Field Emission*

For a negatively charged surface, the current density J_{fp} (in analogy with (2.41a)) on account of tunneling of low energy photoelectrons ($\varepsilon'_x < w_a - \alpha f^{1/2}$) is given by

$$J_{\text{fp}} = -en_{\text{life}} = -A_0 T^2 \beta(v) \Lambda(v) \int_0^{w_a - \alpha f^{1/2}} D(\varepsilon'_x) \ln[1 + \exp(\varepsilon_\upsilon + \varepsilon_F - \varepsilon'_x)] \mathrm{d}\varepsilon'_x \quad \text{for } \upsilon_0 > w_a \tag{2.43a}$$

and

$$J_{\text{fp}} = -en_{\text{life}} = -A_0^2 T^2 \beta(v) \Lambda(v) \int_{w_a - \alpha f^{1/2} - \upsilon_0}^{w_a - \alpha f^{1/2}} D(\varepsilon'_x) \ln[1 + \exp(\varepsilon_\upsilon + \varepsilon_F - \varepsilon'_x)] \mathrm{d}\varepsilon'_x \quad \text{for } \upsilon_0 < w_a \tag{2.43b}$$

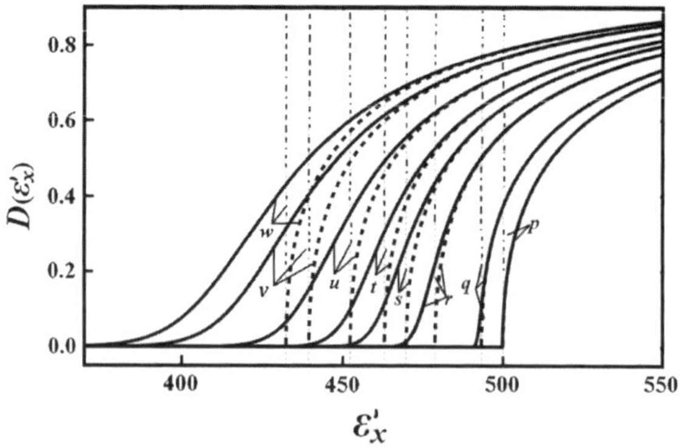

Fig. 2.6 Dependence of $D(\varepsilon_x')$ on ε_x' for the parameters $T = 300\,K$, $w_a = 500$ and $\upsilon_0 = 1.0 \times 10^4$. The labels on the curves *p, q, r, s, t, u, v* and *w* correspond to $f = 0$, 1, 10, 20, 30, 50, 80, and 100, respectively. The *solid curves* correspond to the present analysis while *dashed curves* refer to the step potential barrier with reduced height. The portion of the curves corresponding to *left-* and *right-hand side* of the *vertical lines* refer to the light induced field emission (life) and photoelectric emission, respectively (after Agarwal et al. [1]; curtsey authors and publishers NRC Press)

This phenomenon was predicted by Sodha et al. [41] and verified experimentally by Kher et al. [23] and Iwami et al. [21]. The prediction of (2.43a), (2.43b) that J_{fp} is proportional to the light irradiation $\Lambda(\nu)$ is in accordance with the observations of Iwami et al. [21]; other theories do not explain this observation.

2.3.4 Numerical Results and Discussions

Figures 2.6 and 2.7 illustrate the dependence of $D(\varepsilon_x')$ on ε_x' for a temperature of 300 K; the variation is similar to that in Figs. 2.2 and 2.3, corresponding to a temperature of 1000 K.

The dependence of light-induced field emission (*life*) (J_{fp}/J_{ph0}) and photoemission (J_{ph}/J_{ph0}) currents on the dimensionless field strength (*f*) for different values of the parameters ε_ν and φ has been illustrated in Figs. 2.8 and 2.9. Figure 2.8 indicates that the photoemission current (J_{ph}/J_{ph0}) increases with increasing field strength and parameter ε_ν; this is because of large availability of high energy electrons and corresponding smaller energy barrier height. The *life* current (J_{fp}) displays a trend opposite to that in case of J_{ph} with increasing ε_ν; this is explained on the basis of large availability of low energy electrons for tunneling for small ε_x. The parameters J_{ph} and J_{fp} are indicative of the effect of the electric field f on the ε_x dependence of $D(\varepsilon_x)$. The effect of the work function of the metallic plate (φ) on emission currents has been displayed in Fig. 2.9. It is noticed that the current dependence (J_{ph} and J_{fp}) on φ displays a trend opposite to that in the case of ε_ν; this nature can be understood in terms of $\varsigma(= \varepsilon_\upsilon - \varphi)$, which

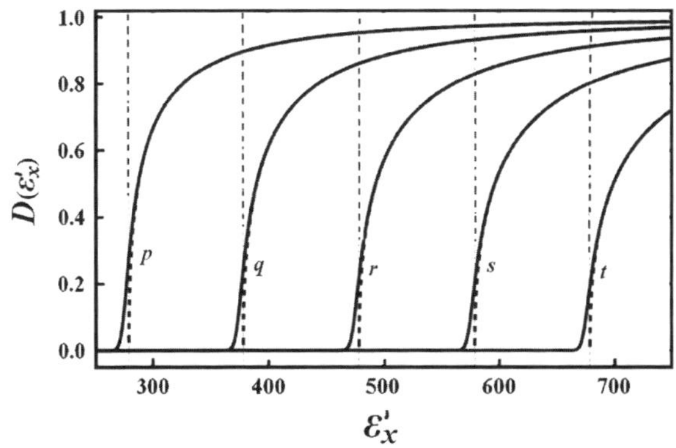

Fig. 2.7 Dependence of $D(\varepsilon_x')$ on ε_x' for the parameters $T = 300$ K, $f = 10$ and $\upsilon_0 = 1.0 \times 10^4$. Curves *p, q, r, s* and *t* correspond to $w_a = 300, 400, 500, 600$, and 700, respectively. The nature of the *curves* (*solid, dashed* and *vertical lines*) is the same as in Fig. 2.6 (after Agarwal et al. [1]; curtsey authors and publishers NRC Press)

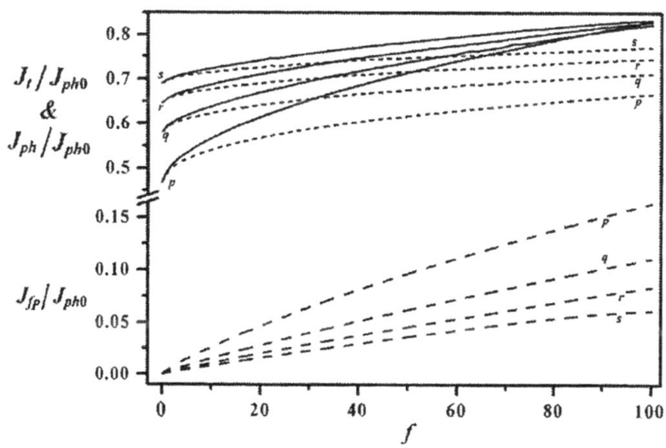

Fig. 2.8 Dependence of (J_{ph}/J_{ph0}), (J_{fp}/J_{ph0}) *and* (J_t/J_{ph0})on dimensionless electric field (f) for the parameters $T = 300$ K $w_a = 500$ and 120 and $\upsilon_0 = 1.0 \times 10^4$. *Dashed, short dashed* and *solid curves* correspond to $\varphi = 200$, The labels on the curves *p, q, r* and *s* correspond to$\varepsilon_\upsilon = 250$, 300, 350, and 400, respectively. *Dashed* (*left-hand* scale), short *dashed* (*right-hand* scale) and *solid curves* (*right hand* scale) correspond to the life (J_{fp}/J_{ph0}), photo (J_{ph}/J_{ph0}) and total (J_t/J_{ph0}) emission current density, respectively (after Agarwal et al. [1]; curtsey authors and publishers NRC Press)

increases with increasing ε_v and decreasing φ. The figures also display the fact that *life* current significantly contributes to the total emission current ($J_t = J_{ph} + J_{fp}$) for large f; it is also interesting to point out that the total current enhances by a factor of about 1.5 from its initial value (at $f = 0$) for the chosen set of parameters.

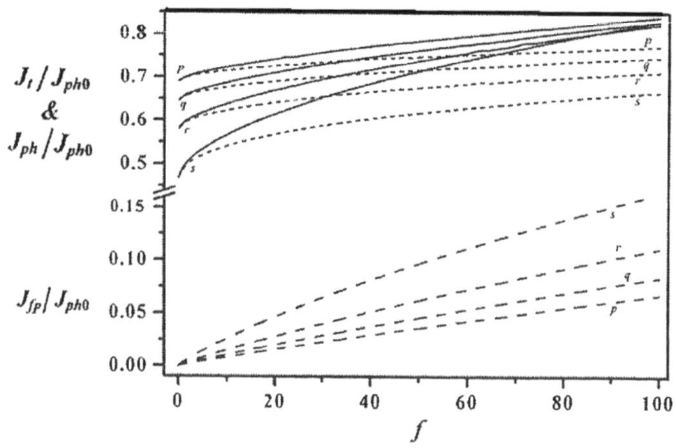

Fig. 2.9 Dependence of (J_{ph}/J_{ph0}), (J_{fp}/J_{ph0}) and (J_t/J_{ph0}) on dimensionless electric field (f) for the parameters $T = 300\ K$ $w_a = 500$ and 120 and $\upsilon_0 = 1.0 \times 10^4$. *Dashed, short dashed* and *solid curves* correspond to $\varepsilon_\upsilon = 300$, The labels on the curves p, q, r and s correspond to $\varphi = 100$, 150, 200 and 250 respectively. *Dashed* (*left hand* scale), *short dashed* (right hand scale) and *solid curves* (*right hand* scale) correspond to the life (J_{fp}/J_{ph0}), photo (J_{ph}/J_{ph0}) and total (J_t/J_{ph0}) emission current density respectively (after Agarwal et al. [1]; curtsey authors and publishers NRC Press)

2.4 Modified Dubridge Theory: Case IIa and other effects

According to this theory, the total (not normal) energy of an electron gets enhanced by $h\nu$ on absorption of a photon of frequency ν. Hence if p and p' denote the momentum of an electron before and after the absorption of a photon and E' is the energy of the electron after absorption of the photon [11].

$$E' = (p'^2/2m) = h\nu + (p^2/2m). \tag{2.44}$$

Hence from (2.8b) and (2.44) the momentum distribution of electrons after absorption of a photon is

$$n(p')\mathrm{d}p' = (2/h^3)\beta(\nu)A(\nu)F_D(\varepsilon' - \varepsilon_\upsilon - \varepsilon_F)4\pi[1 - (\varepsilon_\upsilon/\varepsilon')]^{1/2}p'^2\mathrm{d}p',$$

which is equivalent to

$$n(p')\mathrm{d}p'_x\mathrm{d}p'_y\mathrm{d}p'_z = (2/h^3)\beta(\nu)A(\nu)F_D(\varepsilon' - \varepsilon_\upsilon - \varepsilon_F)[1 - (\varepsilon_\upsilon/\varepsilon')]^{1/2}\mathrm{d}p'_x\mathrm{d}p'_y\mathrm{d}p'_z,$$

where $\varepsilon' = E'/k_BT$.

Out of the electrons having $\varepsilon' > w_a$, the fraction having $\varepsilon'_x > w_a$ is $[1-(w_a/\varepsilon')]^{1/2}/2$.[1] Hence, the number of electrons emitted per unit area per unit time is

$$n_{\text{ph}} = \int_{\sqrt{2mW_a}}^{\infty}\int_{-\infty}^{\infty}\int_{-\infty}^{\infty} \bar{D}(\varepsilon'_x)[1-(w_a/\varepsilon')]^{1/2}(p'_x/m)n(p')\mathrm{d}p'_x\mathrm{d}p'_y\mathrm{d}p'_z.$$

As in the case of *Fowler*'s theory, the above integral may be simplified as

$$\begin{aligned} n_{\text{ph}} &= (-J_{\text{ph}}/e) \\ &= (A_0/2e)T^2\beta(v)\Lambda(v) \times \int_0^{\infty}\int_0^{\infty} D(\varepsilon''_x + w_a)\{1-[w_a/(\varepsilon''_x+\varepsilon''_t+w_a)]^{1/2}\} \\ &\quad \{1-[\varepsilon_v/(\varepsilon''_x+\varepsilon''_t+w_a)]\}^{1/2}F_D(\varepsilon''_x+\varepsilon''_t-\xi)d\varepsilon''_x d\varepsilon''_x, \end{aligned}$$

As an approximation $D(\varepsilon''_x + w_a)$ can be replaced by an average value $\bar{D}(\varepsilon'' + w_a)$,

where $\varepsilon'' = \varepsilon''_x + \varepsilon''_t$ and $\bar{D}(\varepsilon'' + w_a) = \frac{1}{\varepsilon''}\int_0^{\varepsilon''} D(\varepsilon''_x + w_a)\mathrm{d}\varepsilon''_x$.

Thus using the identity $\int_0^{\infty}\int_0^{\infty} f(x_1+x_2)\mathrm{d}x_1\mathrm{d}x_2 = x\int_0^{\infty} xf(x)$

One obtains

$$\begin{aligned} n_{\text{ph}} = (-J_{\text{ph}}/e) = (A_0/2e)T^2\beta(v)\Lambda(v) \times \int_0^{\infty} \bar{D}(\varepsilon''+w_a)\{1-[w_a/(\varepsilon''+w_a)]^{1/2}\} \\ \times\{1-[\varepsilon_v/(\varepsilon''+w_a)]\}^{1/2}\{1+\exp(\varepsilon''-\xi)\}^{-1}\varepsilon\mathrm{d}\varepsilon \end{aligned} \tag{2.45}$$

The energy distribution of emitted photoelectrons, based on this theory is in a little better agreement with experiments than that corresponding to *Fowler's* theory. In view of the mathematically untreatable expressions occurring in *Dubridge's* theory, it has not been used to an appreciable extent in the study of the kinetics of complex plasmas. Hence, no further discussions of this theory has been made later in this book.

---

[1] The possible values of u_x, u_y, u_z corresponding to an electron speed u are the coordinates on the surface of a sphere of radius u. The area of the surface corresponding to $u_x > u_c$ is $2\pi(u - u_c)u$, while the area of the whole spherical surface is $4\,\pi u^2$. Hence the fraction of electrons with speed u having $u_x > u_c$ is simply

$$[2\pi u(u-u_c)/4\pi u^2] = 1/2(1-u_c/u) = 1/2[1-(w_a/\varepsilon')^{1/2}],$$

where $w_a = (m_e u_c^2/2)$ and $\varepsilon' = (m_e u^2/2)$.

2.5 *Spicer's* Three Step Model

No discussion of the photoelectric effect is complete, without a reference to the widely cited three step model, advanced by Spicer [52] and formalized by Bergland and Spicer [2]; a simple account of the model has been given by Spicer and Herrara-Gomez [53]. According to this model, photoemission of electrons is a bulk (rather than a surface) process and consists of three steps, viz

(i) Generation of photoelectrons, (deep in material) having enough energy to overcome the surface barrier.
(ii) Transport of these photoelectrons to the surface, taking into account the scattering of the electrons and
(iii) Transmission through the surface.

Consider the normal incidence of light of frequency v and irradiance $I_0(v)$ incident normally on the surface of a material with reflectivity $R(v)$. The irradiance $I(v)$ at the depth x in the material is given by

$$I(v) = I_0(v)[1 - R(v)]\exp[-\alpha(v)x], \tag{2.46}$$

where $\alpha(v)$ is the attenuation constant of light in the material and $R(v)$ is the reflection coefficient at the surface.

The number of photoelectrons with enough energy to escape from the surface, which is generated from photoexcitation per unit volume per unit time, is proportional to the irradiance. Hence, the number $\mathrm{d}n_{\mathrm{exc}}$ of such photoelectrons generated per unit area in a thickness $\mathrm{d}x$ of the material is given by

$$\mathrm{d}n_{\mathrm{exc}} = \alpha_{\mathrm{exc}} \cdot I \cdot \mathrm{d}x, \tag{2.47}$$

where α_{exc} is a coefficient indicative of the efficiency of photoexcitation

The probability of traversing a distance x without significant loss of energy is given by

$$P_T(x) = \exp(-x/L). \tag{2.48}$$

where L is the mean free path of the electrons.

Hence, the number of photoelectrons emitted per unit area per unit time from the surface is

$$n_{\mathrm{ph}} = \int_0^{\infty} P_{em}^{(v)} P_T(x)\mathrm{d}n_{\mathrm{exc}},$$

where $P_{em}^{(v)}$ is the probability of transmission through the surface barrier.

Substituting for $I(v)$, $\mathrm{d}n_{\mathrm{exc}}$ and $P_T(x)$ from (2.46), (2.47) and (2.48) in the above equation and integrating the R.H.S. one obtains

$$n_{\mathrm{ph}} = I_0(v)[1 - R(v)]\alpha\alpha_{\mathrm{exc}}\{\alpha_{\mathrm{exc}} + 1/L\}^{-1}. \tag{2.49}$$

The predictions for *Spicer*'s model agree with experiments; however the inputs viz. absorption of the radiation in the solid, electron scattering/transport data and band structure are not available for many photoelectric surfaces. Further the computation, necessary for the application of *Spicer*'s model is formidable. Hence phenomenological models, in good agreement with experiments are still relevant. In fact *Fowler's* expression for the photoelectric emission is still in vogue for interpretation of photoelectric data and models for physical processes, where photoelectric emission is important.

2.6 Size Effect

Watson [57, 58] adopted a simple model and predicted that the photoelectric efficiency χ of the surface of a particle of radius a is given by

$$(\chi/\chi_b) = \left(\frac{\beta_w}{\alpha_w}\right)^2 \frac{\alpha_w^2 - 2\alpha_w + 2 - 2\exp(-2\alpha_w)}{\beta_w^2 - 2\beta_w + 2 - 2\exp(-2\beta_w)} \tag{2.50}$$

where $\alpha_w = a\alpha + a/L, \beta_w = a\alpha$.

2.7 Secondary Electron Emission

2.7.1 Secondary Electron Emission by Electron Impact

When a high energy primary electron is incident on the surface of a material, it may either be reflected or enter the material. Once it is in the material, it may collide with scattering centers and ultimately get out of the material; this process is known as back scattering. However, some of the energy of the electron may be utilized in the excitation of electrons, which may escape from the material; this process is known as secondary electron emission.

The reflection coefficient is of the order of 0.05 at low energies of primary electrons and it rapidly decreases with increasing energy of primary electrons [19]. Both reflection and back scattering are not of importance in the kinetics of complex plasmas and therefore not considered hereinafter.

In what follows a phenomenological theory of secondary electron emission from (i) plane surface of a semi infinite solid and (ii) spherical grain has been presented. The case of a cylindrical surface has been discussed in Appendix B of this chapter.

2.7.1.1 Plane Surface of Semi-infinite Solid (After Jonker [20])

Jonker [20] has given a theory of secondary electron emission from the plane surface of a semi-infinite solid on account of the incidence of primary high energy electrons. The following four assumptions made by him constitute the basis of his theory as well as subsequent theories, including this presentation: (*i*) The primary electrons loose energy according to Whiddington's [59] law and that the absorption and scattering of electrons is negligible, (*ii*) The number of secondary electrons generated by a primary electron is proportional to the rate of energy loss in the solid, (*iii*) The absorption of secondary electrons in the material follows the exponential law and (*iv*) The distribution of generated secondary electrons is isotropic. Further for the sake of simplicity, it has also been implicitly assumed that V_p, the energy of the incident electron is much larger than the surface energy barrier.

The energy $V(x)$ in eV of a primary electron in a solid is given by [59]

$$V^2(x) = V_p^2 - \alpha x \tag{2.51}$$

where

V_p is the electron energy at $x = 0$,
x is the distance traversed in cm and
α is a constant dependent on the material.

The rate of secondary electron production n_s is proportional to the energy loss per unit length; thus the rate of secondary electron production in a distance dx is as per Bruining [3]

$$n_s\mathrm{d}x = -KI_p\frac{\mathrm{d}V(x)}{\mathrm{d}x}\mathrm{d}x = (K\alpha I_p/2)(V_p^2 - \alpha x)^{-1/2}\mathrm{d}x, \tag{2.52}$$

where I_p is the primary beam current, i.e., number of primary electrons, incident on the surface per unit time and K is a material specific constant.

Consider the normal (along x direction) incidence of a beam of primary (high energy) electrons on the plane surface ($x = 0$) of a semi-infinite solid substance. The secondary electrons, generated at x (the distance from the surface) and moving in a direction making an angle between θ and $\theta + \mathrm{d}\theta$ with the axis (corresponding to a solid angle $2\pi\sin\theta\mathrm{d}\theta$), traverse a distance $x\sec\theta$, before hitting the surface and getting emitted. Hence the number of such secondary electrons getting emitted from the surface is from 2.51 and 2.52 given by

$$\mathrm{d}i_s = (K\alpha I_p/2)(V_p^2 - \alpha x)^{-1/2}\mathrm{d}x\int_0^{\pi/2}\frac{2\pi\sin\theta\mathrm{d}\theta}{4\pi}\exp(-\beta x\sec\theta)\mathrm{d}\theta, \tag{2.53}$$

where β is the attenuation constant for secondary electrons and ($2\pi\sin\theta d\theta/4\pi$) is the fraction of electrons emitted, making an angle between θ and $\theta + d\theta$ with the normal.

Integrating the above equation in the limit $x = 0$ to $x = x_m = V_p^2/\alpha$ (corresponding to $V(x) = 0$) one obtains.

$$I_s = (K\alpha I_p/4)\int_0^{x_{\max}} (V_p^2 - \alpha x)^{-1/2}\mathrm{d}x \int_0^{\pi/2} \exp(-\beta x\sec\theta)\sin\theta \mathrm{d}\theta. \tag{2.54a}$$

where $x_m = V_p^2/\alpha$ is the depth of penetration of the primary electron in the material and corresponds to $V(x_m) = 0$.

It is common to express the results in terms of secondary yield $\delta(V_p)$, defined as

$$\delta(V_p) = I_s/I_p. \tag{2.54b}$$

In case of oblique incidence at an angle λ to the normal, the distance of the point x from the surface is $x\cos\lambda$ and hence

$$I_s = (K\alpha I_p/4)\int_0^{x_{\max}} (V_p^2 - \alpha x)^{-1/2}\mathrm{d}x \int_0^{\pi/2} \exp(-\beta' x)\sin\theta \mathrm{d}\theta, \tag{2.54c}$$

where $\beta' = \beta\cos\lambda$.

If $I_p'(V_p)\mathrm{d}V_p$ is the energy distribution of the primary beam the secondary current i_s is

$$i_s = \int \delta(V_p) I_p'(V_p)\mathrm{d}V_p \tag{2.54d}$$

The various approaches to the evaluation of secondary electron emission essentially differ in the expressions for the rate of energy loss by primary electrons. A popular empirical expression for $\delta(V_p)$, number of secondary electrons emitted by the incidence of primary electrons of energy V_p is Sternglass [55], Jonker [20], Meyer-Vernet [26] is

$$\delta(V_p) = 7.4\delta_m(V_p/V_m)\exp[-2(V_p/V_m)^{1/2}], \tag{2.55}$$

where the maximum value of $\delta = \delta_m$ occurs for $V_p = V_m$.

Representative values [60] of δ_m and V_m are given in Table 2.4.

The energy distribution of secondary electrons is approximately Maxwellian with a characteristic energy of 2 eV (Hachenberg and Brauer [18].

The electron emission on account of electron beams has been discussed in literature.

Table 2.4 Values of δ_m and V_m

Material	Sio$_2$	MgO	Teflon	Kapton	Al_2O_3	Mg	Al
δ_m	2.4	4.0	3.0	2.1	1.5–1.9	0.92	0.97
V_m (keV)	0.4	0.4	0.3	0.15	0.35–1.3	0.25	0.3

2.7.2 Spherical Particle (After Misra et al. [31])

On the basis of rather simple analysis Draine and Salpeter [65] have emphasized the importance of secondary electron emission in the charging of small grains ($\approx$ 10 nm or less in radius) when the penetration depth x_m of the primary electrons exceeds the diameter of the grains. In view of the existence of such small particles in the interstellar medium (Puget and Leger [66]) and space, e.g., *Hailey's* comet environment (Sagdeev et al. [67]), this phenomenon is of much interest in the charging of dust. A theory of the secondary electron emission and charging of a dust particle, exposed to high energy electrons (from the plasma or elsewhere) should also be of interest to situations other than those in space environment.

Meyer-Vernet [26] developed an elaborate theory for the charging of a dust particle in space, taking into account the phenomenon of secondary electron emission; however this theory was based on the widely used empirical relation by Sternglass [55]. However the validity of the results is severely limited by the fact that this relation is applicable only for the plane surface of the semi-infinite solid and hence completely ignores the effect of the size and shape of the particle.

Chow et al. [5] modified the theory by Jonker [20] to make it applicable to spherical particles and applied it to investigate the charging of the particles in *Maxwellian and Lorentzian* plasmas. The theory distinguishes between the cases when the diameter of the particle is larger or smaller than the penetration depth of the electrons. However the relevance of the theory is limited because it overlooks the following points: (*i*) The length of an electron path (*AB*) in the spherical particle depends on the perpendicular distance (ρ) from the center of the particle; hence it varies from *zero* to the diameter or x_m depending upon ρ (see Fig. 2.10), (*ii*) As a corollary to (*i*), when the diameter exceeds the penetration depth x_m there is an optimum perpendicular distance ρ_m from the center (corresponding to the path length equal to the penetration depth x_m) such that electrons corresponding to $\rho \geq \rho_m$ will pass through the particle while the electrons, corresponding to $\rho < \rho_m$ will get stuck in the particle and (*iii*) The electrons, stuck in the particle and primary electrons sticking to the surface contribute to the charging of the particles.

Here we have modified the theory of *Chow et al* by incorporating the points, enumerated above. Specifically the parameter δ (the number of secondary electrons, emitted by the particle per primary electron) has been evaluated as a function of primary electron energy, radius, and electric potential of the particle. The parameter δ for *Maxwellian distribution* of the primary electron energy has also been evaluated.

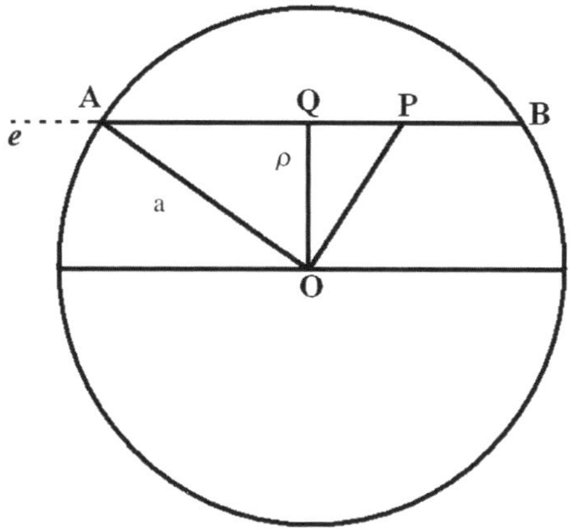

Fig. 2.10 Path of incident primary electron in a particle

Referring to Fig. 2.10, considering the four basic assumptions [20] stated earlier and using (2.52), the rate of secondary electron generation induced by a primary electron in traversing a distance dx in the substance is given by [5]

$$\begin{aligned} dn_s &= (K\alpha/2)(V_p^2 - \alpha x)^{-1/2}\langle \exp[-\beta l(x,\rho,\theta,\varphi)]\rangle dx \\ &= (K\alpha/2)(V_p^2 - \alpha x)^{-1/2} f(x,\rho)]dx, \end{aligned} \tag{2.56}$$

where K is a constant for a given material, β is the attenuation constant for the secondary electrons in the substance, l is the distance of any point (θ, φ) on the spherical surface $(r = a)$ from P and $< >$ denotes the average over all points on the sphere. To determine the distance l of point P from any point on the sphere viz. $(a\cos\theta\cos\varphi,\ a\cos\theta\sin\varphi,\ a\sin\theta)$, one may without loss of generality consider $z = 0$, to be the plane containing the center of particle and the path. Thus one has

$$l^2(x,\theta,\varphi,\rho) = \left[\left[a\cos\theta\cos\varphi - [x - (a^2-\rho^2)^{1/2}]\right]^2 + (a\cos\theta\sin\varphi - \rho)^2 + a^2\sin^2\theta\right].$$

Hence

$$f(x,\rho) = \langle \exp(-\beta l)\rangle = \frac{1}{4\pi}\int_{\theta=0}^{\pi}\int_{\varphi=0}^{2\pi} \exp[-\beta l(x,\theta)]\sin\theta\, d\theta d\varphi. \tag{2.57}$$

It is convenient to define a value of $\rho = \rho_m$ so that the path length of the electron in the particle is x_m, the penetration depth; thus $\rho_m^2 = a^2 - (x_m/2)^2$. It may be noted that for $x_m < 2a$, when $\rho < \rho_m$, $AB > x_m$ and the primary electron gets stuck in the particle; when $\rho > \rho_m$, $AB < x_m$ and the electron passes through the particle. In case $x_m > 2a$, all the electrons pass through the particle.

Consider a mono-energetic beam of primary electrons with n_e electrons incident per unit area, per unit time, normal to the direction of the beam. The number of primary electrons incident per unit cross section of the dust surface normal to the beam, per unit time n_p is given by (Abbas et al. [68])

$$n_p = n_e[1 - (V_s/V_p)] \tag{2.58}$$

where V_s is the potential energy of the electron at the surface of the particle and $V_s << V_p$.

Hence, the number of secondary electrons produced per unit time is

$$n_s = (1-\eta_s)\left(\frac{K\alpha}{2}\right)\left[\int_0^{\rho_m} 2\pi n_p I_1(V_p,\rho)\rho\mathrm{d}\rho + \int_{\rho_m}^{a} 2\pi n_p I_2(V_p,\rho)\rho\mathrm{d}\rho\right] \quad \text{for } x_m < 2a \tag{2.59a}$$

and

$$n_s = (1-\eta_s)\left(\frac{K\alpha}{2}\right)\int_0^{a} 2\pi n_p I_2(V_p,\rho)\rho\mathrm{d}\rho \quad \text{for } x_m \geq 2a \tag{2.59b}$$

where

$$I_1(V_p,\rho) = \left(\int_0^{x_\mathrm{m}} (V_p^2 - \alpha x)^{-1/2} f(x,\rho)\mathrm{d}x\right),$$

$$I_2(V_p,\rho) = \left(\int_0^{2\sqrt{a^2-\rho^2}} (V_p^2 - \alpha x)^{-1/2} f(x,\rho)\mathrm{d}x\right)$$

and η_s is the electron sticking (at the surface) coefficient.

The secondary electron emission yield δ can be expressed as: $\delta = \left(n_s/\pi a^2 n_p\right)$. In the dimensionless form δ viz. $\delta' = \delta/(1-\eta_s)K\sqrt{a\alpha}$ can be expressed as a function of $\beta' = \beta a$, and $x'_m = (x_m/a)$; to do so one has to use $\boldsymbol{r} = a\boldsymbol{r}'$

For a charged particle, the yield $\delta(V_s)$ is given by[6]

$$\delta'(V_s) = \delta'(V_s = 0) \quad \text{for } V_s \geq 0 \tag{2.60a}$$

$$\delta'(V_s) = \kappa\delta'(V_s = 0) \quad \text{for } V_s < 0 \tag{2.60b}$$

where $\kappa = (1 - V_s/kT_s)\exp(V_s/k_B T_s)$ and T_s is the temperature of secondary electrons.

The dependence of δ' on x'_m has been illustrated for different values of β' in Fig. 2.11. It is seen that the yield $\left(\delta'_m\right)$ takes a maximum value for an optimum value of x'_m and it decreases with increasing β' (i.e., particle size).

When diameter of the particle exceeds the penetration distance, the number of primary electrons which gets stuck in the dust particle is given by

Fig. 2.11 Dependence of δ' on x'_m; the labels *p, q, r, s* and *t* refer to $\beta' = 0.01, 0.1, 0.5, 1.0$ and 3.0 respectively for $b = 5a$. *Solid* and *broken curves* correspond to uniform and Gaussian distribution of electrons in the beam (after Misra et al. [31], curtsey authors and publishers AIP)

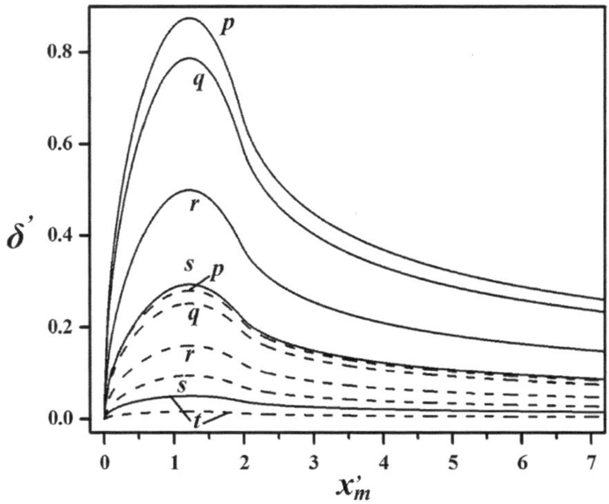

$$n_{\text{stuck}} = (1 - \eta_s) \int_0^{\rho_m} 2\pi n_p \rho \mathrm{d}\rho = (\pi \rho_m^2)(1 - \eta_s) n_p. \tag{2.60c}$$

For a Gaussian beam of primary electrons of width b, n_p is not constant and may be expressed as

$$n_p = n_{p0} \exp(-\rho^2/b^2). \tag{2.61a}$$

With this modification (2.59a) and (2.59b) are valid, while (2.60c) changes to

$$n_{\text{stuck}} = n_{p0} \pi b^2 (1 - \eta_s)[1 - \exp(-\rho_m^2/b^2)] \tag{2.61b}$$

The secondary electron emission yield may be obtained by substituting $\pi b^2[1 - \exp(-a^2/b^2)]$ instead of πa^2 in (60). Figure 2.11 indicates that the Gaussian profile of primary electron beam significantly suppresses the maximum value of the yield (δ'_m).

2.7.3 Spherical Particle in Maxwellian Plasma

Consider a spherical particle immersed in a plasma with high energy (primary) electrons having a *Maxwellian distribution* of energy. Following the well-established *Orbital Motion Limited* (*OML*) *approach,* the net electron flux incident on the dust particle can be expressed as (Sodha and Guha [46])

$$J_p = n_e (k_B T_e / 2\pi m_e)^{1/2} (1 - V_s / k_B T_e) \quad \text{for } V_s < 0 \tag{2.62a}$$

and

$$J_p = n_e(k_BT_e/2\pi m_e)^{1/2}\exp(-V_s/k_BT_e) \quad \text{for } V_s \geq 0 \tag{2.62b}$$

Following the approach similar to the earlier analyses [5, 26], the net electron flux associated with secondary electron emission is given by

$$J_s = n_e(2\pi/m_e^2)(2\pi k_BT_e/m_e)^{-3/2}\kappa\int_0^\infty \varsigma_s(V_p)(V_p - V_s)\exp[-V_p/k_BT_e]\mathrm{d}V_p \quad \text{for } V_s < 0 \tag{2.63a}$$

and

$$J_s = n_e(2\pi/m_e^2)(2\pi k_BT_e/m_e)^{-3/2}\int_{V_s}^\infty \varsigma_s(V_p)(V_p - V_s)\exp[-V_p/k_BT_e]\mathrm{d}V_p, \quad \text{for } V_s \geq 0, \tag{2.63b}$$

where

$$\varsigma_s = (1-\eta_s)(K\alpha/2)\left[\int_0^{\rho_m} 2\pi I_1(V_p,\rho)\rho\mathrm{d}\rho + \int_{\rho_m}^a 2\pi I_2(V_p,\rho)\rho\mathrm{d}\rho\right] \quad \text{for } x_m < 2a, \tag{2.64a}$$

$$\varsigma_s = (1-\eta_s)(K\alpha/2)\int_0^a 2\pi I_2(V_p,\rho)\rho\mathrm{d}\rho \quad \text{for } x_m \geq 2a, \tag{2.64b}$$

and

$$\kappa = 1 - (V_s/k_BT_e).$$

The secondary emission yield can be expressed as: $\delta = (J_s/J_p)$.

The effect of particle size and electron temperature on secondary emission yield (δ) for a particle in a Maxwellian plasma has been illustrated in Fig. 2.3a, b, for a standard set of parameters, viz. $a = 50$ nm, $k_BT_s = 3$ eV, $k_BT_e = 50$ eV, $\alpha = 10^{12}(\text{eV})^2/\text{cm}$, $\beta = 10^5/\text{cm}$, $K = 0.01/\text{eV}$ and $\mu_s = 0.1$. The figures indicate that δ linearly increases with increasing negative potential on the surface while it sharply decays with increasing positive potential. Increasing the size of the particle significantly reduces the generation of secondaries because of decreasing function $f(x, \rho)$, which leads to decrease in the secondary emission yield (δ); this nature has been displayed in Fig. 2.12. Further, the increase in the temperature of primary electrons (k_BT_e) enhances x_m and the accretion over dust surface, resulting in the increasing secondary yield (this behavior has been illustrated in Fig. 2.12).

(iii) Cylindrical Particles (see Appendix B).

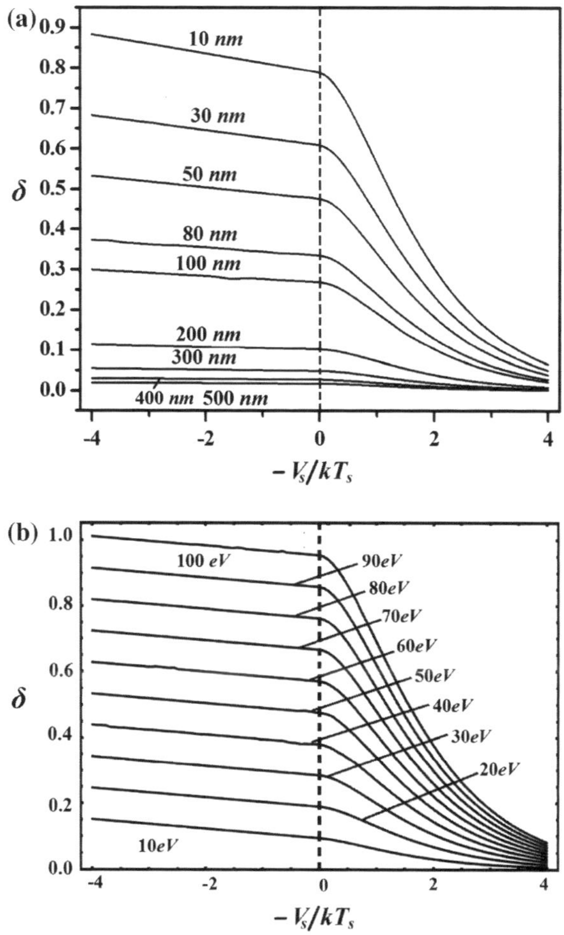

Fig. 2.12 Dependence of δ on (V_s/kT_s) for different values of **a** radius (a, in nm) of the particle and **b** mean temperature (kT_e, in eV) of the incident primary electrons for standard set of parameters in the text; the magnitude of varying parameter (kT_e) has been indicated on the *curves* (after Misra et al. [31], curtsey authors and publishers AIP)

2.8 Electron Emission from Charged Spherical and Cylindrical Surfaces of Metals

2.8.1 What is Different About Electron Emission from Curved Surfaces?

Consider an electron just outside the surface (after overcoming the surface barrier) of a metal, charged to a potential $-(V_0/e)$.

The conditions for the electron to overcome the surface potential and thus get emitted are

I. **Negatively Charged surface** (positive electron potential energy V_0 after emission of an electron): All such electrons will escape (or get emitted). Hence, the expressions for thermionic and photoelectric emission current and the mean energy, just outside the surface are identical to the case of uncharged plane surface.

II. **Positively Charged surface** (negative electron potential energy $-V_0$ after emission of an electron just outside the surface):

1. *Plane surface*:

$$E_x(= m_e u_x^2/2) > V_0, \tag{2.65a}$$

2. *Spherical surface*:

$$E(= m_e u^2/2) > V_0, \tag{2.65b}$$

3. *Cylindrical surface*:

$$[E_x(= m_e u_x^2/2) + E_y(= m_e u_y^2/2)] > V_0, \tag{2.65c}$$

where $\boldsymbol{u}$ is the velocity of the electron just outside the metal, $\boldsymbol{x}$ is the direction, normal to a surface element and z is the direction along the axis (in case 3 only).

Further since the expressions for the potential energy of an electron outside a metal, taking into account the image force are dependent on the nature of the surface, the reduction in the work function on account of a negative electric potential on the surface also depends on its nature. The expressions for the transmission coefficient also depend on the variation of electron potential energy with distance, which is determined by the shape of the surfaces.

2.8.2 Reduction of Work Function by Negative Electric Potential on a Spherical Surface

The potential energy $V(r)$ of an electron near the surface of a metallic spherical particle of radius a is given by [50]

$$V(r) = (-e^2/a)[2(r^2 - a^2) + e^2 a/W_a]^{-1} + (e^2 a/2r^2) + (Ze^2/r) \quad \text{for } r > a, \tag{2.66a}$$

and

$$V(r) = -[W_a - e^2/2a] + (Ze^2/a) \quad \text{for } r < a, \tag{2.66b}$$

where a is the radius of the particle, and $-Ze$ is the charge on the particle.

When e^2a/W_a is negligible as compared to $(r^2 - a^2)$ (2.66a) reduces to the usual expression, in text books (e.g., Page and Adams [33].

Equation (2.66b) is the usual expression, except for a surface barrier $-[W_a - e^2/2a]$ instead of $-W_a$.

Based on (2.66a), Sodha and Sharma [50] tabulated the reduction in the work function $\Delta\varphi/(e^2/a)$ as a function of the charge on the particle. An identical analysis was also made much later by [8]. Recently Sodha and Srivatsava [51] have made a similar analysis, taking into account the *Debye* shielding; thus (2.66a) gets replaced by

$$U(\rho) = -\{2(\rho^2 - 1) + (e^2/aW_a)\}^{-1} + (1/2\rho^2) + (Z/\rho)\exp[(1-\rho)/\mu] \quad \text{for } \rho > 1 \tag{2.66c}$$

and

$$U(\rho) = (Z + 1/2) - W_a/(e^2/a) \quad \text{for } \rho < 1, \tag{2.66d}$$

where $\rho = r/a$ and $U(\rho) = V(r)/(e^2/a)$.

The value of ρ corresponding to the maximum value of U viz. ρ_m can be obtained by putting by $(\mathrm{d}U/\mathrm{d}\rho) = 0$; thus

$$\mu\rho_m^4 - \mu(\rho_m^2 - 1)^2 + Z\rho_m(\rho_m^2 - 1)^2(\rho_m + \mu)\exp[(1-\rho_m)/\mu] = 0.$$

It is seen that $(\rho_m^2 - 1)$ is in general much larger than e^2/aW_a.

Hence, the maximum value of the potential energy $U_m(Z, \mu)$ is given by

$$U_m(Z, \mu) = -[1/2(\rho_m^2 - 1)] + [1/2\rho_m^2] + (Z/\rho_m)\exp[(1-\rho_m)/\mu] \tag{2.67}$$

In view of (2.66d), the energy E_F corresponding to *Fermi* level of a charged particle is given by

$$[E_F/(e^2/a)] = [E_{F0}/(e^2/a)] + (Z + 1/2) - [W_a/(e^2/a)],$$

where E_{F0} refers to an uncharged plane surface.

Hence the work function φ, corresponding to the charged particle is given by

$$\begin{aligned}[\varphi/(e^2/a)] &= [(W_a - E_{F0})/(e^2/a)] + [U_m(Z,\mu) - (Z + 1/2)] \\ \text{or } (\varphi - \varphi_0)/(e^2/a) &= [\Delta\varphi/(e^2/a)] = [U_m(Z,\mu) - (Z + 1/2)],\end{aligned} \tag{2.68}$$

where φ_0 refers to the work function of the uncharged plane surface and $\Delta\varphi(= \varphi - \varphi_0)$ is the change in work function due to the charge and curvature of the surface of the particle.

Table 2.5 illustrates the dependence of $\Delta\varphi/(e^2/a)$ on Z and μ. The *Coulomb limit* is approached when μ tends to infinity as given in Table 2.5.

Table 2.5 Dependence of $[-\Delta\varphi/(e^2/a)] on\ Z\ and\ \mu$

$Z \rightarrow$ $\mu \downarrow$	1	2	3	4	7	10	20	40	60	80	100	200	400
2	1.16	1.59	1.94	2.24	2.99	3.60	5.17	7.42	9.14	10.61	11.89	16.95	24.12
3	1.11	1.51	1.84	2.12	2.83	3.40	4.86	6.99	8.61	9.99	11.21	15.98	22.73
4	1.09	1.47	1.79	2.06	2.74	3.30	4.72	6.76	8.34	9.67	10.84	15.46	22.00
5	1.07	1.45	1.75	2.02	2.68	3.23	4.62	6.62	8.17	9.47	10.62	15.15	21.55
6	1.06	1.43	1.73	1.99	2.65	3.18	4.56	6.53	8.05	9.34	10.47	14.93	21.25
7	1.05	1.42	1.72	1.98	2.62	3.15	4.51	6.46	7.97	9.24	10.36	14.78	21.03
8	1.04	1.41	1.70	1.96	2.60	3.12	4.47	6.41	7.90	9.16	10.28	14.66	20.86
9	1.04	1.40	1.69	1.95	2.58	3.10	4.44	6.37	7.85	9.11	10.21	14.57	20.73
10	1.03	1.39	1.68	1.94	2.57	3.08	4.42	6.33	7.81	9.06	10.16	14.49	20.63
50	1.01	1.35	1.62	1.87	2.47	2.97	4.25	6.09	7.51	8.71	9.78	13.9	19.9
100	1.00	1.34	1.62	1.86	2.46	2.96	4.23	6.06	7.48	8.67	9.73	13.9	19.8
∞	1.00	1.34	1.61	1.85	2.45	2.94	4.20	6.03	7.44	8.63	9.68	13.8	19.7

After Sodha and Srivastava [23], curtsey authors and publishers Elseiver

2.8.3 Simple Theory of Electron Emission from Curved Surfaces

The theory presented in this section assumes that the transmission coefficient of an electron across the surface is unity when the normal energy is greater than W_a and is zero otherwise.

2.8.3.1 Negatively Charged Surfaces: ($V_s > 0$)

In the approximation, stated before, the rate and mean energy of emitted electrons (just outside the surface) are the same as in planar uncharged case. The mean energy far away from the surface gets enhanced by an amount equal to the potential energy of the electron at the surface. We have used the symbol V_0 for potential energy of an electron in case of a plane surface but it is convenient to use symbol V_s in case of curved surface.

2.8.3.2 Positively Charged Spherical Surfaces: ($V_s < 0$)

References [38, 39, 46] are relevant.

2.8.3.3 Thermionic Emission

The energy distribution of the number of electrons incident per unit area per unit time on the surface is given by

$$d^2 n_1 = (A_0/e)T^2 \exp(-\varepsilon_x - \varepsilon_t + \varepsilon_F) d\varepsilon_x d\varepsilon_t. \tag{2.9}$$

The energy distribution just outside the surface (after the electrons have crossed the surface barrier $w_a (= W_a/k_B T)$) is

$$\begin{aligned} d^2 n_1' &= (A_0/e)T^2 \exp(-\varepsilon_x' - \varepsilon_t - w_a + \varepsilon_F) d\varepsilon_x' d\varepsilon_t \\ &= (A_0/e)T^2 \exp(-\varphi) \exp(-\varepsilon_x' - \varepsilon_t) d\varepsilon_x' d\varepsilon_t, \end{aligned} \tag{2.69}$$

where $\varepsilon_x' = \varepsilon_x - w_a$ denotes the normal energy of an electron just outside the surface and $\varphi = (w_a - \varepsilon_F) = (W_a - E_F)/k_B T$.

If the surface has an electrical potential $-V_s/e(V_s < 0)$, only electrons having a total energy larger than $-V_s$ or $(\varepsilon_x' + \varepsilon_t) > -\upsilon_s (= -V_s/k_B T)$ can escape from the surface and get emitted; in many publications instead of $-\upsilon_s$, $Ze^2/ak_B T = Z\alpha$ is used.

Hence from (2.69) the number of electrons n_{th} emitted from the surface per unit area per unit time is given by

$$\begin{aligned} (-J_{\text{th}}/e) = n_{\text{th}} &= (A_0/e)T^2 \exp(-\varphi) \int_0^\infty \int_0^\infty \exp(-\varepsilon_x' - \varepsilon_t) d\varepsilon_x' d\varepsilon_t. \\ &\qquad (\varepsilon_x' + \varepsilon_t) > -\upsilon_s \\ &= -(A_0/e)T^2 \exp(-\varphi) \int_{-\upsilon_s}^\infty \varepsilon \exp(-\varepsilon) d\varepsilon \\ &= -(A_0/e)T^2 (1 - \upsilon_s) \exp[-(\varphi - \upsilon_s)]; \end{aligned} \tag{2.70a}$$

In writing the single integral from the double integral identity (2.37) has been used.

In many investigations the term $(1 - \upsilon_s)$ is missing, which may be responsible for unacceptably large errors.

The mean energy of electrons, just outside the surface is from (2.69), given by

$$\begin{aligned} \varepsilon_{\text{th0}} &= [(A_0/e)T^2 \exp(-\varphi)/n_{\text{th}}]. \int_0^\infty \int_0^\infty (\varepsilon_x' + \varepsilon_t) \exp(-\varepsilon_x' - \varepsilon_t) d\varepsilon_x' d\varepsilon_t \\ &\qquad (\varepsilon_x' + \varepsilon_t) > -\upsilon_s \\ &= [\exp(-\upsilon_s)/(1 - \upsilon_s)] \int_{-\upsilon_s}^\infty \varepsilon^2 \exp(-\varepsilon) d\varepsilon \\ &= -\upsilon_s + (2 - \upsilon_s)/(1 - \upsilon_s) \end{aligned}$$

Hence, the mean energy far away from the surface is given by

$$\varepsilon_{\text{th}} = \varepsilon_{\text{th0}} + \upsilon_s = (2 - \upsilon_s)/(1 - \upsilon_s) \tag{2.70b}$$

For the evaluation of the integrals, occurring in the expressions for n_{th} and ε_{th0} the identity from (2.37) has been used.

If we consider a particle of radius a with charge $(Z - 1)e$, it acquires a charge Ze after an electron gets outside the surface and hence the corresponding value of $-\upsilon_s$ is Ze^2/ak_BT.

2.8.3.4 Photoelectric Emission

From (2.39), the energy distribution of photoelectrons crossing the surface of the metal per unit area per unit time just outside the surface is given by.

$$\mathrm{d}^2 n_{\text{ph}} = (A_0 T^2/e)\beta(\nu)\Lambda(\nu)F_D(\varepsilon_x'' + \varepsilon_t - \xi)\mathrm{d}\varepsilon_x''\mathrm{d}\varepsilon_t, \tag{2.71}$$

where $\xi = \varepsilon_\upsilon - [w_a - \varepsilon_F] = (h\nu - \Phi)/k_BT$.

Of these only those electrons can escape or be emitted for which $(\varepsilon_x'' + \varepsilon_t) > -\upsilon_s$. Hence

$$\begin{aligned}(-J_{\text{ph}}/e) = n_{\text{ph}} &= (A_0/e)T^2\beta(\nu)\Lambda(\nu)\underset{(\varepsilon_x'' + \varepsilon_t) > -\upsilon_s}{\int_0^\infty\int_0^\infty} F_D(\varepsilon_x'' + \varepsilon_t - \xi)\mathrm{d}\varepsilon_x''\mathrm{d}\varepsilon_t,\\ &= (A_0/e)T^2\beta(\nu)\Lambda(\nu)I_1(-\upsilon_s, \xi)\end{aligned} \tag{2.72a}$$

and

$$\begin{aligned}\varepsilon_{\text{ph},0} &= [(A_0/e)T^2\beta(\nu)\Lambda(\nu)/n_{\text{ph}}]\underset{(\varepsilon_x'' + \varepsilon_t) > -\upsilon_s}{\int_0^\infty\int_0^\infty}(\varepsilon_x'' + \varepsilon_t - \xi)F_D(\varepsilon_x'' + \varepsilon_t - \xi)\mathrm{d}\varepsilon_x''\mathrm{d}\varepsilon_t.\\ &= (A_0/e)T^2\beta(\nu)\Lambda(\nu)I_2(-\upsilon_s, \xi)/n_{\text{ph}}\end{aligned} \tag{2.73a}$$

where using identity (2.37)

$$\begin{aligned}
I_1 &= \int_{-\upsilon_s}^{\infty} \varepsilon[1+\exp(\varepsilon-\xi)]^{-1}\mathrm{d}\varepsilon \\
&= \{-\varepsilon\ln[1+\exp(\xi-\varepsilon)]\}_{-\upsilon_s}^{\infty} + \int_{-\upsilon_s}^{\infty} \ln[1+\exp(\xi-\varepsilon)]\mathrm{d}\varepsilon, \\
&= -\upsilon_s\ln[1+\exp(\xi+\upsilon_s)] + \int_0^{\infty} \ln[1+\exp(\xi+\upsilon_s-\varepsilon'')]\mathrm{d}\varepsilon'', \\
&= -\upsilon_s\ln[1+\exp(\xi+\upsilon_s)] + \Phi(\xi+\upsilon_s),
\end{aligned} \tag{2.72b}$$

$$\begin{aligned}
\Phi(\xi+\upsilon_s) &= \int_0^{\exp(\xi+\upsilon_s)} \frac{\ln[1+\Omega]}{\Omega}d\Omega, \quad I_2 = \int_{-\upsilon_s}^{\infty} \varepsilon^2[1+\exp(\varepsilon-\xi)]^{-1}\mathrm{d}\varepsilon \\
&= \{-\varepsilon^2\ln[1+\exp(\xi-\varepsilon)]\}_{-\upsilon_s}^{\infty} + 2\int_{-\upsilon_s}^{\infty} \varepsilon\ln[1+\exp(\xi-\varepsilon)]\mathrm{d}\varepsilon \\
&= \upsilon_s^2\ln[1+\exp(\xi+\upsilon_s)] + 2\int_{-\upsilon_s}^{\infty} \varepsilon\ln[1+\exp(\xi-\varepsilon)]\mathrm{d}\varepsilon \\
&= \upsilon_s^2\ln[1+\exp(\xi+\upsilon_s)] + 2\int_0^{\infty} (-\upsilon_s+\varepsilon_1)\ln[1+\exp(\xi-\varepsilon_1+\upsilon_s)]\mathrm{d}\varepsilon_1 \\
&= \upsilon_s^2\ln[1+\exp(\xi+\upsilon_s)] - 2\upsilon_s\Phi(\xi+\upsilon_s) + 2I_3(\xi+\upsilon_s).
\end{aligned} \tag{2.73b}$$

and

$$I_3(\Omega) = \int_0^{\infty} \eta\ln[1+\exp(\Omega-\eta)]\mathrm{d}\eta. \tag{2.73c}$$

Thus, the mean energy far away from the surface is given by

$$\varepsilon_{\mathrm{ph}} = \varepsilon_{\mathrm{ph0}} + \upsilon_s = \frac{-\upsilon_s\Phi(\xi+\upsilon_s) + 2I_3(\xi+\upsilon_s)}{\Phi(\xi+\upsilon_s) - \upsilon_s\ln[1+\exp(\xi+\upsilon_s)]}.$$

If n_p denotes the rate of photoelectron emission from an uncharged surface $(-\upsilon_s = 0)$

$$[n_{\mathrm{ph}}/n_p] = I_1(\xi, -\upsilon_s)/\Phi(\xi). \tag{2.72c}$$

Many investigators have used the intuitive and erroneous relation

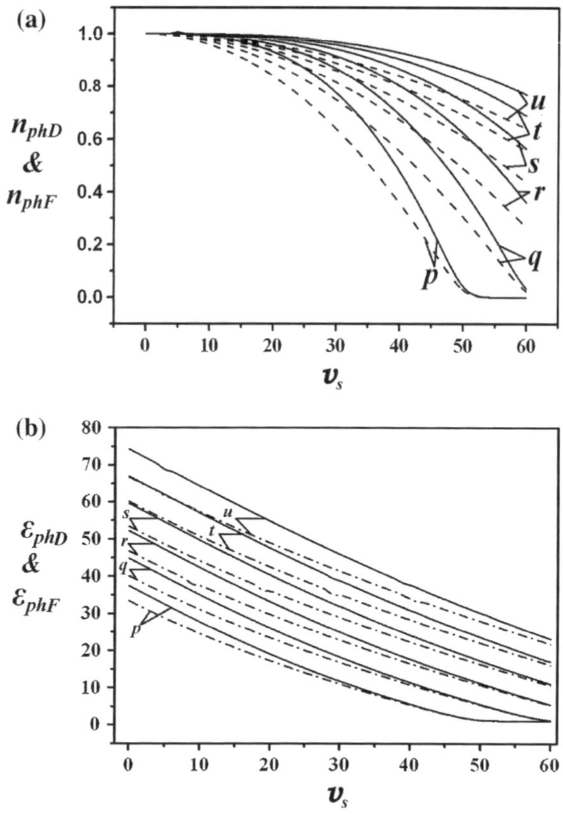

Fig. 2.13 Dependence of rate of emission (n_{phD}/n_{ph0}), (n_{phF}/n_{ph0}) and mean energy ε_{phD}, ε_{phF} of emitted of photoelectrons on the electric potential $\upsilon_s(= -eV_s/kT)$ at the surface of an emitting spherical particle; the continuous and *broken curves* refer to Du-Bridge and Fowler theories respectively, while the letters p, q, r, s, t and u refer to equal to $\xi = 50, 60, 70, 80, 90$, and 100, respectively (after Misra et al. [30], curtsey authors and publishers NRC Press)

$$(n_{ph}/n_p) = \exp(\upsilon_s). \tag{2.72d}$$

Misra and Sodha [30] have evaluated (n_{ph}/n_p) and ε_{ph} on the basis of modified DuBridge theory; the υ_s dependence of the both the parameters corresponding to Fowler's and modified DuBridge theory is shown in Fig. 2.13a, b.

It is seen from Fig. (2.14) that the disagreement between the correct (2.72c) and erroneous (2.72d) increases with increasing ξ and $-\upsilon_s$. Since for most situations of interest ξ is large, the use of (2.72d) can lead to unacceptably large errors.

In Tables 2.6a, b, c, and 2.7a, b, c n_{ph}/n_p and ε_{ph} have been tabulated as functions of ξ and $-\upsilon_s$.(after Sodha et al. [44], curtsey authors and publishers APS)

2.8.3.5 Cylindrical Surface

We consider emission from an element of the surface, normal to x direction and assume the z axis to be along the axis of the cylindrical surface; the potential energy of an electron at the surface is $-V_s$ (Sodha et al. [49]).

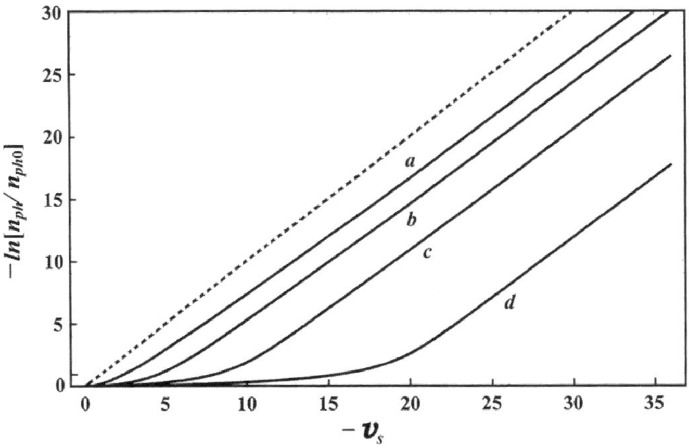

Fig. 2.14 Dependence of n_{ph}/n_p on $-\upsilon_s$ for different values of ξ. The *solid curves a, b, c, d* refer to (2.59c) and $\xi = 5, 10, 15$, and 20. The *broken straight line* refer to the erroneous (2.59d) (after Misra et al. [29], curtsey authors and publishers Springer)

2.8.3.6 Thermionic Emission

From (2.8a) and (2.8c), the number of electrons incident per unit area per unit time on an element of the surface, normal to the x direction is

$$d^3n_1 = (p_x/m_e)(2/h^3)F_D[(p^2/2m_ek_BT) - (E_F/k_BT)]dp_xdp_ydp_z. \tag{2.8d}$$

For electrons of interest to thermionic emission $p_x^2/2m_e > W_a$ and the above equation reduces to (see the argument after (2.33).

$$d^3n_1 = (p_x/m_e)(2/h^3)\exp(E_F/k_BT)\exp[-(p_x^2 + p_y^2 + p_z^2/2m_ek_BT)]dp_xdp_ydp_z \tag{2.8e}$$

If $\boldsymbol{p}'$ denotes the electron momentum just outside the surface

$$\left(p_x'^2/2m_e\right) = \left(p_x^2/2m_e\right) - W_a, p_y' = p_y \text{ and } p_z' = p_z.$$

The momentum distribution of electrons just outside the surface element is thus

$$d^3n_{\text{th}} = (p_x'/m_e)(2/h^3)\exp(-\Phi/k_BT)\exp[-(p_x'^2 + p_y'^2 + p_z'^2/2m_ek_BT)]dp_x'dp_y'dp_z' \tag{2.8f}$$

where Φ is the work function of the metal.

As discussed before (47c) only those electrons get emitted for which $\left(p_x'^2 + p_y'^2\right)/2m_e > -V_s$; thus the number of electrons, emitted per unit area from the surface can be obtained by integrating (3f), subject to this inequality. Hence,

Table 2.6 n_{ph}/n_p for $1 < \xi < 300$

$\xi \rightarrow$ $\downarrow (-\upsilon_S)$	1	2	3	4	5	6	7	8
1	1.73234	1.96651	2.30933	2.74097	3.261	3.77477	4.34342	4.93313
2	1.43231	1.56106	1.79139	2.12938	2.55559	3.04494	3.57764	4.14028
3	1.28821	1.34624	1.47161	1.69632	2.02678	2.44437	2.92464	3.44821
4	1.21417	1.23733	1.29404	1.41665	1.63662	1.9605	2.37022	2.84195
5	1.17184	1.18058	1.20332	1.25902	1.37946	1.59565	1.91414	2.31726
6	1.14474	1.14796	1.15657	1.17899	1.23388	1.35262	1.56579	1.87989
7	1.12569	1.12686	1.13004	1.13855	1.16071	1.21497	1.33234	1.54306
8	1.11136	1.11179	1.11296	1.11611	1.12453	1.14648	1.20022	1.31647
9	1.10009	1.10025	1.10067	1.10183	1.10496	1.11331	1.13509	1.1884
10	1.09094	1.091	1.09116	1.09158	1.09273	1.09583	1.10413	1.12576

$\xi \rightarrow$ $\downarrow (-\upsilon_S)$	9	10	15	20	30	40	50	60
1	5.53803	6.15411	9.33055	12.5829	19.1673	25.7922	32.4375	39.0876
2	4.72408	5.32328	8.44792	11.6733	18.2294	24.8397	31.4746	38.1173
3	4.00191	4.57712	7.62838	10.8145	17.3283	23.916	30.533	37.1653
4	3.35671	3.9016	6.86264	10.0005	16.4609	23.0191	29.617	36.2363
5	2.78163	3.2886	6.1435	9.22611	15.6242	22.1474	28.7248	35.33
6	2.27754	2.73567	5.46544	8.487	14.8159	21.2993	27.8512	34.4406
7	1.85357	2.24665	4.82446	7.7796	14.0335	20.4731	26.9949	33.5633
8	1.52518	1.8327	4.21803	7.10093	13.2752	19.6676	26.1574	32.6998
9	1.30372	1.51075	3.64525	6.4486	12.5391	18.8816	25.338	31.8541
10	1.17871	1.29325	3.1074	5.82069	11.8237	18.138	24.5343	31.0262

$\xi \rightarrow$ $\downarrow (-\upsilon_S)$	70	80	90	100	150	200	250	300
1	45.7381	52.3544	58.9834	65.8947	98.2935	131.018	166.614	198.15
2	44.7672	51.3854	58.0506	64.8368	97.2761	130.446	165.765	199.011
3	43.812	50.4415	57.1345	63.8029	96.352	129.744	164.54	199.884
4	42.8736	49.5158	56.2234	62.7946	95.4268	128.576	162.82	200.395
5	41.9518	78.5982	55.3038	61.8123	94.5055	127.428	161.362	199.966
6	41.0457	47.6824	54.369	60.8548	93.6061	126.829	160.658	197.953
7	40.154	46.7712	53.4304	59.9208	92.7189	126.483	160.325	194.689
8	39.2759	45.8728	52.4999	59.009	91.8461	125.745	159.643	191.947
9	38.4125	44.9915	51.587	58.1178	90.981	124.464	158.073	190.827
10	37.564	44.1268	50.6945	57.2456	90.1328	123.327	155.97	190.958

Table 2.7 ε_{ph} for $1 < \xi < 300$

$\xi \rightarrow$ $\downarrow (-\upsilon_S)$	1	2	3	4	5	6	7	8
1	0.839078	0.887760	0.925374	0.949896	0.965100	0.974670	0.980910	0.985150
2	0.534340	0.628570	0.727196	0.806901	0.862410	0.899220	0.92379	0.940640
3	0.283342	0.363810	0.476054	0.059689	0.699856	0.775940	0.829300	0.866660
4	0.134824	0.181770	0.261115	0.373447	0.499309	0.612023	0.699590	0.763850
5	0.060334	0.083130	0.125604	0.197883	0.303306	0.426250	0.541180	0.634240
6	0.026030	0.036170	0.05589	0.09272	0.156890	0.253600	0.370480	0.483750
7	0.010965	0.015290	0.023833	0.040439	0.072110	0.128878	0.217050	0.326920
8	0.004541	0.006340	0.009916	0.016977	0.030970	0.058365	0.108810	0.189260
9	0.001856	0.002590	0.004061	0.006977	0.012843	0.024766	0.048710	0.093860
10	0.00075	0.001050	0.001645	0.002829	0.00523	0.010168	0.020470	0.041620

$\xi \rightarrow$ $\downarrow (-\upsilon_S)$	9	10	15	20	30	40	50	60
1	0.988129	0.99033	0.995620	0.99752	0.998890	0.999400	0.999460	0.999660
2	0.952560	0.96128	0.982480	0.990080	0.995570	0.997520	0.998300	0.998880
3	0.893337	0.91290	0.960580	0.977680	0.990040	0.994400	0.996430	0.997630
4	0.810651	0.84524	0.929910	0.960330	0.982290	0.990030	0.993650	0.995700
5	0.705120	0.75848	0.890490	0.938010	0.972320	0.984420	0.989970	0.993080
6	0.578649	0.65322	0.842320	0.910730	0.960150	0.977560	0.985550	0.959920
7	0.436643	0.53124	0.785380	0.878500	0.09458	0.969450	0.980400	0.986350
8	0.292142	0.39751	0.719710	0.841310	0.929150	0.960100	0.974430	0.982320
9	0.167535	0.26383	0.645360	0.799150	0.910330	0.949490	0.967600	0.977660
10	0.082364	0.15014	0.562490	0.752040	0.889290	0.937640	0.960000	0.972290

$\xi \rightarrow$ $\downarrow (-\upsilon_S)$	70	80	90	100	150	200	250	300
1	0.999711	1.00000	0.99929	1.00100	0.999950	0.999120	1.001030	0.998600
2	0.999060	0.99944	0.99815	1.001800	0.999720	0.998130	1.002010	0.997160
3	0.998056	0.99834	0.99666	1.002380	0.999320	0.997050	1.002910	0.995680
4	0.996600	0.99684	0.99502	1.002550	0.998740	0.995870	1.003750	0.994150
5	0.994800	0.99513	0.99351	1.002340	0.997980	0.994590	1.004530	0.992570
6	0.992550	0.99337	0.99227	1.001730	0.997050	0.993210	1.005240	0.990950
7	0.989900	0.99152	0.99117	1.000720	0.995940	0.991730	1.005890	0.989290
8	0.986878	0.98937	0.98993	0.999310	0.994660	0.990150	1.006470	0.987580
9	0.983410	0.98681	0.98836	0.997510	0.993200	0.988471	1.006990	0.985830
10	0.979486	0.98382	0.98636	0.995300	0.991560	0.986710	1.007450	0.984030

$$n_{\text{th}} = (2/m_e h^3)\exp(-\varphi)\left[\int_{-\infty}^{\infty} \exp(-p_z'^2/2m_e k_B T)\mathrm{d}p_z'\right]$$
$$\times 2\left\{\left\{\int_{p_y'=0}^{p_0} \int_{p_x'=(p_0^2-p_y'^2)^{1/2}}^{\infty} p_x' \exp[-(p_x'^2+p_y'^2)/2m_e k_B T]\mathrm{d}p_x'\mathrm{d}p_y'\right\}\right\}$$
$$+\left\{\left\{\int_{p_y'=p_0}^{\infty} \int_{p_x'=0}^{\infty} p_x' \exp[-(p_x'^2+p_y'^2)/2m_e k_B T]\mathrm{d}p_x'\mathrm{d}p_y'\right\}\right\}$$
$$= (4\pi m_e k_B^2 T^2/h^3)\exp[-\varphi + \upsilon_s].\{(2/\sqrt{\pi})(-\upsilon_s)^{1/2} + \exp[-\upsilon_s]\mathrm{erfc}((-\upsilon_s)^{1/2})\} \tag{2.73e}$$

where $\varphi = (W_a - E_F)/k_B T$, $p_0^2/2m_e = -V_s$, $\upsilon_s = V_s/k_B T$ and $\mathrm{erfc}(\eta) = (2/\sqrt{\pi}) \int_{\eta}^{\infty} \exp[-t^2]\mathrm{d}t$.

The factor 2 in (2.73e) is on account of the fact that p_y' varies between $-\infty$ to $+\infty$. The corresponding mean energy of the electrons far away from the surface of the particle can be shown [43] to be given by

$$\varepsilon_{\text{th,far away}} = 2 + \frac{\upsilon_s \exp(-\upsilon_s)\mathrm{erfc}((-\upsilon_s)^{1/2})}{(2/\sqrt{\pi})(-\upsilon_s)^{1/2} + \exp(-\upsilon_s)\mathrm{erfc}((-\upsilon_s)^{1/2})} \tag{2.73f}$$

The dependence of the rate of thermionic emission $n_{\text{th}}/n_{\text{th0}}$ (subscript 0 refers to the uncharged surface) and mean electron energy ε_{th} far away from the surface on $-\upsilon_s$ is shown in Figs. 2.15 and 2.16 respectively.

It is seen that for a given value of the surface potential $-\upsilon_s$, $n_{\text{th}}/n_{\text{th0}}$ is higher for the spherical surface case, as compared to the cylindrical one. This is because the inequality (47c) allows more electrons to be emitted as compared to the case of inequality (47b).

It may also be noted that for a given value of $-\upsilon_s$ the mean energy of emitted electrons is higher in case of the cylindrical surface than in the case of spherical surface. This is because of the inequalities (47b) and (47c), which restrict three and two components of the momentum in the two cases.

2.8.3.7 Photoelectric Emission

If Λ photons of frequency ν are incident per unit area per unit time on the surface element and β is the probability of absorption of a photon, which would increase the normal energy of an electron incident on the surface, then the distribution of momenta $\boldsymbol{p}'$ of electrons incident on the surface per unit area per unit time which

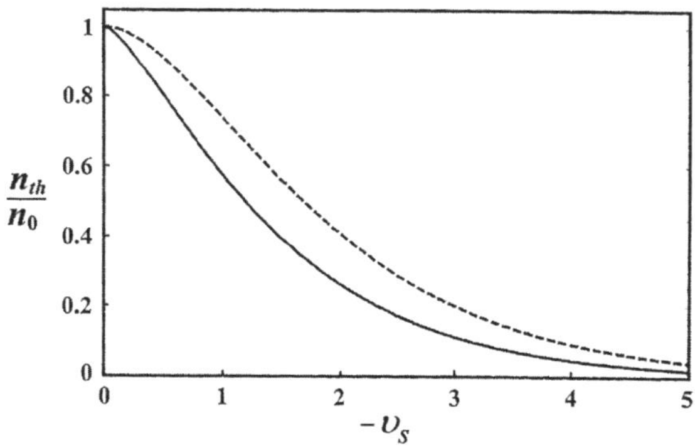

Fig. 2.15 Dependence of the rate of thermionic emissions (n_{th}/n_{th0}) on the surface potential ($-\upsilon_s$) *solid* and *broken curves* refer to cylindrical and spherical particles, respectively (after Sodha et al. [49], curtsey authors and publishers AIP)

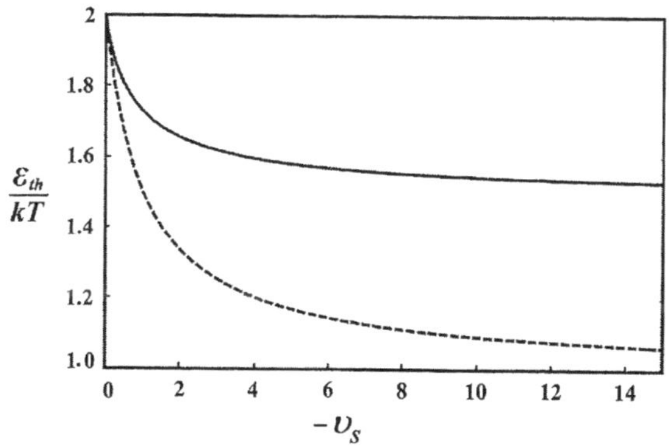

Fig. 2.16 Dependence of the mean energy of thermionically emitted electrons at a large distance from the dust particle, (ε_{th}) on the surface potential ($-\upsilon_s$); *solid* and *broken curves* corresponds to cylindrical and spherical particles respectively (after Sodha et al. [49], curtsey authors and publishers AIP)

have the normal energy enhanced by absorption of a photon can using (3d) be written down as:

$$d^3 n_1 = \beta A(2/h^3)(p'_x/m_e)F_D[(p'^2/2m_e k_B T) - (h\nu/k_B T) - E_F/k_B T]dp'_x dp'_y dp'_z,$$

where $p'^2/2m_e = p^2/2m_e + h\nu$, $p'^2_x/2m_e = p^2_x/2m_e + h\nu$, $p'_y = p_y$, $p'_z = p_z$.

Hence, the momentum $\boldsymbol{p}''$ distribution of the electrons, outside the surface is given by

$$\begin{aligned} d^3 n_{ph} &= \beta A\left(2/h^3\right)\left(p''_x/m_e\right)F_D\left[\left(p''^2/2m_e k_B T\right) + \left(W_a/k_B T\right) - \left(h\nu/k_B T\right) - E_F/k_B T\right]dp''_x dp''_y dp''_z \\ &= \beta A\left(2/h^3\right)\left(p''_x/m_e\right)F_D\left[\left(p''^2/2m_e k_B T\right) - \xi\right]dp''_x dp''_y dp''_z, \end{aligned} \tag{2.74}$$

where $p''^2_x = p'^2_x - W_a$, $p''_y = p_y$, $p''_z = p_z$ and $\xi = (h\nu - \Phi)/k_B T$.

Proceeding as in the case of thermionic emission, the number of emitted photoelectrons per unit area per unit time from the surface and corresponding mean energy far away from the particle is given by

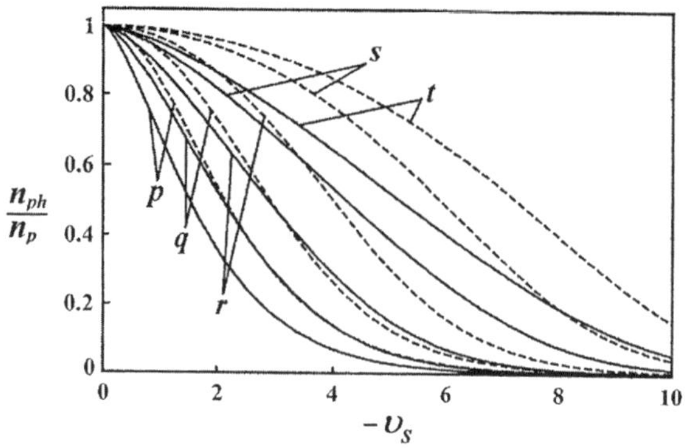

Fig. 2.17 Dependence of photoelectric emission rate of electrons (n_{ph}/n_p) on the surface potential ($-\upsilon_s$). The letters p, q, r, s and t on the *curves* refer to $\xi = 1, 3, 5, 8$ and 10 respectively; *solid* and *broken curves* correspond to cylindrical and spherical particles, respectively (after Sodha et al. [49], curtsey authors and publishers AIP)

$$n_{\text{ph}} = \iiint \mathrm{d}^3 n_{\text{ph}} = \beta \Lambda (8\, m_e k_B^2 T^2/h^3) \times \{((-\upsilon_s)^{1/2}) \int\limits_{\xi_3=0}^{\infty} (\xi_3)^{-1/2} \times \ln\{1 + \exp[\xi + \upsilon_s - \xi_3]\}\mathrm{d}\xi_3$$
$$+ \frac{1}{2} \int\limits_{\xi_3=0}^{\infty} \int\limits_{\xi_2=-\upsilon_s}^{\infty} (\xi_2 \xi_3)^{-1/2} \times \ln\{1 + \exp[\xi - \xi_2 - \xi_3]\}\mathrm{d}\xi_2 \mathrm{d}\xi_3]\}, \tag{2.75a}$$

where

$$\xi_2 = p_x''^2/2m_e k_B T \quad \text{and} \quad \xi_3 = p_z''^2/2m_e k_B T. \tag{2.75b}$$

$$\frac{\varepsilon_{\text{ph}}^l}{k_B T} = \frac{\varepsilon_{\text{ph}}}{k_B T} + \upsilon_s, \tag{2.75c}$$

and

$$n_{\text{ph}}(\varepsilon_{\text{ph}}/k_B T) = \beta \Lambda (8 m_e k_B^2 T^2/h^3) \times \{(-\upsilon_s)^{1/2} \int\limits_{\xi_3=0}^{\infty} (\xi_3)^{-1/2}[\xi_3 - \upsilon_s] \times \ln\{1 + \exp[-\{-\upsilon_s + \xi_3 - \xi\}]\}\mathrm{d}\xi_3$$
$$+ (1/2) \int\limits_{\xi_3=0}^{\infty} \int\limits_{\xi_2=0}^{e\upsilon} \int\limits_{\xi_1=-\upsilon_s-\xi_2}^{\infty} (\xi_2\xi_3)^{-1/2} \times \ln\{1 + \exp[-(\xi_1 + \xi_2 + \xi_3 - \xi)]\}\mathrm{d}\xi_1 \mathrm{d}\xi_2 \mathrm{d}\xi_3$$
$$+ (1/2) \int\limits_{\xi_3=0}^{\infty} \int\limits_{\xi_2=-\upsilon_s}^{\infty} (\xi_2\xi_3)^{-1/2}(\xi_2 + \xi_3) \times \ln\{1 + \exp[-(\xi_2 + \xi_3 - \xi)]\}\mathrm{d}\xi_2 \mathrm{d}\xi_3$$
$$+ (1/2) \int\limits_{\xi_3=0}^{\infty} \int\limits_{\xi_2=-\upsilon_s}^{\infty} \int\limits_{\xi_1=0}^{\infty} (\xi_2\xi_3)^{-1/2} \times \ln\{1 + \exp[-(\xi_1 + \xi_2 + \xi_3 - \xi)]\}\mathrm{d}\xi_1 \mathrm{d}\xi_2 \mathrm{d}\xi_3\} \tag{2.76}$$

The dependence of the rate of photoelectric emission and the mean energy of photoelectrons far away from the surface on the surface potential is illustrated in Figs. 2.17 and 2.18.

(i) at $\rho = 1$,

$$A + B = C_1 Ai[\eta_1] + D_1 Bi[\eta_1] \tag{2.83a}$$

and

$$ik_1(A - B) = f_1^{(1/3)}(C_1 Ai'[\eta_1] + D_1 Bi'[\eta_1]); \tag{2.83b}$$

(ii) at $\rho = 1 + (q + 1)\delta$ where $(0 \leq q \leq (n - 1))$,

$$C_q Ai[\eta_q] + D_q Bi[\eta_q] = C_{q+1} Ai[\eta_{q+1}] + D_{q+1} Bi[\eta_{q+1}] \tag{2.83c}$$

and

$$f_q^{1/3}(C_q Ai'[\eta_q] + D_q Bi'[\eta_q]) = f_{q+1}^{1/3}(C_{q+1} Ai'[\eta_{q+1}] + D_{q+1} Bi'[\eta_{q+1}]). \tag{2.83d}$$

(iii) at $\rho = \rho_n (= 1 + n\delta)$,

$$C_{\rho_n} Ai[\eta_{\rho_n}] + D_{\rho_n} Bi[\eta_{\rho_n}] = E + F, \tag{2.83e}$$

and

$$f_{\rho_n}^{1/3}(C_{\rho_n} Ai'[\eta_{\rho_n}] + D_{\rho_n} Bi'[\eta_{\rho_n}]) = ik_3(E - F). \tag{2.83f}$$

Thus, the transmission coefficients, corresponding to emission/accretion of electrons from/on a spherically negatively charged particle can be evaluated from (2.82a) and (2.82b), making use of (2.83a) to (2.83f) to obtain A, B, E, F.

In this section, the transmission coefficient for an electron having an arbitrary energy has been evaluated. This is in contrast to earlier work, applicable to electron energy e_r, which is less than half of the surface energy barrier.

For a better understanding of underlying physics and numerical appreciation of the results, three cases viz. (i) a spherical metallic particle at a temperature $T = 1500$ K, (thermionic and field emission) (ii) spherical metallic particle at a temperature $T = 300$ K, illuminated by a continuous source of monochromatic radiation causing photoemission, characterized by a parameter ξ and light induced field emission and (iii) accretion (classical as well as tunneling) of electrons at the surface of the spherical metallic particle, have been considered. Corresponding dependence of $D_e(e_r)$ and $D_a(e_r)$ on the radial energy of electrons e_r, potential energy at the surface and height of the surface energy barrier has been graphically illustrated. For computational purpose, the following standard sets of parameters have been used.

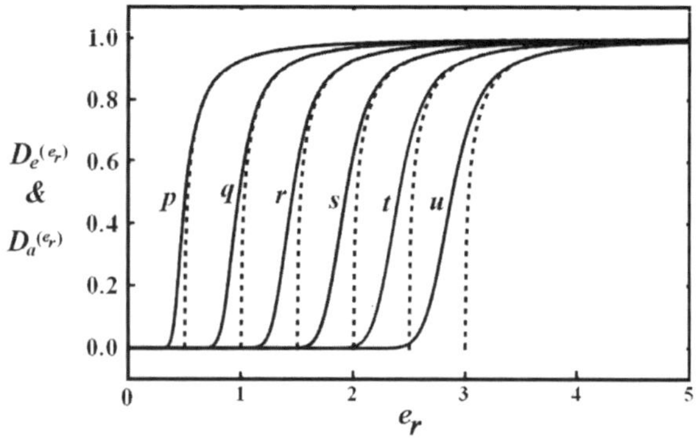

Fig. 2.19 Dependence of transmission coefficient [$D_e(e_r)$ or $D_a(e_r)$] on e_r for $\mu_d = (\lambda_D/a) = 5$ and $\beta = 1{,}000$. The labels on the curves *p, q, r, s* and *t* correspond to $\upsilon_s = 0.5, 1.0, 1.5, 2.0$ and 3.0 respectively. The *solid curves* correspond to the present analysis while the *dashed curves* refer to the step potential barrier of height $(1 - \upsilon_s)$ (after Mishra et al. [28], curtsey authors and publishers AIP)

2.8.4.6 For the Evaluation of Transmission Coefficient

$\upsilon_s = 0.5$, $\mu_d = (\lambda_D/a) = 5$ and $\beta = 1{,}000$.

2.8.4.7 For the Evaluation of Electron Currents

Case-I: $a = 10$ nm, $W_a = 10$ eV, $\mu_d = 5$ with $\phi = 5$ eV and $T = 1500$ K.
Case-II: $a = 10$ nm, $W_a = 10$ eV, $\mu_d = 5$ with $T = 300$ K.
Case-III: $a = 10$ nm, $W_a = 10$ eV, $\mu_d = 5$.

Parametric dependences have been investigated by varying one of these parameters, keeping others unchanged.

The set of Fig. 2.19 illustrates the dependence of the transmission coefficient [$D_e(e_r)$ and $D_a(e_r)$] on the radial energy of electrons e_r and e'_r corresponding to far away from and inside the particle as a function of υ_s (surface potential) (Figs. 2.19, 2.20), parameter β (Fig. 2.21) and *Debye* length (μ_d) (Fig. 2.22).

It is seen from Fig. 2.19 that the transmission coefficient [i.e., $D(e_r)$] increases monotonically with increasing e_r and υ_s, and approaches unity asymptotically for large e_r. In the curves the region $e_r < \upsilon_s$ is indicative of the contribution to the electron emission through tunneling (electric field emission) while the rest (i.e., $e_r > \upsilon_s$) corresponds to classically allowed emission; the broken curves correspond to transmission coefficient for step potential barrier [$D_0(e_r)$] corresponding to υ_s (when W_a is substituted by $(W_a - V_s)$). The transmission coefficient corresponding to irradiated metallic particles (causing photoemission) may be obtained by shifting e_r-*axis* by $e_\nu(= h\nu)$. Further it is seen that the numerical values of the transmission coefficient, corresponding to accretion [$D_a(e_r)$] and emission [$D_e(e_r)$] of electrons are very close to each other and almost indistinguishable on the same graph.

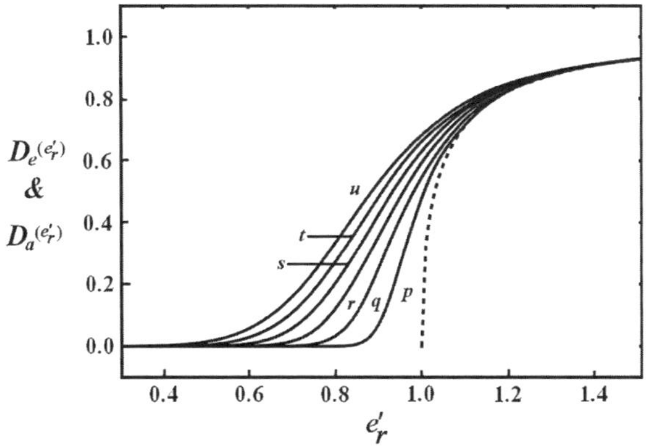

Fig. 2.20 Dependence of transmission coefficient $\left[D_e\left(e'_r\right) \text{ or} D_a\left(e'_r\right)\right]$ on e'_r for $\mu_d = (\lambda_D/a) = 5$ and $\beta = 1{,}000$. The labels on the *curves p, q, r, s* and *t* correspond to $\upsilon_s = 0.5$, 1.0, 1.5, 2.0 and 3.0 respectively. The *solid curves* correspond to the present analysis while the *dashed curves* refer to the step potential barrier of height $(1 - \upsilon_s)$ (after Mishra et al. [28], curtsey authors and publishers AIP)

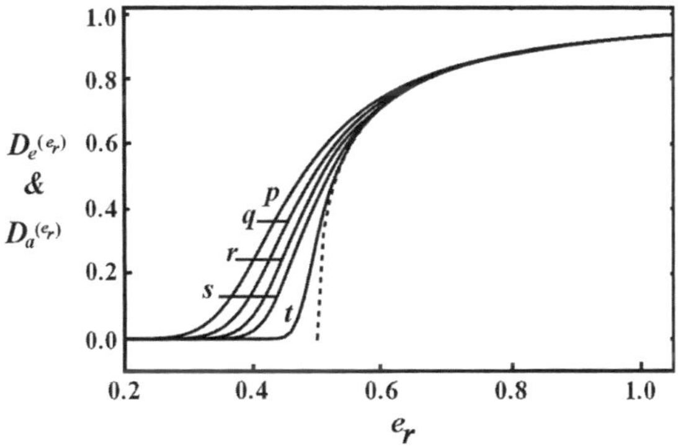

Fig. 2.21 b Dependence of transmission coefficient $[D_e(e_r)$ or $D_a(e_r)]$ on e_r for $\upsilon_s = 0.5$, and $\mu_d = (\lambda_D/a) = 5$. The labels on the curves *p, q, r, s*, and *t* correspond to $\beta = 250$, 500, 1,000, 2,000 and 26,000 respectively. The *solid curves* correspond to the present analysis while the *dashed curves* refer to the step potential barrier of height $(1 - \upsilon_s)$ (after Mishra et al. [28], curtsey authors and publishers AIP)

The graphical representation for transmission of the set of Fig. 2.19 is well applicable to the latter two cases viz. irradiated particles and accretion. The fact that $D_e(e_r) \approx D_a(e_r)$ is the necessary and sufficient condition [48] for the validity of *Saha's equation* in thermal equilibrium of a system of electron emitting dust and electrons accreting on the surface of the particles. Since *Saha's equation* is based on statistical thermodynamics, without regard to process details, this result is a source of satisfaction. By the way this result relaxes the previous condition [48] for

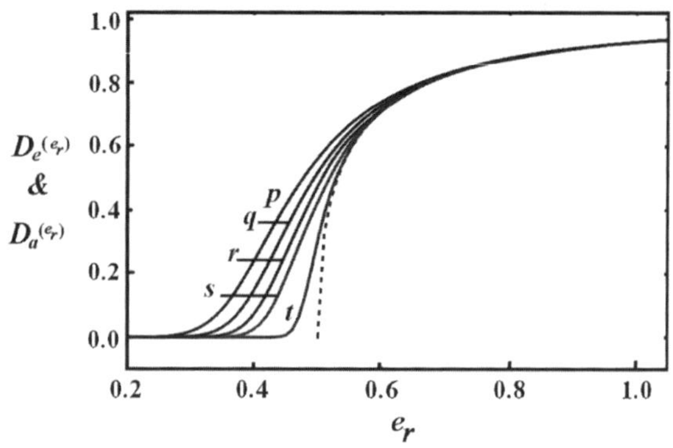

Fig. 2.22 Dependence of transmission coefficient $[D_e(e_r)$ or $D_a(e_r)]$ on e_r for $\upsilon_s = 0.5$ and $\beta = 1{,}000$. The labels on the curves p, q, r, s and t correspond to $\mu_d = 5$, 10, 20, 50 and 100 respectively. The *solid curves* correspond to the present analysis while the *dashed curves* refer to the step potential barrier of height $(1 - \upsilon_s)$ (after Mishra et al. [28], curtsey authors and publishers AIP)

the validity of *Saha's equation* viz the applicability of *Born's approximation*. In what follows, the discussion is in terms of $D_e(e_r)$ and $D_a(e_r)$.

It is interesting to notice that contribution of the field emission and departure of $D(e_r)$ from $D_0(e_r)$ increases with increasing υ_s while the converse is true with increasing $\beta(a\ \&\ W_a)$ and μ_d; the nature of dependence may be understood in terms of changing barrier width and has been displayed in Figs. 2.21 and 2.22. This behavior also underlines the fact that the contributions of pure thermionic or photoelectric emission currents get significantly enhanced with increasing υ_s.

2.8.5 Electron Emission

2.8.5.1 Electron Emission Current

The number of electrons incident per unit area per unit time on an element of surface normal to x (as explained earlier) is

$$d^2 n_1 = (A_0 T^2/e) F_D[(\varepsilon'_x + \varepsilon'_t - \varepsilon'_f)] d\varepsilon'_x d\varepsilon'_t. \tag{2.9a}$$

where $(A_0/e) = (4\pi\ m_e k_B^2/h^3)$ and $A_0 = 120\ \text{A/cm}^2\text{K}^2$.

Equation (2.9) is valid for all elements of a curved surface when the x direction is normal to it. For a spherical surface, it is convenient to replace ε'_x by ε'_r where r refers to radial. Thus,

$$d^2 n_1 = (A_0 T^2/e) F_D[(\varepsilon'_r + \varepsilon'_t - \varepsilon'_f)] de'_r de'_t. \tag{2.9b}$$

It is important to remember that integration over ε'_t is equivalent to sum over l the orbital quantum number.

2.8.5.2 Dark Electron Emission Current (Thermionic and Electric Field Emission)

From (2.9b), the electron emission current n_t per unit area is given by

$$n_t = \int\int D_e(e_r)(A_0T^2/e)F_D[(\varepsilon'_r + \varepsilon'_t - \varepsilon_f)]\mathrm{d}\varepsilon'_r\mathrm{d}\varepsilon'_t$$

and using (2.67)

$$n_t = \int\int D_e(\varepsilon'_r + \upsilon_s - 1)(A_0T^2/e)F_D[(\varepsilon'_r + \varepsilon'_t - \varepsilon_f)]\mathrm{d}\varepsilon'_r\mathrm{d}\varepsilon'_t. \quad (2.84a)$$

The lower limit for e'_r is given by $e_r = 0$ and hence it is $e'_{rm} = 1 - \upsilon_s = 1 - (V_s/W_a)$ for $(V_s < W_a)$ and zero for $(V_s > W_a)$.

Hence, the electron emission current per unit area of the surface is given by

$$n_t = n_{\mathrm{th}} + n_{\mathrm{fe}},$$

where

$$n_{\mathrm{th}} = \left(A_0T^2/e\right)\int_g^\infty\int_0^\infty D_e\left(e'_r + \upsilon_s - 1\right)F_D[(\varepsilon'_r + \varepsilon'_t - \varepsilon_f)]\mathrm{d}\varepsilon'_r\mathrm{d}\varepsilon'_t \quad (2.84b)$$

and

$$n_{\mathrm{fe}} = (A_0T^2/e)\int_{ge'_{rm}}^{g}\int_0^\infty D_e(e'_r + \upsilon_s - 1)F_D[(\varepsilon'_r + \varepsilon'_t - \varepsilon_f)]\mathrm{d}\varepsilon'_r\mathrm{d}\varepsilon'_t, \quad (2.84c)$$

where n_{th} and n_{fe} correspond to thermionic $(E'_r > W_a)$ and field emission $(E'_r < W_a)$ and $g = (W_a/k_BT)$.

Substituting $\varepsilon'_r = ge'_r$, $\varepsilon'_t = ge'_t$ and $\varepsilon_f = ge_f$ in the above two equations one obtains

$$n_{\mathrm{th}} = (A_0/e)T^2g^2\int_1^\infty\int_0^\infty D_e(e'_r + \upsilon_s - 1)F_D[g(e'_r + e'_t - e_f)]\mathrm{d}e'_r\mathrm{d}e'_t \quad (2.85a)$$

$$n_{\mathrm{fe}} = (A_0/e)T^2g^2 \int_{e'_{rm}}^{1} \int_{0}^{\infty} D_e(e'_r + \upsilon_s - 1)F_D[g(e'_r + e'_t - e_f)]\mathrm{d}e'_r\mathrm{d}e'_t. \tag{2.85b}$$

In case of thermionic emission one is interested only in electrons having an energy higher that W_a, which itself is much higher than the Fermi energy ($>> k_BT$). Hence $\gamma = g(e'_x + e'_t - e'_f) >> 1$ and $F_D(\gamma) = \exp(-\gamma)$.

Thus (2.72a) simplifies to

$$\begin{aligned} n_{\mathrm{th}} &= (A_0/e)T^2g^2 \exp(ge_f) \int_{0}^{\infty} \exp(-ge'_t)\mathrm{d}\varepsilon'_t \int_{1}^{\infty} D_e(e'_r + \upsilon_s - 1)\exp(-ge'_r)\mathrm{d}e'_r \\ &= (A_0/e)T^2g \exp(ge_f) \int_{1}^{\infty} D_e(e'_r + \upsilon_s - 1)\exp(-ge'_r)\mathrm{d}e'_r. \end{aligned} \tag{2.86}$$

If it is assumed as in many books and papers that $D_e = 1$, the above equation reduces to

$$n_{\mathrm{th}0} = (A_0/e)T^2 \exp(-g + ge_f) = (A_0/e)T^2 \exp(-\phi/k_BT) \tag{2.87}$$

where $\Phi = W_a - E_F$ is the work function.

Equation (2.85b) can also be simplified by integrating with respect to e'_t. Thus,

$$n_{\mathrm{fe}} = (A_0/e)T^2g \int_{e'_{rm}}^{1} D_e(e'_r + \upsilon_s - 1)\ln[1 + \exp[-g(e'_r - e_f)]]\mathrm{d}e'_r \tag{2.88}$$

The mean energy of emitted electron (inside the metal) e'_{th} and e'_{fe} are given by

$$e'_{\mathrm{th}} = (A_0/en_{\mathrm{th}})T^2g^2 \textstyle\int_1^\infty \int_0^\infty (e'_r + e'_t)D_e(e'_r + \upsilon_s - 1)F_D\left[g(e'_r + e'_t - e_f)\right]\mathrm{d}e'_r\mathrm{d}e'_t \tag{2.89}$$

$$e'_{\mathrm{fe}} = (A_0/en_{\mathrm{fe}})T^2g^2 \int_{e'_{rm}}^{1} \int_{0}^{\infty} (e'_r + e'_t)D_e(e'_r + \upsilon_s - 1)F_D[g(e'_r + e'_t - e_f)]\mathrm{d}e'_r\mathrm{d}e'_t \tag{2.90}$$

The mean energy of electrons far away from the particle is using (2.80) given by $e_{\mathrm{th}} = e'_{\mathrm{th}} + 1 - \upsilon_s$ and $e_{\mathrm{fe}} = e'_{\mathrm{fe}} + 1 - \upsilon_s$.

2.8.5.3 Photoelectric Current

(Photoelectric effect and light-Induced field emission)

The most widely used quantitative theory of photoelectric emission was formulated by [13], who assumed that the normal energy of a fraction $\beta(v)\Lambda(v)\mathrm{d}v$ of the electrons incident on the surface from inside gets enhanced by an amount (hv), when $\Lambda(v)\mathrm{d}v$ photons of frequency between (v) and $(v + \mathrm{d}v)$ are incident on the surface per unit area per unit time. Hence, the energy distribution of photoelectrons [whose normal energy has been enhanced by (hv)], incident per unit area per unit time from the inside is, using (2.9b), given by

$$\mathrm{d}^2 n_{\mathrm{ph}} = \beta(v)\Lambda(v)\mathrm{d}v(A_0 T^2/e)F_D[(\varepsilon_r'' + \varepsilon_t'' - \varepsilon_v - \varepsilon_f)]\mathrm{d}\varepsilon_r''\mathrm{d}\varepsilon_t'', \tag{2.39}$$

where $\varepsilon_t'' = \varepsilon_t'$, $\varepsilon_r'' = \varepsilon_r' + (hv/k_B T)$ is the enhanced (after absorption of a photon) dimensionless radial energy and $e_v = hv/k_B T$.

Proceeding in the same way, as in the evaluation of dark electron current the total photoelectric current is using (2.79b), given by

$$n_{\mathrm{phT}} = n_{\mathrm{ph}}(v) + n_{\mathrm{life}}(v), \tag{2.91a}$$

where

$$n_{\mathrm{ph}}(v) = \int_{v_1}^{v_2} \beta(v)\Lambda(v)\mathrm{d}v(A_0/e)T^2 g \int_1^{\infty} D_e(e_r'' + \upsilon_s - 1)\ln[1 + \exp[-g(e_r'' - e_v - e_f)]]\mathrm{d}e_r'' \tag{2.91b}$$

and

$$\begin{aligned} n_{\mathrm{life}}(v) = &\int_{v_1}^{v_2} \beta(v)\Lambda(v)dv(A_0/e)T^2 g \int_{\varepsilon_{rm}''}^{1} D_e(e_r'' + \upsilon_s - 1) \\ &\times \ln[1 + \exp[-g(e_r'' - e_v - e_f)]]\mathrm{d}e_r'' \end{aligned} \tag{2.91c}$$

In case $D_e(e_r'') = 1$, (2.90) reduces to (2.41b) viz.

$$n_{\mathrm{ph0}} = (A_0/e)T^2[\beta(v)\Lambda(v)]\Phi(\xi).$$

The corresponding mean energy of the emitted electrons while inside the particle is given by

$$\begin{aligned} n_{\mathrm{ph}}\varepsilon_{\mathrm{ph}}'' = &\int_{v_1}^{v_2} \beta(v)\Lambda(v)\mathrm{d}v(A_0/e)T^2 g \int_1^{\infty}\int_0^{\infty} (e_r'' + e_t'')D_e(e_r'' + \upsilon_s - 1) \\ &(1 + \exp[g(e_r'' + e_t'' - e_v - e_f)])^{-1}\mathrm{d}e_r''\mathrm{d}e_t \end{aligned} \tag{2.91d}$$

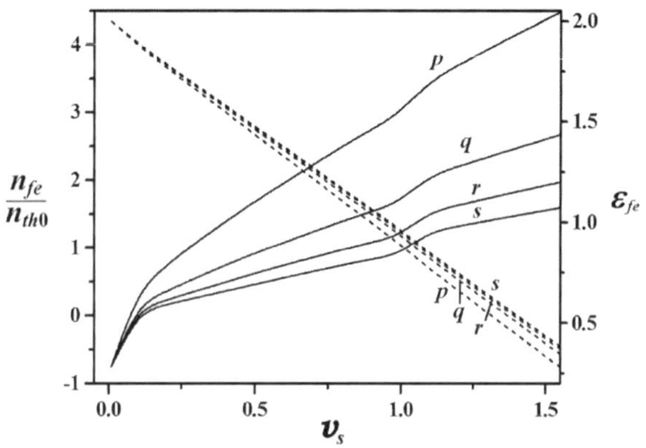

Fig. 2.23 Dependence of (n_{fe}/n_{th0}) and ε_{fe} with dimensionless surface potential (υ_s), for Case-I $a = 10$ nm, $W_a = 10$ eV, $\mu_d = 5$ and $\Phi = 5$ eV. The labels on the curves p, q, r and s correspond to $T = 800$, 1000, 1500 K and 2000 K respectively. *Solid* and *broken lines* correspond to left and right hand scale respectively (after Mishra et al. [28], curtsey authors and publishers AIP)

$$n_{\text{life}}\varepsilon''_{\text{life}} = \int\limits_{v_1}^{v_2} \beta(v)\Lambda(v)\mathrm{d}v(A_0/e)T^2 g \int\limits_{\varepsilon''_{rm}}^{1}\int\limits_{0}^{\infty} (e''_r + e''_t)D_e(e''_r + \upsilon_s - 1)$$
$$(1 + \exp[g(e''_r + e''_t - e_v - e_f)])^{-1}\mathrm{d}e''_r \mathrm{d}e_t \qquad (2.91e)$$

The mean energy of electrons far away from the particle is using (2.80) given by $\varepsilon_{\text{ph}} = \varepsilon''_{\text{ph}} + 1 - \upsilon_s$ and $\varepsilon_{\text{life}} = \varepsilon''_{\text{life}} + 1 - \upsilon_s$.

2.8.5.4 Numerical Results and Discussion

The dependence of the electric field (n_{fe}/n_{th0}) and thermionic (n_{th}/n_{th0}) emission currents and associated respective mean energy (ε_{fe} and ε_{th}, far away from the particle surface) on the dimensionless surface potential (υ_s) has been illustrated in the set of Fig. 2.23.

The figures indicate that the field emission currents are strongly influenced by surface potential and can largely contribute to the emission current (Figs. 2.23, 2.25) while thermionic emission also increases monotonically with υ_s but with much slower rate than field emission (Figs. 2.24, 2.25 2.26). This behavior may be understood in terms of the availability of electrons for field emission inside metallic particles which is more than the availability of high energy electrons, corresponding to thermionic emission. The Figs. (2.23, 2.24) also display the fact that the thermionic emission current increases with increasing temperature of the spherical particle while field emission current displays the opposite trend; this may be ascribed to the large availability of electrons at high temperature for thermionic emission.

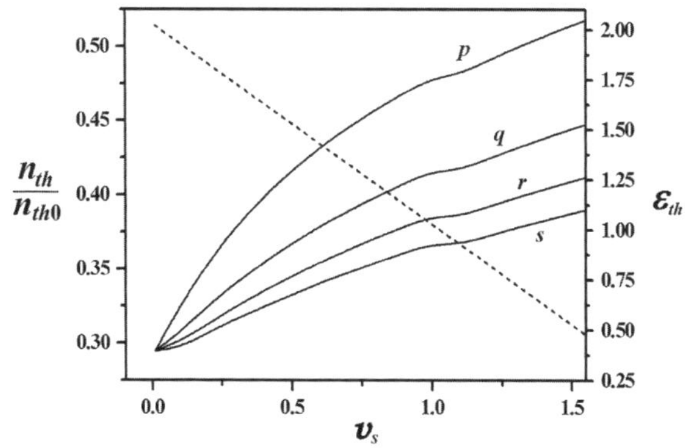

Fig. 2.24 Dependence of (n_{th}/n_{th0}) and ε_{th} with dimensionless surface potential (υ_s), for Case-I $a = 10$ nm, $W_a = 10$ eV, $\mu_d = 5$ and $\Phi = 5$ eV. The labels on the *curves* *p, q, r* and *s* correspond to $T = 800$, 1000, 1500 and 2000 K respectively. *Solid* and *broken lines* correspond to *left* and *right hand* scale respectively (after Mishra et al. [28], curtsey authors and publishers AIP)

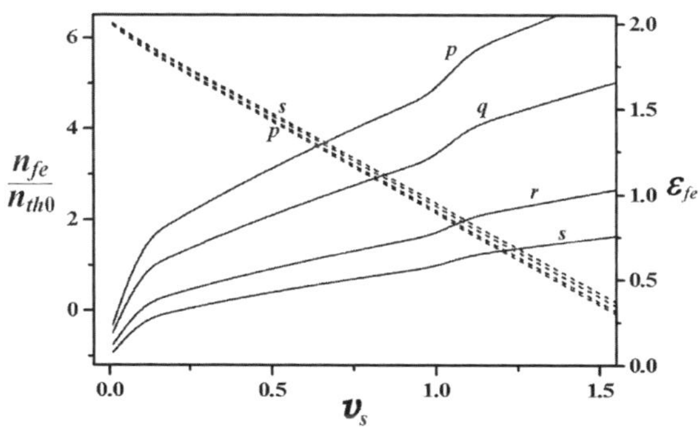

Fig. 2.25 Dependence of (n_{fe}/n_{th0}) and ε_{fe} with dimensionless surface potential (υ_s), for Case-I, $T = 1500$ K, $W_a = 10$ eV, $\mu_d = 5$ and $\Phi = 5$ eV. The labels on the *curves* *p, q, r* and *s* correspond to $a = 5$, 10, 15 and 20 nm respectively. *Solid* and *broken lines* correspond to *left* and *right hand* scale respectively (after Mishra et al. [28], curtsey authors and publishers AIP)

Further, the size dependence has been illustrated in Figs. (2.25, 2.26) and can be understood in terms of $D(\varepsilon_r)$ dependence on β which ensures the large contribution from field emission with smaller size (larger electric field for given (υ_s)). Correspondingly the mean energy decreases with (υ_s) on account of larger emission of low energy electrons with increasing surface potential (υ_s); $\varepsilon_{fe}/\varepsilon_{th}$ display very weak dependence on size and temperature of the metallic particle.

The electron emission from spherical metallic particles in the presence of monochromatic radiation resulting in photoemission has been illustrated in Figs. 2.27, 2.28, 2.29. Figures 2.27 displays the dependence of light-induced field emission (n_{life}/n_{ph0}) and corresponding mean energy ε_{life} on surface potential (υ_s)as

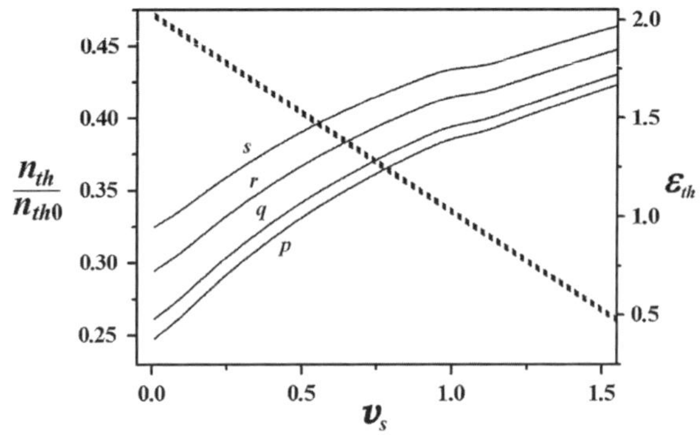

Fig. 2.26 Dependence of $(n_{\mathrm{th}}/n_{\mathrm{th0}})$ and $\varepsilon_{\mathrm{th}}$ with dimensionless surface potential (υ_s), for Case-I, $T = 1500$ K, $W_a = 10$ eV, $\mu_d = 5$ and $\Phi = 5$ eV. The labels on the curves p, q, r and s correspond to $a = 5$, 10, 15 and 20 nm respectively. *Solid* and *broken lines* correspond to *left* and *right hand* scale respectively (after Mishra et al. [28], curtsey authors and publishers AIP)

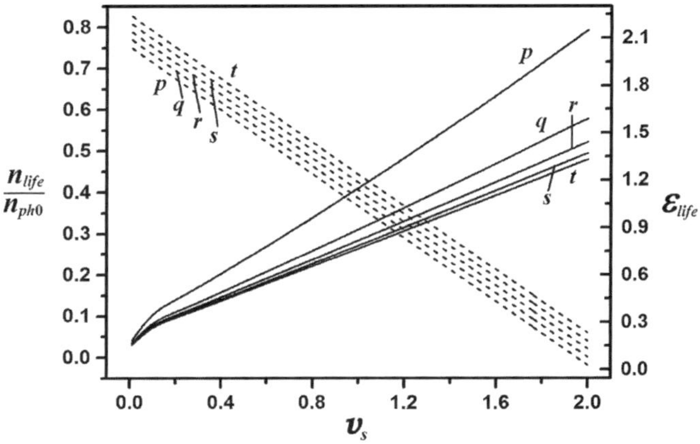

Fig. 2.27 Dependence of $(n_{\mathrm{life}}/n_{\mathrm{ph0}})$ and $\varepsilon_{\mathrm{life}}$ with dimensionless surface potential (υ_s), for Case-II, $a = 10$ nm, $W_a = 10$ eV, $\mu_d = 5$ with $T = 300$ K. The labels on the *curves* p, q, r, s, and t correspond to $\xi = 0.1$, 0.2, 0.3, 0.4, and 0.5, respectively. *Solid* and *broken lines* correspond to *left* and *right hand* scale, respectively (after Mishra et al. [28], curtsey authors and publishers AIP)

a function of the parameter $\xi[= g(\varepsilon_v - \varphi)]$, which usually characterizes the energy of the incident radiation and work function of the material. The figure suggests that $n_{\mathrm{life}}/n_{\mathrm{ph0}}$ decreases with increasing ξ; this behavior may be explained on the basis of the large availability of low energy electrons for the electric field emission. On the other hand, large availability of high energy electrons for photoemission (Fig. 2.28) explains the reverse trend for the ξ dependence of $n_{\mathrm{ph}}/n_{\mathrm{ph0}}$. Further the mean energy in both the cases increases with increasing ξ and may be understood in terms of emission of high energy electrons with increasing ξ. The

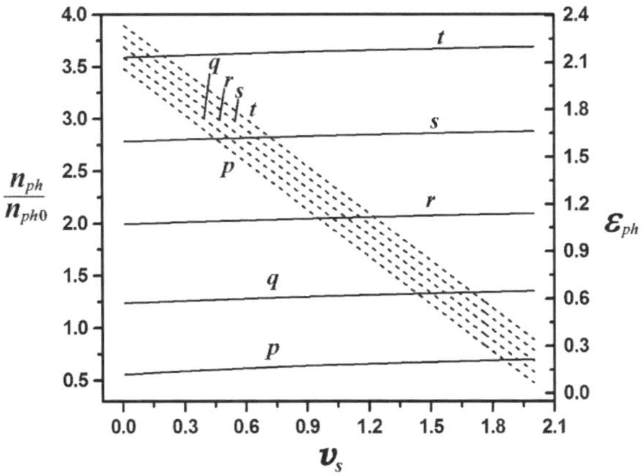

Fig. 2.28 Dependence of $(n_{\text{ph}}/n_{\text{ph0}})$ and ε_{ph} with dimensionless surface potential (υ_s), for Case-II, $a = 10$ nm, $W_a = 10$ eV, $\mu_d = 5$ with $T = 300$ K. The labels on the *curves p, q, r, s,* and *t* correspond to $\xi = 0.1$, 0.2, 0.3, 0.4, and 0.5, respectively. *Solid* and *broken lines* correspond to *left* and *right hand* scale respectively (after Mishra et al. [28], curtsey authors and publishers AIP)

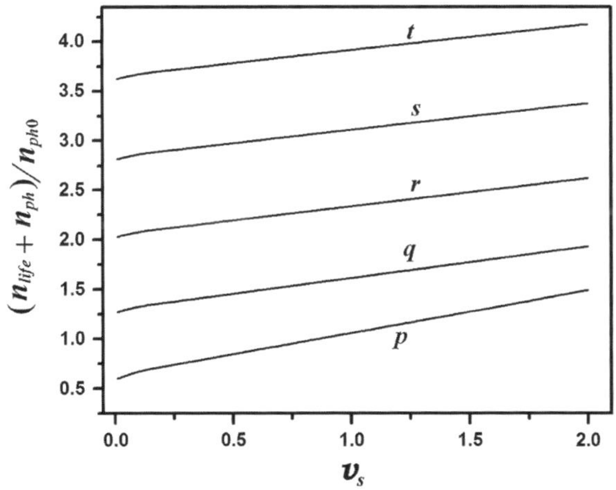

Fig. 2.29 Dependence of $((n_{\text{life}} + n_{\text{ph}})/n_{\text{ph0}})$ with dimensionless surface potential (υ_s), for Case-II, $a = 10$ nm, $W_a = 10$ eV, $\mu_d = 5$ with $T = 300$ K. The labels on the *curves p, q, r, s* and *t* correspond to $\xi = 0.1$, 0.2, 0.3, 0.4, and 0.5, respectively (after Mishra et al. [28], curtsey authors and publishers AIP)

results are in conformance with the transmission coefficient curves illustrated in Fig. 2.19. Figure 2.29 displays the fact that light induced field emission (*life*) current significantly contributes to the total emission current $(n_t = n_{\text{ph}} + n_{\text{life}})$ with increasing υ_s; it is interesting to point out that the total current enhances by a factor of about 2.1 from its initial value (at $\upsilon_s = 0$) for the chosen set of parameters.

2.9 Mie's Theory of Light Scattering by Spherical Particles

For large particles, the power incident on a sphere of radius a from a beam of irradiation I is simply

$$P_a = \pi a^2 I.$$

On account of diffraction and scattering, the above equation gets modified to

$$P_a = M_f(a, N, \lambda)\pi a^2 I$$

where the Mie factor M_f depends on a, and the complex refractive index $N = N_1 + iN_2$ of the sphere for the radiation of wavelength λ.

The theory for the interaction of the spherical particles with electromagnetic radiation has been given by *Mie* [27] and elucidated in the excellent books by Stratton [54], Born and Wolf [62], Goody [63] and Van de Hulst [56]. In this section only the results, obtained by Mie [27] have been stated; for a derivation the reader is referred to the cited sources.

The power S of light incident per unit area, normal to the direction of incidence is

$$S = \left(E_0^2/2\right)\left(\varepsilon_m/\mu_m\right)^{1/2}, \tag{2.92}$$

where E_0 is the amplitude of the electric vector, and $\varepsilon_m/\varepsilon_m\mu_m.\mu_m$ are the dielectric and magnetic permittivity of the medium.

The scattered power W_s is given by

$$W_s = \pi \frac{E_0^2}{k_m^2}\left(\varepsilon_m/\mu_m\right)^{1/2} \sum_{n=1}^{n=\infty} (2n+1)\left(|a_n|^2 + |b_n|^2\right) \tag{2.93}$$

where $k_m = 2\pi/\lambda_m$.

λ_m is the wavelength of the incident radiation in the medium in which the sphere is situated,

$$a_n = -\frac{\mu_s j_n(N\rho)[\rho j_n(\rho)]' - \mu_m j_n(\rho)[N\rho j_n(N\rho)]'}{\mu_s j_n(N\rho)[\rho h_n^{(1)}(\rho)]' - \mu_m h_n^{(1)}(\rho)[N\rho j_n(N\rho)]'}, \tag{2.94a}$$

$$B_n = -\frac{\mu_s j_n(\rho)[N\rho j_n(N\rho)]' - \mu_m N^2 j_n(N\rho)[\rho j_n(N\rho)]'}{\mu_s h_n^{(1)}(\rho)[N\rho j_n(N\rho)]' - \mu_m N^2 j_n(N\rho)[\rho h_n^{(1)}(\rho)]'}, \tag{2.94b}$$

$\rho = k_m a$, N is the complex refractive index of the sphere, relative to the medium

$$j_n(x) = (\pi/2x)^{1/2} J_{n+1/2}(x),$$

$J_m(x)$ is the *Bessel function* of order m,

$$h_n^{(1)}(x) = (\pi/2x)^{1/2} H_{n+1/2}^{(1)}(x),$$

and $H_{n+1/2}^{(1)}(x)$ is *Henkel function*.

The sum of absorbed and scattered power W_t is given by

$$W_t = \frac{\pi E_0^2}{k_m^2}\left(\frac{\varepsilon_m}{\mu_m}\right)^{1/2} \mathrm{Re}\left\{\sum_{n=1}^{n=\infty}(2n+1)(a_n+b_n)\right\} \tag{2.95}$$

The efficiency factors for scattering Q_S and for scattering cum absorption (total) Q_t are defined as

$$Q_S = (W_s/\pi a^2 S) \tag{2.96a}$$

and

$$Q_t = (W_t/\pi a^2 S) \tag{2.96b}$$

The efficiency factor corresponding to absorption Q_a is given by

$$Q_a = Q_t - Q_s \tag{2.96c}$$

The parameters Q_a, Q_s and Q_t have been tabulated in two books. (*National Bureau of Standards* 1940 and Wickramasinghe [61]).

A user friendly *Mie Calculator*, made available by *Scott Prahl* (on website: http://omlc.ogi.edu/calc/mie_calc.html); may be used for computation of Q_a, Q_s and Q_t. Values corresponding to ice and silicates have been tabulated by Dorschner [6].

Appendix A

Electron Transmission Coefficient Across a Negatively Charged Cylindrical Surface (After Mishra et al. [28], Sodha and Dubey [45])

Sodha and Dixit [42] have evaluated the tunneling probability of an electron in the case of a negatively charged cylindrical surface by following a formalism, proposed by Ghatak et al. [16] and [35]. This approximation is better than *Born's approximation*, but it is not valid when $E_r > 0.7\ V_s$. In this section, the case of the cylindrical surface is analyzed on the same lines as in the case of spherical surface.

Assuming that the electron potential energy is zero at $r \geq \lambda_D$, the electron potential energy in and out of the surface is given by

$$\upsilon(\rho) = (\upsilon_s - 1) \quad \text{for}\ \rho < 1 \quad \text{(Region-I)} \tag{2.97a}$$

$$\upsilon(\rho) = \upsilon_s\left(\frac{\ln[\mu_d/\rho]}{\ln[\mu_d]}\right)\exp[-(\rho-1)/\mu_d] \quad \text{for}\ 1 < \rho < \rho_n \quad \text{(Region-II)} \tag{2.97b}$$

and

$$\upsilon(\rho) = 0 \quad \text{for}\, \rho > \rho_n \quad (\text{Region-III}) \tag{2.97c}$$

where $\upsilon(\rho_n) = 0.05\ \upsilon_s$ and the symbols are the same as in the case of a spherical surface.

Substituting in

$$\Psi = r^{1/2}\psi(r)\exp(\pm in\theta)\exp(icz),$$

time independent *Schrödinger's* equation, interpreting n and c in term of the θ and z component of the linear momentum of an electron, one obtains [45]

$$\frac{d^2\psi}{dr^2} + \frac{8\pi^2 m_e}{h^2}(E_r - V(r))\psi = 0$$

or

$$\frac{d^2\psi}{d\rho^2} + \beta(\varepsilon_r - \upsilon(\rho))\psi = 0$$

where the nomenclature is the same as in the spherical case.

Substituting for $\upsilon(\rho)$ from (2.97) in the above equation one obtains

$$\frac{d^2\psi}{d\rho^2} + \beta(\varepsilon_r - \upsilon_s + 1)\psi = 0 \quad \text{for}\, \rho < 1 \quad (\text{Region-I}), \tag{2.98a}$$

$$\frac{d^2\psi}{d\rho^2} + \beta\left\{\varepsilon_r - \upsilon_s\left(\frac{ln[\mu_d/\rho]}{ln[\mu_d]}\right)\exp[-(\rho-1)/\mu_d]\right\}\psi = 0 \quad \text{for}\, 1 < \rho < \rho_n \quad (\text{Region-II}) \tag{2.98b}$$

and

$$\frac{d^2\psi}{d\rho^2} + \beta\varepsilon_r\psi = 0 \quad \text{for}\, \rho > \rho_n \quad (\text{Region-III}). \tag{2.98c}$$

It may be noticed that the dependences of $\upsilon(\rho)$ on ρ, are very close (*for* $\mu_d > 5$) for the spherical and cylindrical surfaces, corresponding to the regions when $1 < \rho < 3$ and $\upsilon_s < \upsilon(\rho) < 0.2\ \upsilon_s$. Since these are the main regions, contributing to the transmission probability the ε_r dependence of $D_e(\varepsilon_r)$ and $D_a(\varepsilon_r)$ should also be almost the same in the two cases. Hence, the electron current for given υ_s, μ_d and a will also be the same in both cases. Computations for the two cases support this conclusion.

Appendix B

Secondary Emission from Cylindrical Particles

Chow et al. [4] have analyzed secondary electron emission from a cylindrical grain on account of the incidence of a high energy beam of (primary) electrons, propagating in a direction parallel to the axis of the grain. This restriction puts a severe limitation on the application of this model to complex plasma kinetics.

The phenomenon of secondary electron emission from the surface of an infinitely long cylindrical grain for oblique incidence of primary electrons has been analyzed herein. Specifically, the ratio δ_0 of the number of emitted secondary electrons to that of accreting incident primary (high energy) electrons has been evaluated for an electrically neutral particle. The multiplying factor for a charged particle has been given. For typical parameters, numerical results have been obtained and discussed.

Consider the incidence of a uniform electron beam on an infinitely long cylinder, with its axis at an angle λ with the direction of propagation of the beam. Fig. A.1 represents the cross section of the cylinder at an angle λ to the axis, which is an ellipse with semi major axis $a$cosec λ and semi minor axis a, where a is the radius of the cylinder; the primary electrons travel along chords, parallel to the major axis. Thus, the typical path of a primary electron is AB, parallel to the major axis at a distance ρ from the center O.

Using simple coordinate geometry

$$AB = 2(a^2 - \rho^2)^{1/2} \mathrm{co\,sec}\,\lambda = 2x_1. \tag{A1}$$

The distance r_1 of point P from the axis of the cylinder is given by

$$r_1^2 = (x_1 - x)^2 \sin^2\lambda + \rho^2 \tag{A2}$$

where $(x_1 - x)\sin\lambda$ is the projection of $(x_1 - x)$on a plane, normal to the axis.

If one chooses a system of coordinates with P (at a distance r_1 from the axis) as origin, $\Phi = 0$ as the perpendicular from P on the axis and $\theta = 0$ in the plane having P and normal to the axis, the distance l of any point on the cylinder to P can be shown to be given by

$$l = (r_1^2 + a_1^2 + 2r_1 a\cos\varphi)^{1/2} \mathrm{co\,sec}\,\theta \tag{A3}$$

Equations (2.56) and (2.57) are valid in this case also, where l is given by (A3).

The number of secondary electrons generated by a primary electron in motion along AB is

$$F(\rho) = (1 - \eta_{se}) \int_0^X \mathrm{d}n_s = (1 - \eta_{se}) \frac{K\alpha^{1/2}}{2} \int_0^X (x_m - x)^{-1/2} f(x, \rho) \mathrm{d}x, \tag{A4}$$

where

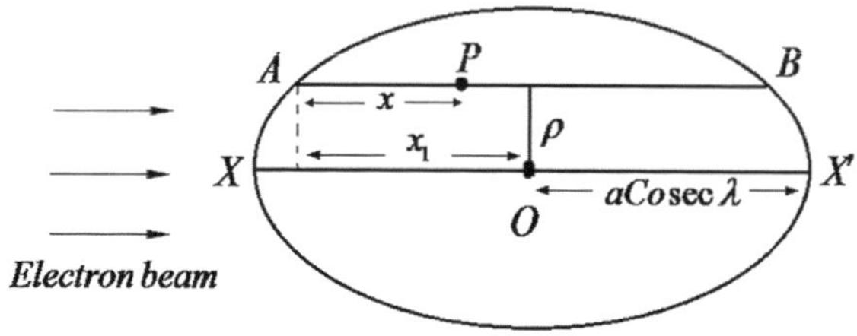

Fig. A.1 Oblique incidence of electrons on a cylindrical particle

$$X = 2x_1 \quad \text{if } x_m > 2x_1 \text{or } \rho > \rho_m, \tag{A5a}$$

$$X = x_m \quad \text{if } x_m < 2x_1 \text{ or } \rho < \rho_m, \tag{B5b}$$

$2x_1$ is given by (A1).

η_{se} is sticking coefficient of electrons on the particle surface and

$$\rho_m^2 = a^2 - (x_m \sin\lambda/2)^2 \tag{A5c}$$

If $x_m < 2a\text{cosec } \lambda$, ρ_m is real and positive and the limits of x are obtained from (A5a) and (B5b) and if $x_m > 2a\text{cosec } \lambda$, the limits are given by (A5a).

Hence, the mean number of secondary electrons $\delta_0(\lambda)$ generated by the motion of a primary electron in the cylindrical particle

$$\delta_0(\lambda) = (1/2a)\int_{-a}^{a} F(\rho)\mathrm{d}\rho = (1/a)\int_0^a F(\rho)\mathrm{d}\rho \tag{A6}$$

where $F(\rho)$ is given by (A4).

The corresponding mean number of electrons stuck ($x_m < 2x_1$) per primary electron is given by

$$\delta_{st}(\lambda) = (1 - \eta_{se})(2\rho_m/2a) = (1 - \eta_{se})(\rho_m/a) \tag{A7}$$

In case the particle is at an electric potential V_s with respect to the surroundings one has (Sodha et al. 2005)

$$\delta(\upsilon_s, \lambda) = \delta_0(\lambda) \quad \text{for } \upsilon_s \geq 0 \tag{A8a}$$

and

$$\delta(\upsilon_s, \lambda) = \exp(-\upsilon_s)\{(2/\pi^{1/2})\upsilon_s^{1/2} + \exp(\upsilon_s)\text{erfc}(\upsilon_s^{1/2})\}\delta_0(\lambda) \quad \text{for } \upsilon_s < 0 \tag{A8b}$$

where $\upsilon_s = \text{eV}_s/\text{kT}_s$, e is the electronic charge, k is Boltzmann's constant and T_s is the temperature of the particle.

In general, the angle λ, between the incoming electron direction and the axis of the cylinder varies at random. Hence

$$\overline{\delta_0} = \int_0^{\pi/2} \delta_0(\sin\lambda/2)\mathrm{d}\lambda \tag{A9a}$$

and

$$\overline{\delta_{st}} = \int_0^{\pi/2} \delta_{st}(\sin\lambda/2)\mathrm{d}\lambda = ((1-\eta_{se})/a)\int_0^{\pi/2} \rho_m(\sin\lambda/2)\mathrm{d}\lambda, \tag{A9b}$$

where $(\sin\lambda/2)\mathrm{d}\lambda$ is the solid angle contained between cones making angle between λ and $(\lambda + \mathrm{d}\lambda)$ with the axis divided by 4π.

From (8Aa) and (8Ab), one can write

$$\bar{\delta}(\upsilon_s) = \overline{\delta_0} \quad \text{for } \upsilon_s \geq 0 \tag{A8c}$$

and

$$\bar{\delta}(\upsilon_s, \lambda) = \exp(-\upsilon_s)\{(2/\pi^{1/2})\upsilon_s^{1/2} + \exp(\upsilon_s)\mathrm{erfc}(\upsilon_s^{1/2})\}\overline{\delta_0}(\lambda) \quad \text{for } \upsilon_s < 0. \tag{A8d}$$

The energy (kT_e) distribution of the electrons incident on the particle per unit time is given by

$$n_e(\upsilon_p)\mathrm{d}\upsilon_p = \pi al(2kT_e/m\pi)^{1/2} n_e(\upsilon_p - \upsilon_s)\exp(-\upsilon_p)\mathrm{d}\upsilon_p, \tag{2.10}$$

where $\upsilon_p = (V_P/kT_e)$.

Hence, the number of secondary electrons emitted by the particle per unit time is

$$n_{\text{see}} = \int_0^{\infty} n_e(\upsilon_p)\bar{\delta}(\upsilon_s, \upsilon_p)\mathrm{d}\upsilon_p \quad \text{for } \upsilon_s < 0 \tag{2.11a}$$

and

$$n_{\text{see}} = \int_{\upsilon_s}^{\infty} n_e(\upsilon_p)\bar{\delta}(\upsilon_s, \upsilon_p)\mathrm{d}\upsilon_p \quad \text{for } \upsilon_s \geq 0. \tag{2.11b}$$

The number of electrons stuck in the particles per unit time is

$$n_{\text{Stuck}} = \int_0^{\upsilon'} n_e(\upsilon_p)\bar{\delta}_{st}(\upsilon_p)\mathrm{d}\upsilon_p \quad \text{for } \upsilon_s < 0 \tag{2.12a}$$

$$n_{\text{Stuck}} = \int_{\upsilon_s}^{\upsilon'} n_e(\upsilon_p)\bar{\delta}_{st}(\upsilon_p)\mathrm{d}\upsilon_p \quad \text{for } \upsilon_s \geq 0 \tag{2.12b}$$

where

$$\upsilon' = V'/kT_e = \upsilon_s(T_s/T_e) + \left[(2a\alpha\mathrm{co}\sec\lambda)^{1/2}/kT_e\right], \tag{2.12c}$$

corresponding to the primary electron energy needed to traverse a distance $2a\mathrm{cosec}\lambda$; in other words,

$$x_m(V') = 2a\mathrm{cosec}\lambda.$$

References

1. S. Agarwal, S. Misra, S.K. Mishra, M.S. Sodha, Can. J. Phys. **90**, 265 (2012)
2. C.N. Berglend, W.E. Spicer, Phys. Rev. **4A**, A1030 (1964)
3. H. Bruining, Thesis, Leydew, Die Sekandar-*Electron Emission Festa Korper*, (Springer Berlin, 1942). p. 60
4. V.W. Chow, M. Rosenberg, Planet. Space Sci. **43**, 613 (1995)
5. V.W. Chow, D.A. Mendis, M. Rosenberg, J. Geophys. Res. **98**, 19065 (1993)
6. J. Dorschner, Astron. Nachr. Bd. **292**, H. 2 (1970)
7. B.T. Draine, Astrophys. J. Suppl. Ser. **36**, 595 (1978)
8. B.T. Draine, B. Sutin, Astrophys. J. **320**, 803 (1987)
9. J.W. Dewdney, Phys. Rev. **125**, 399 (1962)
10. P.K. Dubey, J. Phys. D **3**, 145 (1970)
11. L.A. Dubridge, Phys. Rev. **43**, 727 (1933)
12. R.G. Forbes, J.H.B. Deane, Proc. Roy. Soc. Lond. **A467**, 2927 (2011)
13. R.H. Fowler, Phys. Rev. **38**, 45 (1931)
14. R.H. Fowler, *Statistical Mechanics: The Theory of the Properties of Matter in Equilibrium* (Cambridge University Press, London, 1955)
15. R.H. Fowler, L.W. Nordheim, Proc. Roy. Soc. Lond. **A112**, 781 (1928)
16. A. Ghatak, R.L. Gallawa, I.C. Goyal, IEEE J. Quat. Electr. **28**, 400 (1992)
17. A. Ghatak, S. Lokanathan, *Quantum Mechanics, Theory and Applications* (Macmillan, New Delhi, 2005)
18. O. Hachenberg, W. Brauer, Adv. Electron Phys. **11**, 413 (1959)
19. C. Herring, M.H. Nichols, Rev. Mod. Phys. **21**, 185 (1949)
20. J.H. Jonker, Phillips Res. Repts. **7**, 1 (1952)
21. K. Iwami, A. Iuzuka, N. Umeda, J. Vac. Sci. Technol. B, **29**, 028103 (2011)
22. R.O. Jenkins, W.G. Trodden, *Electron and Ion Emission from Solids* (Dover Publications, New York, 1965)
23. S. Kher, A. Dixit, D. N. Rawat, M.S. Sodha Appl. Phys. Lett. **96**, 044101 (2010)
24. B.A. Klumov, S.I. Popel, R. Bingham, JETP Lett. **72**, 364 (2000)
25. D.R. Lide, Editor-in-chief, *CRC Handbook of Chemistry and Physics,* 89th edn. (CRC Press, New York, 2008–2009)
26. N. Meyer Vernet, Astron. Astrophys. **105**, 98 (1982)
27. G. Mie, Ann. Physik [4], **25**, 377 (1908)
28. S.K. Mishra, S. Misra, M.S. Sodha, Phys. Plasmas **19**, 073705 (2012)
29. S.K. Mishra, M.S. Sodha, S. Srivastava, Astrophys. Space Sci. **344**, 193 (2013)
30. S. Misra, M.S. Sodha, Can. Phys. Under publication
31. S. Misra, S.K. Mishra, M.S. Sodha Phys. Plasma **20**, 013702 (2013)
32. L.W. Nordheim, Proc. Roy. Soc. Lond. **A121**, 788 (1928)
33. L. Page, N.I. Adams, *Principles of Electricity* (D Van Nostranel, New York, 1931). p. 103
34. G. Prakash, Can. J. Phys. **8**, 617 (2010)
35. S. Roy, A.K. Ghatak, I.C. Goyal, IEEE J. Quant. Electr. **29**, 340 (1993)
36. W. Schottky, Z. Phys. **14**, 63 (1923)
37. F. Seitz, *Modern Theory of Solids* (Mc Graw Hill Book Co., New York, 1940)
38. M.S. Sodha, J. Appl. Phys. **32**, 2059 (1961)
39. M.S. Sodha, Brit. J. Appl. Phys. **14**, 172 (1963)
40. M.S. Sodha, A. Dixit, J. Appl. Phys. **104**, 064909 (2008)
41. M.S. Sodha, A. Dixit, S. Srivastava, Appl. Phys. Lett. **94**, 251501 (2009a)
42. M.S. Sodha, A. Dixit, J. Appl. Phys. **105**, 034909 (2009b)
43. M.S. Sodha, A. Dixit, S.K. Agarwal, Can. J. Phys. **87**, 175 (2009c)
44. M. S. Sodha, A. Dixit, S. Srivastava, Phys. Rev. E **79**, 046407 (2009d); erratum E **80**, 06990 (2010)
45. M.S. Sodha, P.K. Dubey, Brit. J. Appl. Phys. D **2**, 1617 (1969)

46. M.S. Sodha, S. Guha, *Physics of Colloidal Plasmas*, vol. 4 eds. by A. Simon, W.B. Thompson. *In Advances Plasma Physics* (InterScience New York, 1971). pp. 219–369
47. M.S. Sodha, P.K. Kaw, Brit. J. Appl. Phys. D **1**, 1303 (1968)
48. M.S. Sodha, S.K. Mishra, Phys. Plasmas **18**, 083708 (2011)
49. M.S. Sodha, S.K. Mishra, S. Misra, Phys. Plasmas **16**, 123701 (2009e)
50. M.S. Sodha, S. Sharma, Brit. J. Appl Phys. D **18**, 1127 (1967)
51. M.S. Sodha, S. Srivastava, Phys. Lett. A **374**, 4733 (2010)
52. W.E. Spicer, Phys. Rev. **112**, 114 (1958)
53. W.E. Spicer, A Herrara-Gomez, Modern Theory and Applications of Photocathodes. *Paper Presented at SPIE's International Symposium on Optics, Imaging and Instrumentation*, San Diego, CA, July 11–16, 1993; SLAC-PUB-6306;SLAC/SSRL-0042, August 1993 (A-SSRL-H)
54. J.A. Stratton, *Electromagnetic Theory* (Mc Graw Hill, New York, 1941), pp. 563–573
55. E.J. Sternglass, *Scientific Paper 1773* (Westinghouse Research lab, Pittsburgh, 1954)
56. H.C. Van de Hulst, *Light Scattering by Small Particles* (Wiley, New York, 1957)
57. W.D. Watson, Astrophys. J. **176**, 103 (1972)
58. W.D. Watson, J. Opt. Soc. Am. **63**, 164 (1973)
59. R. Whiddington, Proc. Roy. Soc. **A86**, 360 (1912)
60. E.C. Whipple, Rep. Prog. Phys. **44**, 1197 (1981)
61. N.C. Wikramsinghe, *Light Scattering Functions for Small Particles* (Wiley, New York, 1973)
62. M. Born, E. Wolf, *Principles of Optics*, Chapter XIII, McMillan, New York (1964)
63. R.M. Goody, *Atmospheric Radiation I. Theoretical Basis*, Chapter 7, Oxford University Press (Clarendon, London and New York , 1964)
64. L. Spitzer, Astrophys. J. **93**, 369 (1948)
65. B.T. Draine, E.E. Salpeter, Astrophys. J. **231**, 77 (1979)
66. J.L. Puget, A. Leger, Ann. Rev. Astron. Astrophys. **27**, 161 (1989)
67. R.Z. Sagdeev, E.N. Evlanov, M.N. Formenkova, O.F. Prilutskii, B.V. Zubov, Adv. Space Res. **9**, 263 (1989)
68. M.M. Abbas, D. Tankosio, P.D. Craven, A.C. McClair, J.F. Spann, Astrophys. J. **T18**, 795 (2010)

Chapter 3
Accretion of Electrons/Ions on Dust Particles

Orbital Model
Orbital model is the simplest model of accretion of electrons/ions on dust particles and the results have been used extensively in complex plasma kinetics. The model assumes that the number of collisions with gaseous species, suffered by ions/electrons is negligible as compared to the number of accretions on the dust particles. It is also of interest to realize that from classical considerations the results for a monotonically varying electric potential around a charged particle do not depend on the nature of the variation but only on the electric potential of the particle with reference to the free plasma. Quantum effects, charge exchange ion collisions with neutral atoms and ion trapping also play an important role in accretion, which has been highlighted.

3.1 Classical Rate of Accretion of Electrons/Ions on Spherical and Cylindrical Particles (After Mott-Smith and Langmuir [13])

3.1.1 General Considerations

Consider (1) a spherical particle of radius a and (2) a cylindrical particle of radius a and length $l(\gg a)$ in a plasma, with electron/ion density n, charge q ($+e$ for ions and $-e$ for electrons), temperature T with suffixes e and i referring to electrons and ions; the particle is at an electric potential V_s/q with respect to the plasma. Also consider a concentric sphere or coaxial cylinder of radius b $(b \gg a)$, such that the electric field and electric potential due to the charged particle is zero for $r > b$.

Let u and υ denote the radial and tangential components of the electron/ion velocity and let u be taken as positive when directed toward the center/axis of the particle. The number of electrons/ions, having radial velocity between u and $u + \mathrm{d}u$ and tangential velocity between υ and $\upsilon + \mathrm{d}\upsilon$, which cross the surface $r = b$, per unit time is

M. S. Sodha, *Kinetics of Complex Plasmas*, Springer Series on Atomic, Optical, and Plasma Physics 81, DOI: 10.1007/978-81-322-1820-3_3,

$$S_b\, u n f(u, \upsilon) \mathrm{d}u\, \mathrm{d}\upsilon \tag{3.1}$$

where $nf(u, \upsilon)\mathrm{d}u\,\mathrm{d}\upsilon$ is the number of electrons/ions per unit volume having radial and tangential velocity components between u and $u + \mathrm{d}u$ and υ and $\upsilon + \mathrm{d}\upsilon$, at $r = b$, where the electric field/potential due to the particle is zero and S_b is the area of the surface at $r = b$; $S_b = 4\pi b^2$ for the sphere and $2\pi bl$ for the cylinder. The nature of $f(u, \upsilon)$ has been discussed later (refer to 3.8a, 3.8b and 3.9).

If u_a and υ_a denote the radial/tangential components of the electron/ion velocity at the surface of the particle $r = a$, the conservation of energy and angular momentum requires

$$V_s + \frac{m}{2}(u_a^2 + \upsilon_a^2) = \frac{m}{2}(u^2 + \upsilon^2) \tag{3.2a}$$

and

$$a\upsilon_a = b\upsilon \tag{3.2b}$$

where m is the mass of the electron/ion.

From (3.2a) and (3.2b) one obtains

$$u_a^2 = u^2 - (b^2/a^2 - 1)\upsilon^2 - (2V_s/m). \tag{3.3}$$

It may be noted that only those electrons/ions will reach the surface of the particle $r = a$ for which

$$u > 0 \tag{3.4a}$$

and

$$u_a^2 > 0. \tag{3.4b}$$

The minimum value of u, corresponding to $u_a^2 > 0$ and $\upsilon = 0$ is from (3.3) given by

$$u_m^2 = (2V_s/m). \tag{3.5a}$$

When V_s is positive, u_m is the positive square root of $(2V_s/m)$; in case V_s is negative, $u_m = 0$ in view of the inequality (3.4a).

For a given value of u, the maximum value of υ^2 *viz.* υ_1^2 corresponds to $u_a^2 = 0$ and is from (3.3) given by

$$\upsilon_1^2 = (u^2 - (2V_s/m))(b^2/a^2 - 1)^{-1}. \tag{3.6a}$$

By choice of $b \gg a$, $|\upsilon_1| \ll u$.

Without detriment to the evaluation of the rate of accretion of electrons and ions one can choose a large value of $b \gg a$ such that

$$\upsilon_1^2 = (a^2/b^2)(u^2 - (2V_s/m)) \tag{3.6b}$$

and hence

$$mv_1^2/2k_BT_e) \ll 1 \tag{3.6c}$$

From (3.1), the number of electrons/ions incident on the surface of the particle per unit time or the accretion current is given by

$$n_c = S_b n \int_{u_m}^{\infty} \int_{v^2=0}^{v^2=v_1^2} uf(u,v)\mathrm{d}u\,\mathrm{d}v \tag{3.7}$$

3.1.2 Function $f(u, v)$ for Maxwellian Distribution of Velocities

The Maxwellian distribution of electron/ion velocities $\boldsymbol{c}$ may be expressed as

$$f(c_x, c_y, c_z)\mathrm{d}c_x\mathrm{d}c_y\mathrm{d}c_z = (m/2\pi k_BT_e)^{3/2}\exp[-(m/2k_BT)(c_x^2 + c_y^2 + c_z^2)]\mathrm{d}c_x\mathrm{d}c_y\mathrm{d}c_z \tag{3.8a}$$

Since in the case of a cylindrical particle with axis along z axis c_z is not relevant, the distribution function may be expressed in terms of u and v by integrating the above distribution with respect to c_z in the limits $\pm\infty$; thus

$$f(u,v)\mathrm{d}u\mathrm{d}v = (m/2\pi k_BT_e)\exp[-(m/2k_BT_e)(u^2 + v^2)]\mathrm{d}u\mathrm{d}v \tag{3.8b}$$

where $u = c_x$ and $v = c_y$.

For the spherical case, both the tangential components are effective and we may replace $\mathrm{d}c_x$ by $\mathrm{d}u$ and $\mathrm{d}c_y\mathrm{d}c_z$ by $2\pi v\mathrm{d}v$; hence

$$f(u,v)\mathrm{d}u\mathrm{d}v = 2\pi(m/2\pi k_BT_e)^{3/2}v\exp[-(m/2k_BT_e)(u^2 + v^2)]\mathrm{d}u\mathrm{d}v \tag{3.9}$$

where $u = c_x$ and v is the resultant of c_y and c_z.

Equation (3.9) has been obtained by putting $2\pi v\mathrm{d}v$ as the first integral of $\mathrm{d}c_x\mathrm{d}c_y$ where c_x and c_y are the two tangential components of the electron/ion velocity; this implies that v is positive.

3.1.3 Spherical Particles

From (3.7) and (3.9), one obtains

$$n_c = n \cdot 4\pi b^2 \cdot 2\pi(m/2\pi k_B T_e)^{3/2} \int_{u_m}^{\infty} u \exp[-(mu^2/2k_B T_e)]\mathrm{d}u \int_{\upsilon^2=0}^{\upsilon^2=\upsilon_1^2} \upsilon \exp[-(m\upsilon^2/2k_B T_e)]\mathrm{d}\upsilon \quad (3.10)$$

Since $(m\upsilon_1^2/2k_B T_e) \ll 1$ (by the choice of $b \gg a$), $\exp(-m\upsilon_1^2/2k_B T) \approx 1$ and (3.10) reduces to

$$n_c = n \cdot 8\pi^2 b^2 (m/2\pi k_B T_e)^{3/2} \int_0^{\infty} u(\upsilon_1^2/2) \exp[-(mu^2/2k_B T_e)]\mathrm{d}u.$$

Substituting for υ_1^2 from (3.6b) in the above equation

$$\begin{aligned} n_c &= n \cdot 4\pi^2 a^2 (m/2\pi k_B T_e)^{3/2} \int_{u_m}^{\infty} u(u^2 - 2V_s/m) \exp[-(mu^2/2k_B T_e)]\mathrm{d}u \\ &= n \cdot \pi a^2 (8k_B T_e/\pi m)^{1/2} \int_{\varepsilon_{\mathrm{rm}}}^{\infty} (\varepsilon_r - \upsilon_s) \exp(-\varepsilon_r) d\varepsilon_r \end{aligned} \quad (3.11a)$$

where $\varepsilon_r = (mu^2/2k_B T_e)$ is the dimensionless radial energy of the electrons/ions and $\upsilon_s = V_s/k_B T_e$.

Further for $\upsilon_s > 0$, $\varepsilon_{\mathrm{rm}} = \upsilon_s$ and (3.11a) reduces to

$$n_c = n \cdot \pi a^2 (8k_B T_e/\pi m)^{1/2} \exp(-\upsilon_s) \quad (3.11b)$$

and for $\upsilon_s < 0$, $\varepsilon_{\mathrm{rm}} = 0$ and (3.11b) reduces to

$$n_c = n \cdot \pi a^2 (8k_B T_e/\pi m)^{1/2} (1 - \upsilon_s) \quad (3.11c)$$

Remembering that when υ is positive, the mean energy of the accreting electrons/ions, when these are far away $r \gg b$ can from (3.7) and (3.9) be expressed as

$$n_c E_c = n \cdot 4\pi b^2 \cdot 2\pi(m/2\pi k_B T_e)^{3/2} \int_{u_m}^{\infty} u \exp[-(mu^2/2k_B T_e)]\mathrm{d}u \int_{\upsilon^2=0}^{\upsilon^2=\upsilon_1^2} \upsilon \frac{m}{2}(u^2 + \upsilon^2) \exp[-(m\upsilon^2/2k_B T_e)]\mathrm{d}\upsilon.$$

Remembering that $u \gg \upsilon_1$, $(m\upsilon^2/2k_B T_e) < (m\upsilon_1^2/2k_B T_e) \ll 1$, and $\exp(-m\upsilon_1^2/2k_B T_e) \approx 1$; the above equation reduces to

$$n_c E_c = n \cdot 8\pi^2 b^2 (m/2\pi k_B T_e)^{3/2} \int_{u_m}^{\infty} u^3 \exp[-(mu^2/2k_B T_e)] \frac{m}{2} \frac{\upsilon_1^2}{2} \mathrm{d}u.$$

Substituting for v_1^2 corresponding to $b \gg a$ from (3.6a)

$$\begin{aligned} n_c E_c &= n \cdot 8\pi^2 a^2 (m/2\pi k_B T_e)^{3/2} \int_{u_m}^{\infty} u^3 \exp[-(mu^2/2k_B T_e)] \frac{m}{4} (u^2 - 2qV_s/m) \mathrm{d}u \\ &= n \cdot \pi a^2 k_B T (8k_B T_e/m\pi)^{1/2} \int_{\varepsilon_{\mathrm{rm}}}^{\infty} \varepsilon_r (\varepsilon_r - \upsilon_s) \exp(-\varepsilon_r) \mathrm{d}\varepsilon_r \end{aligned} \tag{3.12a}$$

For $\upsilon_s > 0$ $\varepsilon_{\mathrm{rm}} = \upsilon_s$ and substituting for n_c from (3.11b) in the above equation one obtains

$$E_c = (2 + \upsilon_s) k_B T_e \tag{3.12b}$$

For $\upsilon_s < 0$, $\varepsilon_{\mathrm{rm}} = 0$ and

$$E_c = [(2 - \upsilon_s)/(1 - \upsilon_s)] k_B T_e \tag{3.12c}$$

The rates of accretion of electrons and ions and the corresponding mean energy are given by putting the electron/ion mass and temperature and density for m, T and n. Further $\upsilon_s = V_s/k_B T_e$ and is positive if the dust particle and the accreting particle (electron/ion) have like charges and negative for a combination of unlike charges.

3.1.4 Alternate Derivation for Spherical Particles

Consider the grazing incidence of an electron/ion on the dust particle and let r_0 be the perpendicular distance of the path of electron/ion at large distance $(V = 0)$ from the center of the particle. The conservation of energy and angular momentum requires

$$V_s + (mc_a^2/2) = mc^2/2$$

and $ac_a = r_0 c$, where c is the speed of electron/ion at larger distance and c_a is the grazing speed.

From the above equations the accretion cross-section $Q(c)$ is given by

$$Q(c) = \pi r_0^2 = \pi a^2 [1 - 2V_s/mc^2]$$

and the rate of accretion is given by

$$n_c = n \int\int\int cQ(c)f(c)\mathrm{d}c_x\mathrm{d}c_y\mathrm{d}c_z$$
$$= n\int_{u_m}^{\infty} 4\pi c^3 Q(c)f(c)\mathrm{d}c \tag{3.11d}$$

where $f(c) = (m/2\pi k_B T_e)^{3/2}\exp[-(mc^2/2k_B T_e)]$
Similarly

$$n_c E_c = n\int_{u_m}^{\infty} 4\pi c^3 (mc^2/2)Q(c)f(c)\mathrm{d}c \tag{3.12d}$$

Substituting for $f(c)$ and $Q(c)$ one obtains (3.11b), (3.11c), (3.12d) and (3.12c).

3.1.5 Flowing Plasma

There are some situations, when the electrons/ions flow past the dust particles with a velocity $\boldsymbol{C}$, much less than the mean electron speed but comparable to the mean ion speed. This affects the ion flux n_{ic} on the particle and E_{ic} the mean energy of the accreting ions; these parameters have been evaluated by Mishra et al. [8] from (3.11d) and (3.12d) by substituting the following expression for $f(c)$, corresponding to a displaced Maxwellian distribution:

$$f(c) = (m_i/2\pi k_B T_i)^{3/2}\exp[-(m_i/2k_B T_i)(\boldsymbol{c}-\boldsymbol{C})^2]$$

Expressions for n_{ic} and E_{ic} thus obtained corresponding to $V_s < 0$ and $V_s > 0$ are in terms of error functions. The distribution function averaged over all values of the angle between $\boldsymbol{c}$ and $\boldsymbol{C}$ can be shown to be given by

$$f(c) = (m_i/2\pi k_B T_i)^{3/2}(k_B T_i/m_i cC)\{\exp[-(m_i/2k_B T_i)(c-C)^2] - \exp[-(m_i/2k_B T_i)(c+C)^2]\}$$

3.1.6 Cylindrical Particles

From (3.7) and (3.8a, 3.8b), the rate of accretion of charged particles (electrons/ions) is given by

$$n_c = n \cdot (4\pi bl) \cdot (m/2\pi k_B T) \int\limits_{u_m}^{\infty} u \exp[-(mu^2/2k_B T)]\mathrm{d}u \int\limits_{-\upsilon_1}^{\upsilon_1} \exp[-(m\upsilon^2/2k_B T)]\mathrm{d}\upsilon \tag{3.13a}$$

For $b \gg a$, $(m\upsilon_1^2/2k_B T)$ (and hence $m\upsilon^2/2k_B T) \ll 1$ and $\exp(-m\upsilon_1^2/2k_B T)$ may be put as unity; thus the above equation reduces to

$$n_c = n \cdot (2\pi bl) \cdot (m/2\pi k_B T) \int\limits_{u_m}^{\infty} 2u\upsilon_1 \exp[-(mu^2/2k_B T)]\mathrm{d}u. \tag{3.13b}$$

Substituting for υ_1 from (3.13b) in the above equation $b \gg a$ one obtains

$$n_c = n \cdot (2al) \cdot (2k_B T/m)^{1/2} \int\limits_{\varepsilon_{\mathrm{rm}}}^{\infty} (\varepsilon_r - \upsilon_s)^{1/2} \exp(-\varepsilon_r)\mathrm{d}\varepsilon_r \tag{3.13c}$$

For $\upsilon_s > 0$, $\varepsilon_{\mathrm{rm}} = \upsilon_s$, (3.13b) simplifies to

$$n_c = nal \cdot (2\pi k_B T/m)^{1/2} \exp(-\upsilon_s) \tag{3.13d}$$

For $\upsilon_s < 0$, $\varepsilon_{\mathrm{rm}} = 0$ and (3.13b) simplifies to

$$n_c = nal \cdot (2\pi k_B T/m)^{1/2} \left[(2/\sqrt{\pi})\sqrt{-\upsilon_s} + \exp(-\upsilon_s)\mathrm{erfc}(\sqrt{-\upsilon_s})\right] \tag{3.13e}$$

where the complementary error function erfc(x) is defined as

$$\mathrm{erfc}(x) = \int\limits_{x}^{\infty} \exp(-t^2)\mathrm{d}t.$$

The mean energy E_c of the accreting electrons far away from the particle ($r > b$) is (in a way to similar to that for spherical potential) given by

$$\begin{aligned} n_c E_c &= n \cdot (2\pi bl) \cdot (m/2\pi k_B T) \int\limits_{u_m}^{\infty} \frac{mu^2}{2} 2u\upsilon_1 \exp[-(mu^2/2k_B T)]\mathrm{d}u \\ &= n \cdot (2al) \cdot (2k_B T/m)^{1/2} k_B T \int\limits_{\varepsilon_{\mathrm{rm}}}^{\infty} \varepsilon_r (\varepsilon_r - \upsilon_s)^{1/2} \exp(-\varepsilon_r)\mathrm{d}\varepsilon_r \end{aligned} \tag{3.14a}$$

Hence including the axial energy, corresponding to the component of the momentum along the axis of the cylinder (z- axis) viz. $k_B T/2$, which is unaffected by the electric field (normal to the z-axis) the mean energy of the electrons, far away ($r > b$) from the axis is given by

$$\begin{aligned}\varepsilon_c &= [E_c + (k_B T/2)]/k_B T \\ &= 1/2 + \int_{\varepsilon_{rm}}^{\infty} \varepsilon_r(\varepsilon_r - \upsilon_s)^{1/2} \exp(-\varepsilon_r) d\varepsilon_r \Big/ \int_{\varepsilon_{rm}}^{\infty} (\varepsilon_r - \upsilon_s)^{1/2} \exp(-\varepsilon_r) d\varepsilon_r\end{aligned} \tag{3.14b}$$

For $\upsilon_s < 0$, $\varepsilon_{rm} = \upsilon_s$ and (3.14b) simplifies to

$$\begin{aligned}\varepsilon_c &= 1/2 + (1/2)\left[(6/\sqrt{\pi})\sqrt{-v_s} + (3 - 2\sqrt{-v_s})\exp(-v_s)\mathrm{erfc}(\sqrt{-v_s})\right] \\ &\quad /\left[(2/\sqrt{\pi})\sqrt{-v_s} + \exp(-v_s) erfc(\sqrt{-v_s})\right] \\ &= \left[(4/\sqrt{\pi})\sqrt{-v_s} + (2 - \sqrt{-v_s})\exp(-v_s) erfc(\sqrt{-v_s})\right] \\ &\quad /\left[(2/\sqrt{\pi})\sqrt{-v_s} + \exp(-v_s) erfc(\sqrt{-v_s})\right]\end{aligned} \tag{3.14c}$$

Further for $\upsilon_s > 0$, $\varepsilon_{rm} = 0$ and (3.14b) leads to

$$\varepsilon_c = 1/2 + \int_0^{\infty} \varepsilon_r(\varepsilon_r - \upsilon_s)^{1/2} \exp(-\varepsilon_r) d\varepsilon_r \Big/ \int_0^{\infty} (\varepsilon_r - \upsilon_s)^{1/2} \exp(-\varepsilon_r) d\varepsilon_r \tag{3.14d}$$

$$\begin{aligned}\varepsilon_c &= 1/2 + (1/2)(3 + 2\upsilon_s) \\ &= (2 + \upsilon_s)\end{aligned} \tag{3.14e}$$

There are alternate ways to derive (3.11b), (3.11c), (3.12b), (3.12c), (3.13d), (3.13e), (3.14c) and (3.14d), but the present treatment is a rigorous one. These equations are applicable to both electrons and ions, when corresponding values of electron and ion parameters are substituted. One has to remember that $\upsilon_s > 0$, for (i) electrons and negative electric potential on the surface and (ii) ions and positive electric potential on the surface. Further $\upsilon_s < 0$ corresponds to (i) electrons and positive electric potential on the surface and (ii) ions and negative electric potential on the surface.

It may be added that for $\upsilon_s > 0$, the maximum of the potential energy of an electron gets reduced on account of the image force on an electron by an amount $\Delta V_s = \Delta\varphi$, tabulated in Table 2.5. Thus for negatively charged particles one should use $V_s - \Delta V_s$ in place of V_s in the expressions.

3.2 Quantum Effects in Electron Accretion on the Surface of Charged Particles (After Mishra et al. [12])

3.2.1 General Remarks

The classical theory of electron accretion on a charged particle, presented in the preceding sections is based on the implicit assumption that the transmission coefficient of plasma electrons through the surface of the particle is unity when $\varepsilon_r > \upsilon_s$ and zero, when $\varepsilon_r < \upsilon_s$. Such an assumption is valid in classical physics but it breaks down when quantum mechanical considerations are taken into account. Although the quantum effects have been appreciated to some extent in the case of electron emission, particularly in the context of electric field emission, the reverse process viz. the tunneling of plasma electrons into the bulk of a negatively charged particle has been investigated only recently [10, 12, 15, 17, 18].

In contrast to the usual expressions, an expression for the transmission coefficient of electrons across a charged metallic surface, based on the three region model has been derived in Sect. 3.2.2; the results and a commentary thereon is also available in the same section.

3.2.2 Quantum Effects in Electron Accretion

Following the derivations in Sect. 3.1 and introducing the coefficient of tunneling accretion $D_a(\varepsilon_r)$, the rate of accretion on a negatively charged surface is given by

$$n_{\mathrm{ec}} = \pi a^2 (8k_B T_e/m\pi)^{1/2} \int D_a(e_r)\exp[-\varepsilon_r]\mathrm{d}\varepsilon_r,$$

where $e_r = E_r/W_a$, $\varepsilon_r = E_r/k_B T_e$,, $(\varepsilon_r = g' e_r)$ and $(g' = W_a/k_B T_e)$.

The rate of classically allowed $(e_r > \upsilon_s' = (V_S/W_a) = \upsilon_s/g')$ and tunneling accretion are given by

$$n_{\mathrm{ecc}} = \pi a^2 (8k_B T_e/m\pi)^{1/2} g' \int_{\upsilon_s'}^{\infty} D_a(e_r)\exp[-g' e_r]\mathrm{d}e_r$$

$$= n_{c0} g' \int_{\upsilon_s'}^{\infty} D_a(e_r)\exp[-g' e_r]\mathrm{d}e_r \tag{3.15a}$$

$$n_{\mathrm{ect}} = \pi a^2 (8k_B T_e/m\pi)^{1/2} g' \int_{0}^{\upsilon_s'} D_a(e_r)\exp[-g' e_r]\mathrm{d}e_r$$

$$= n_{c0} g' \int_{0}^{\upsilon_s'} D_a(e_r)\exp[-g' e_r]\mathrm{d}e_r \tag{3.15b}$$

and

$$n_{\mathrm{ec}} = n_{\mathrm{ecc}} + n_{\mathrm{ect}} \tag{3.15c}$$

where n_{c0} is the rate of accretion on an uncharged particle.

In semiclassical treatment $D_a(e_r)$ is usually taken to be unity; in this limit (3.15a) reduces to

$$n_{\mathrm{ec0}} = n_{c0}\exp\left[-g'\upsilon_s'\right] = n_{c0}\exp[-\upsilon_s].$$

The corresponding mean radial energies of the accreting electrons are given by

$$n_{\mathrm{ecc}}\varepsilon_{\mathrm{recc}} = n_{c0}g'\int\limits_{\upsilon_s'}^{\infty} e_r D_a(e_r)\exp[-g'e_r]\mathrm{d}e_r \tag{3.16a}$$

and

$$n_{\mathrm{ect}}\varepsilon_{\mathrm{rect}} = n_{c0}g'\int\limits_{0}^{\upsilon_s'} e_r D_a(e_r)\exp[-g'e_r]\mathrm{d}e_r \tag{3.16b}$$

To get the total mean energy of accreting electrons, one has to add the mean transverse energy of electrons k_BT_e to ε_r. Thus

$$\varepsilon_{\mathrm{ec}} = \varepsilon_{\mathrm{ecr}} + k_BT_e, \tag{3.16c}$$

where

$$\varepsilon_{\mathrm{ecr}} = \left[n_{\mathrm{ecc}}\varepsilon_{\mathrm{recc}} + n_{\mathrm{ect}}\varepsilon_{\mathrm{rect}}\right]/n_{\mathrm{ec}}$$

Figure 3.1a illustrates the dependence [12] of the transmission coefficient $[D_e(e_r)\&D_a(e_r)]$ on the radial energy of electrons e_r as a function of υ_s' (surface potential). It is seen from Fig. 3.1a that the transmission coefficient [i.e., $D(e_r)$] increases monotonically with increasing e_r and υ_s', and approaches unity asymptotically for large e_r. The electrons in the curves corresponding to the region $e_r < \upsilon_s'$ contribute to electron emission through tunneling (electric field emission) while the rest (i.e., $e_r > \upsilon_s'$) correspond to classically allowed emission; the broken curves correspond to transmission coefficient for step potential barrier $[D_0(e_r)]$ corresponding to υ_s' (when W_a is substituted by $(W_a - V_s)$).

The effect of parameter $g'(= W_a/kT_e)$ on accretion current has been illustrated in Fig. 3.1b, which suggests that the accretion current through tunneling $(n_{\mathrm{ect}}/n_{\mathrm{ec0}})$ initially increases with increasing υ_s' till $g'\upsilon_s' > 1$ and then decreases with further increase in υ_s'. This can be understood in terms of the larger availability of electrons for accretion through tunneling in case of $g'\upsilon_s' > 1$ and increasing barrier width with increasing surface potential. The figure also indicates that the contribution of tunneling accretion to be accretion current increases with increasing g'; this is primarily because of larger availability of low energy electrons for tunneling through accretion. On the other hand, the accretion current $(n_{\mathrm{ecc}}/n_{c0})$ gradually decreases with increasing υ_s' and g', as indicated in Fig. 3.1c. It is seen that with quantum mechanical considerations that the accretion current gets significantly modified with respect to usual semiclassical treatment [i.e., $D_a(e_r) = 1$]. The

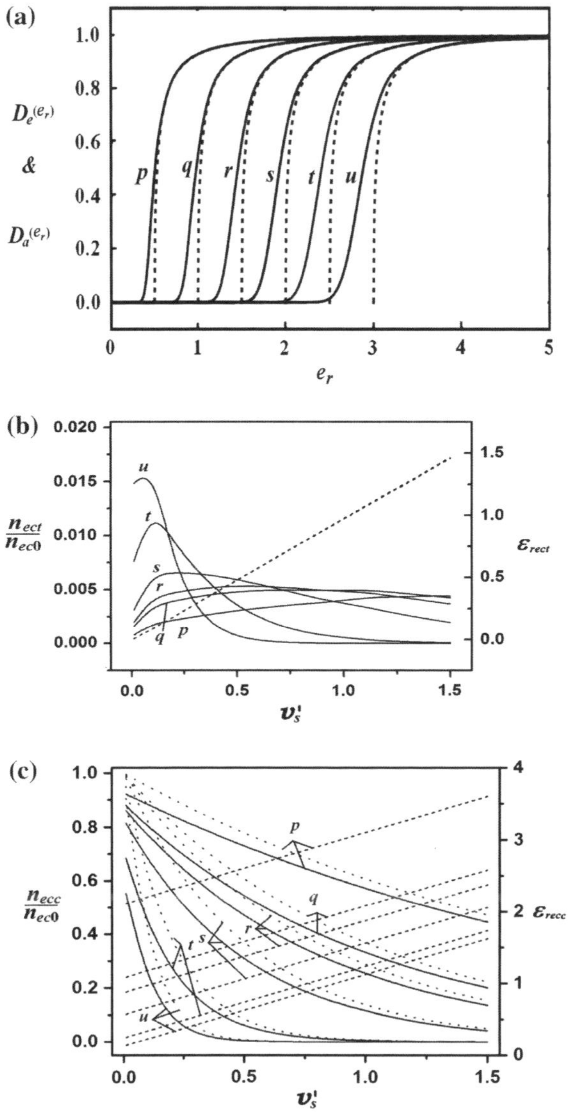

Fig. 3.1 **a** Dependence of transmission coefficient [$D_e(e_r)$ or $D_a(e_r)$] on e_r for $\mu_d = (\lambda_D/a) = 5$ and $\beta = 1000$. The labels on the curves *p, q, r, s,* and *t* correspond to $\upsilon_s' = 0.5$, 1.0, 1.5, 2.0, and 3.0, respectively. The *solid curves* correspond to the present analysis while the *dashed curves* refer to the step potential barrier of height $(1 - \upsilon_s')$ (after Mishra et al. [12], curtsey authors and publishers AIP). **b** Dependence of (n_{ect}/n_{c0}) and ε_{rect} with dimensionless surface potential (υ_s'), for $a = 10$ nm, and $\mu_d = 5$. The labels on the curves *p, q, r, s* and *t* correspond to $g' = 0.5$, 1.0, 1.25, 2.0, 5.0, and 10.0, respectively. *Solid and broken lines* correspond to *left* and *right* hand scale respectively (after Mishra et al. [12], curtsey authors and publishers AIP). **c** Dependence of (n_{ecc}/n_{c0}) and ε_{recc} with dimensionless surface potential (υ_s'), for $a = 10$ nm, and $\mu_d = 5$. The labels on the curves *p, q, r, s* and *t* correspond to $g' = 0.5$, 1.0, 1.25, 2.0, 5.0 and 10.0 respectively. *Solid/dotted lines* on *left hand scale* refer to present/semiclassical approach while *dashed lines* (on *right hand scale*) correspond to mean energy respectively (after Mishra et al. [12], curtsey authors and publishers AIP)

figures also indicate the fact that the mean energy of electrons (away from the surface) associated with usual accretion increases with increasing υ_s' while it displays opposite trend with increasing g' on account of the larger availability of low energy electrons for accretion. The mean energy of the electrons in case of tunneling accretion ($\varepsilon_{\text{rect}}$) increases gradually with surface potential; however it remains almost independent of g'.

3.3 Critique of OML Theory

The previous sections are based on the OML theory, which assumes the absence of electron/ion collisions in the vicinity of the dust particles. This is justified because the mean free path of electrons/ion collisions is much larger than the Debye length in typical complex plasmas in laboratory, industry, and space. However, there are several remaining issues which need to be discussed. First, the potential is of Debye form, only when $\exp(-eV/k_BT_{e,i}) \approx 1 - eV/k_BT_{e,i}$; however it is seen (e.g., [5]) that the Debye form is a useful good approximation for many other situations of interest. Second, the OML theory does not take into account the possible partition [1] of phase space into mutually inaccessible regions by barriers in the radial effective potential $U(r) = V(r) + L^2/2mv^2$, where L denotes the angular momentum and m the electron/ion mass. However, Lampe [4] has shown that the effect of this phenomenon in typical complex plasmas is not significant. Another limitation has been pointed out by Pandey et al. [11] viz. the fact that at low electron temperature, the electron de-Broglie wavelength is comparable or even larger than the size of small dust particles.

3.4 Trapping of Ions

When collisions are neglected, the total (Kinetic + Potential) energy of electrons/ ions in the vicinity of a dust particle is positive and the electrons/ions hit the particle, get reflected back or just fly by. However, Bernstein and Rabinowitz [2] have pointed out that there may be collisions of positive ions with neutral particles (including charge exchange collisions) having low kinetic energy, so that the total energy of the ion after the collision is negative; it may be remembered that the potential energy of a positive ion in the field of a negatively charged particle is negative. Since the collision frequency v is usually small, the production rate of trapped ions is small. But a trapped ion, once created stays trapped for a long time. Hence, the trapped ion density may slowly be large enough to affect accretion. A trapped ion may suffer a collision with a high energy neutral particle and hence acquire positive total energy and escape. The ion may be scattered in a collision to fall on the dust particle and get accreted. Monte Carlo simulations by Goree [3] and Zobnin et al. [19] show that the number density of trapped ions in the vicinity of a dust particle may be quite significant. In what follows an elementary phenomenological theory by Lampe et al. [6] of the effect of ion trapping on the accretion current on the particle has been given.

3.4.1 *Effect of Charge Exchange Ion Collisions with Neutral Atoms on Accretion Current (After Lampe et al. [6])*

In a charge exchange collision of an ion with a neutral atom, the energy of the new ion is less than of that of the old ion, which enhances the probability of accretion on the dust particle. To estimate the rate of accretion, one may define a distance r_0 from the center of a negatively charged particle so that $r = r_0$ is the edge of the sheath, defined by $eV(r_0) = -3k_BT/2$ This newly created ion may fall on the grain surface, immediately if it has a low angular momentum or orbit it if otherwise. The orbiting ion will also get accreted on account of further collisions. Thus, practically all collisions, occurring in the space $r<r_0$ cause accretion of an electron on the surface of the particle. The probability of an ion, suffering a collision inside a sphere of radius r_0 is r_0/λ_m, where λ_m is the mean free path.

Hence, the additional accretion rate of electrons on the dust particle due to collisions is

$$\begin{aligned} n_{\text{ict}} &= 4\pi r_0^2 n_i (k_B T_i/2m_i\pi)^{1/2}(r_0/\lambda_m), \\ &= (r_0^2/a^2) n_{\text{ic0}}(r_0/\lambda_m) \end{aligned} \tag{3.17}$$

where n_{ic0} refers to the rate of accretion on the uncharged particle and equals $4\pi a^2 n_i (k_B T_i/2m_i\pi)^{1/2}$

The usually considered rate of accretion is given by

$$n_{\text{icc}} = n_{\text{ic0}} \exp(-eV_{(r=a)}/k_B T_i) \approx n_{\text{ic0}}(1 - (eV_s/k_B T_i)) \tag{3.18}$$

Hence the total rate of accretion of ions on the dust particle is given by

$$n_{\text{ic}} = n_{\text{ict}} + n_{\text{icc}} = n_{\text{ic0}}[1 - (eV_s/k_B T_i) - (r_0^3/a^2\lambda_m)] \tag{3.19}$$

Elaborate and sophisticated theories for evaluation of $V(r)$ and n_{ic}, taking into account the collision of ions with neutral atoms have been advanced. A discussion of these theories is beyond the scope of this book.

3.5 Schottky Effect and Electron Accretion

As discussed in Sect. 2.3.1, the effective potential energy of an electron near the surface of negatively charged particles is reduced on account of the image force on the electron. Thus [16]

$$V_{\text{eff}} = V_s - \Delta\Phi = (Ze^2/a) - \Delta\Phi \tag{3.20}$$

where $-Ze$ is the charge on the particle, a is the radius of the particle, and $V_s = (Ze^2/a)$ is the potential energy due to the charge on the surface.

Hence from Sect. 3.1 the rate of electron accretion n'_{ec} on the surface of the particle is given by

$$\begin{aligned} n'_{\text{ec}} &= n_e(8k_BT/m_e\pi)^{1/2}\pi a^2\exp(-V_{\text{eff}}/k_BT_e) \\ &= n_{\text{ec}}\exp(-\Delta\Phi/k_BT_e) \end{aligned} \tag{3.21}$$

where n_{ec} is given by (3.11b), $\Delta\Phi$ is tabulated in Table 2.5 as a function of Z and $\mu = D/a$ and D is the Debye length in the plasma.

3.6 Accretion of Electrons/Ions Having Generalized Lorentzian Energy Distribution Function on Dust Particles (After Mishra et al. [9])

In some regions of space, the electrons and ions in the plasma are characterized by a generalized Lorentzian energy distribution function [7, 14]; for such a distribution function (3.9) may be replaced by

$$f_\kappa(u,\upsilon)\mathrm{d}u\mathrm{d}\upsilon = 2\pi\beta(\pi\gamma\upsilon_T^2)^{-3/2}\upsilon\left[1 + (u^2+\upsilon^2)/\gamma\upsilon_T^2\right]^{-(\kappa+1)}\mathrm{d}u\mathrm{d}\upsilon \tag{3.22}$$

where $\upsilon_T = (2k_BT/m)^{1/2}$ is the thermal speed of the plasma electrons/ions, $\gamma = (\kappa - 3/2)$, $\beta = [\Gamma(\kappa+1)/\Gamma(\kappa-1/2)]$ and κ is the spectral index of the distribution.

Following the methodology as outlined in Sect. 3.1, the rate of accretion n_c^κ and the mean energy ε_c^κ of accretion of electrons and ions is given by

$$n_c^\kappa = (4\pi a^2)n\left(\frac{\gamma k_BT}{2\pi m}\right)^{1/2}\left(\frac{\Gamma(\kappa-1)}{\Gamma(\kappa-1/2)}\right)\left(1+\frac{\upsilon_s}{\gamma}\right)^{1-\kappa}, \quad \text{for } \upsilon_s > 0 \tag{3.23a}$$

and

$$n_c^\kappa = (4\pi a^2)n\left(\frac{\gamma k_BT}{2\pi m}\right)^{1/2}\left(\frac{1}{\gamma}\frac{\Gamma(\kappa-1)}{\Gamma(\kappa-1/2)}\right)[\gamma - (\kappa-1)\upsilon_s], \quad \text{for } \upsilon_s \le 0 \tag{3.23b}$$

$$\varepsilon_c^\kappa = (E_c^\kappa/n_c^\kappa) = \left(\frac{2\gamma + \upsilon_s\kappa}{\kappa - 2}\right)k_B T \qquad \text{for } \upsilon_s > 0 \tag{3.24a}$$

and

$$\varepsilon_c^\kappa = (E_c^\kappa/n_c^\kappa) = \left(\frac{\gamma}{(\kappa - 2)}\right)\left(\frac{2\gamma - (\kappa - 2)\upsilon_s}{\gamma - (\kappa - 1)\upsilon_s}\right)k_B T \qquad \text{for } \upsilon_s \leq 0 \tag{3.24b}$$

References

1. J.E. Allen, B.M. Annaratone, U. deAngelis, J. Plasma Phys. **63**, 299 (2000)
2. I.B. Bernstein, I.N. Rabinowitz, Phys. Fluids **2**, 112 (1959)
3. J. Goree, Phys. Rev. Lett. **69**, 277 (1969)
4. M. Lampe, J. Plasma Phys. **65**, 171 (2001)
5. M. Lampe, G. Joyee, G. Ganguli, V. Gavrischaka, Phys. Plasmas **7**, 3851 (2000)
6. M. Lampe, R. Goswami, Z. Sternovsky, S. Robertson, V. Gavrishchaka, G. Ganguli, G. Joyce, Phys. Plasmas **10**, 1500 (2003)
7. D.A. Mendis, M. Rosenberg, Annu. Rev. Astr. Astrophys. **32**, 419 (1994)
8. S.K. Mishra, S. Misra, M.S. Sodha, Phys. Plasmas **18**, 103708 (2011)
9. S.K. Mishra, S. Misra, M.S. Sodha, Europhys. J., *D* , **67**, 210 (2013)
10. S. Misra, S.K. Mishra, M.S. Sodha, Phys. Plasmas **19**, 043702 (2012)
11. B.P. Pandey, S.V. Vladimirov, A.A. Samarian, Phys. Rev. **E83**, 016401 (2011)
12. S.K. Mishra, M.S. Sodha, S. Misra, Phys. Plasmas **19**, 073705 (2012)
13. H.M. Mottsmith, I. Langmuir, Phys. Rev. **28,**727 (1926)
14. M. Rosenberg, D.A. Mendis, J. Geophys. Res. **97**, 14773 (1992)
15. M.S. Sodha, S.K. Mishra, Phys. Plasmas **18**, 044502 (2011)
16. M.S. Sodha, S. Srivastava, Phys. Lett. **A377**, 4773 (2010)
17. M.S. Sodha, S.K. Mishra, S. Misra, Phys. Lett. A **374**, 3376 (2010)
18. Y. Tyshetskiy, S.V. Vladimirov, Phys. Rev. E **83**, 046406 (2011)
19. A.V. Zobnin, A.P. Nefedov, A.P. Sinel'stichikov, V.E. Fortov, J. Exp. Theor. Phys. **91**, 483 (2000)

Chapter 4
Kinetics of Dust-Electron Cloud

4.1 Thermal Equilibrium: Charge Distribution Over Dust

The early experiments by Sugden and Thrush [10] and Shuler and Weber [3] concluded that the electron density in rich hydrocarbon flames (with Carbon dust) is much higher than that predicted by the application of Saha's equation to the gaseous species. Further, it was shown that the electron density could be explained on the basis of thermionic emission from the surface of the graphite dust particles, present in the flames; it was seen that Saha's equation was applicable to the emission from and accretion of electrons on dust (or ionization/recombination in the dust-electron system), when the ionization energy is replaced by the energy needed by an electron to move from inside the dust particle to infinity. The validity of Saha's equation testifies that the dust-electron system is in thermal equilibrium. There are many situations, when the dust-electron system is not in thermal equilibrium; this is particularly true when the dominant electron emission is not of thermal origin. In this section, the statistical mechanics of a dust-electron system in thermal equilibrium has been considered; this analysis is only valid when the gas, in which the dust is suspended, does not directly or indirectly affect the electronic processes.

Consider the following electron emission/accretion process

$$P_{Z-1} \rightleftharpoons P_Z + e^-, \tag{4.1}$$

where P_Z represents a dust particle with charge Ze and e^- represents an electron.

The condition of equilibrium requires

$$n_{Z-1} n_{\text{th}}(Z-1) = n_Z n_{\text{ec}}(Z), \tag{4.2}$$

where n_Z is the number density of particles with a charge Ze, $n_{\text{th}}(Z-1)$ is the rate of thermionic emission from a particle with a charge $(Z-1)e$ and $n_{\text{ec}}(Z)$ (*proportional to* n_e) is the rate of electron accretion on a particle with a charge Ze. Substituting for $n_{\text{th}}(Z-1)$ from Sect. 2.3.2 and for $n_{\text{ec}}(Z)$ from Sect. 3.1 in (4.2) and remembering that the particles and the electrons are at the same temperature T, one obtains an equation, analogous to Saha's equation for thermal ionization of

M. S. Sodha, *Kinetics of Complex Plasmas*, Springer Series on Atomic, Optical, and Plasma Physics 81, DOI: 10.1007/978-81-322-1820-3_4,

atoms, when the ionization potential is replaced by the effective work function $(\varphi + Ze^2/a)$ and the statistical weight of P_Z and P_{Z-1} are equal. Thus

$$K_Z = \frac{n_e n_Z}{n_{Z-1}} = n_s(T)\exp(\alpha/2)\exp(-Z\alpha), \tag{4.3}$$

where K_Z is the equilibrium constant of the reaction represented by (4.1),

$$n_s(T) = (2/h^3)(2\pi m_e k_B T)^{3/2}\exp(-\Phi_0/k_B T),$$

Φ is the work function for an uncharged particle equal to $\Phi_0 - e^2/2a$, Φ_0 is the work function corresponding to a plane surface, $e^2/2a$ arises from the image force, $\alpha = e^2/ak_BT$ and $\upsilon_s = -Z\alpha$.

It can be shown [1, 7] that Saha's equation is valid, even when quantum effects are considered.

Equation (4.3) is valid for positive as well as negative values of Z. From (4.3)

$$\begin{aligned}\frac{n_Z}{n_0} &= \frac{n_Z}{n_{Z-1}}\cdot\frac{n_{Z-1}}{n_{Z-2}}\cdots\cdots\frac{n_1}{n_0} = \prod_{Z=1}^{Z} K_Z/n_e \\ &= \left(\frac{n_s}{n_e}\right)^Z \exp[-\alpha(1+2+\cdots\cdots+Z)]\cdot\exp(Z\alpha/2) \qquad (4.4a)\\ &= \left(\frac{n_s}{n_e}\right)^Z \exp\left[-Z^2\alpha/2\right].\end{aligned}$$

Putting $x = \ln\left(\frac{n_s}{n_e}\right)$, i.e. $n_e = n_s\exp(-x)$ the above equation can be expressed as

$$\ln\frac{n_Z}{n_0} = Zx - Z^2\alpha/2 = -(\alpha/2)\left(Z^2 - 2Z\frac{x}{\alpha} + \frac{x^2}{\alpha^2}\right) + \frac{x^2}{2\alpha} \tag{4.4b}$$

or

$$\frac{n_Z}{n_0} = \exp\left(x^2/2\alpha\right)\exp\left[-\alpha/2(Z - x/\alpha)^2\right]. \tag{4.4c}$$

This is the distribution function of charge on the particles.

The two equations for determination of x and n_0 are

$$n_s\exp(-x) = n_e = \sum_{-\infty}^{\infty} Zn_Z = \overline{Z}n_d \tag{4.5a}$$

and

$$n_d = \sum_{-\infty}^{\infty} n_Z, \tag{4.5b}$$

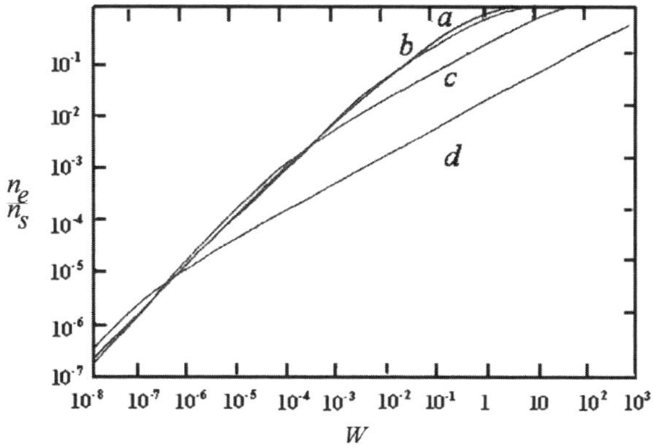

Fig. 4.1 Dependence of (n_e/n_s) on $W = n_d/n_s$ for different values of $\alpha =$ (*a*) 0, (*b*) 2π, (*c*) 4π, (*d*) 24 (after Smith [4], curtsey publisher AIP)

where $\overline{Z}$ is the mean charge on the particles and n_d is the number density of dust particles.

From (4.4a), (4.4b), (4.4c) and (4.5a), (4.5b) one can obtain [4].

$$\overline{Z} = \frac{x}{\alpha} - \frac{h(x, \alpha)}{\alpha} \tag{4.6a}$$

where

$$h(x, \alpha) = \frac{4\pi \sum_{1}^{\infty} n \exp\left(-2\pi^2 n^2/\alpha\right) \sin(2\pi n x/\alpha)}{1 + 2\pi \sum_{1}^{\infty} \exp(-2\pi^2 n^2/\alpha) \cos(2\pi n x/\alpha)}.$$

From (4.5a) and (4.6a) one obtains

$$(n_e/n_s) = \exp[-x] = \left(\overline{Z}\, n_d/n_s\right) = W[x - h(x, \alpha)] \tag{4.6b}$$

where $W = n_d/\alpha n_s$.

The dependence of (n_e/n_s) on W for different values of α is illustrated in Fig. 4.1.

Smith [4] has pointed out that the term $h(x, \alpha)$ is negligible when $\alpha < 6$ (corresponding to particles, larger than 1 nm in radius for a temperature of 2500 K); then (4.6b) reduces to

$$(n_e/n_s) + W \ln(n_e/n_s) = 0. \tag{4.6c}$$

For high values of n_d (i.e. large *W*), $(n_e/n_s) \approx 1$ while for low values of n_d (i.e. small *W*), $n_e \propto n_d^{1/2}$, equivalent to a single charge on some particles. For a specific value of $W(= n_d/\alpha n_s)$, $x = \ln(n_e/n_s)$ and hence n_e can, in general, be obtained from (4.6b) and (4.6c). Knowing x, the charge distribution is given by (4.4c) where n_0 can be evaluated from (4.5b).

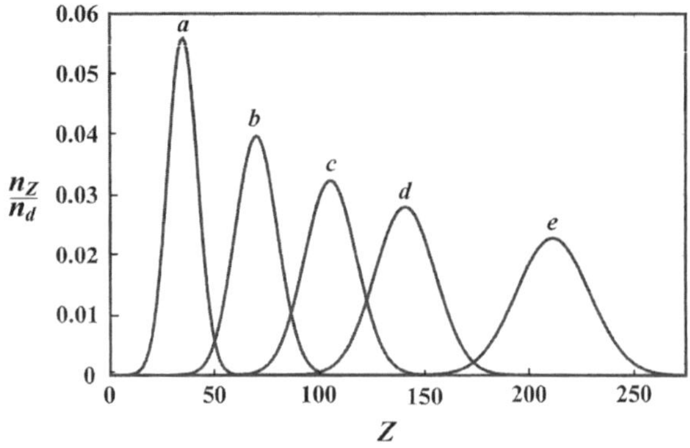

Fig. 4.2 Charge distribution on the particles in a dusty plasma for $n_s/n_e = 2$; the *curves a, b, c, d* and *e* refer to $\alpha = 0.0196$, 0.0098, 0.0065, 0.0049 and 0.00327 respectively (after Misra and Mishra [2])

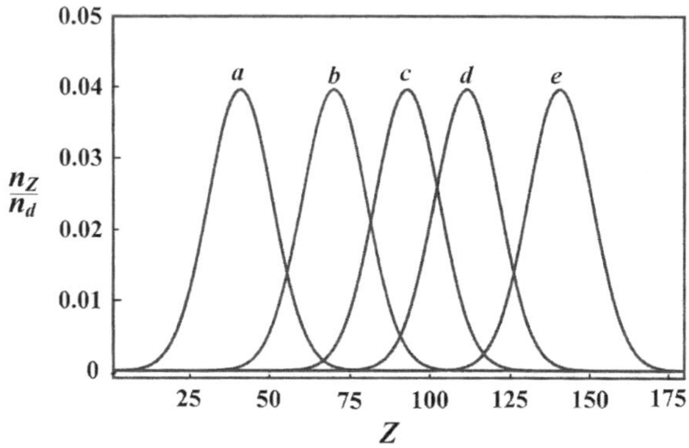

Fig. 4.3 Charge distribution on the particles in a dusty plasma for $\alpha = 0.0098$; the *curves a, b, c, d* and *e* refer to $n_s/n_e = 62.69$, 142.182, 234.54, 337.15 and 566.73 respectively (after Misra and Mishra [2])

To appreciate the charge distribution on particles one may evaluate the mean charge $(\bar{Z} = n_e/n_d)$, and (n_e/n_s) corresponding to set of values of (n_d/n_s) and $\alpha = e^2/akT$. Figures 4.2 and 4.3 illustrate the charge distribution (dependence of n_Z/n_d on Z) for different values of $\alpha = e^2/akT$, corresponding to $(n_s/n_e) = 2$ and for different values of (n_s/n_e), corresponding to $\alpha = 0.0098$. The points on the curves, corresponding to integral values of Z are the only significant points.

4.2 Steady State (Non Equilibrium) Kinetics

4.2.1 Philosophy

Early researchers in the field were interested only in the average charge on the dust particles, which was evaluated from the charge balance on a particle, considering electron emission from and electron/ion accretion on the particle; the effect of dust on the plasma parameters was ignored and hence the theory was applicable only to very low density of dust particles. The next step was to take into account the number balance of electrons/ions, considering electron/ion/dust temperature unchanged. It was, however, realized that since the rates of electron emission from and electron/ion accretion on the dust particles, as well as of the gaseous ionization/deionization processes were strongly temperature dependent, the energy balance of the plasma species and the dust should also be taken into account. To summarize, a satisfactory investigation in the kinetics of complex plasma should include charge balance of the particles and number/energy balance the complex of plasma species. Later considerations included charge distribution on particles having (i) uniform size and (ii) size distribution.

4.2.2 *Charge Distribution in Irradiated Dust Cloud (After Sodha et al. [9])*

Consider a dust cloud, irradiated by light and having dust particles at a temperature T and electrons at a temperature T_e. Following the same logic as in the derivation of (4.3) one obtains

$$K_Z = n_{\text{ph}}(Z-1)/n_Z n_{\text{ec}}, \tag{4.7a}$$

where K_Z is a parameter, corresponding to the present case and $n_{\text{ph}}(Z-1)$ is the rate of photoelectron emission from a particle with charge $(Z-1)e$.

Using the expressions for $n_{\text{ph}}(Z-1)$ from Sect. 2.3.2 and for n_{ec} from Sect. 3.1, one obtains.

$$K_Z = n_p(8k_BT_e/m\pi)^{-1/2}\exp(-Z\alpha_e) \quad \text{for } Z<0 \tag{4.7b}$$

and

$$K_Z = n_p[\psi(\xi, Z\alpha)/\Phi(\xi)](8k_BT_e/m\pi)^{-1/2}(1+Z\alpha_e)^{-1} \quad \text{for } Z>0 \tag{4.7c}$$

where ψ and Φ are given by (2.72b).

n_p is the rate of photoelectron emission per unit area from an uncharged plane surface and $\upsilon_s = -Z\alpha$.

From (4.7b) and (4.7c) one obtains

$$(n_Z/n_0) = \prod_{Z+1}^{0}(n_e/K_Z) \quad \text{for } Z<0 \tag{4.8a}$$

and

$$(n_Z/n_0) = \prod_{1}^{Z}(K_Z/n_e) \quad \text{for } Z>0, \tag{4.8b}$$

Using (4.7b) and (4.7c), (4.8a) and (4.8b) may be expressed as

$$(n_Z/n_0) = \left(YX^{-1/2}\right)^Z \exp[-Z(Z+1)\alpha/2X] \quad \text{for } Z > 0 \tag{4.9a}$$

and

$$(n_Z/n_0) = \left(YX^{-1/2}\right)^Z \prod_1^Z [\psi(\xi, Z\alpha)/(1 + Z\alpha/X)\Phi(\xi)] \exp[-Z(Z+1)\alpha/2X] \quad \text{for } Z > 0, \tag{4.9b}$$

where

$$Y = \left(n_p/n_e\right)\left(8k_BT_e/m\pi\right)^{-1/2}, \quad X = T_e/T$$

$$\psi(\xi, Z\alpha) = Z\alpha \ln[1 + \exp(\xi - Z\alpha)] + \Phi_0(\xi - Z\alpha)$$

and

$$\Phi_0(\Lambda) = \int_0^{\exp\Lambda} \frac{\ln(1+\Omega)}{\Omega} d\Omega.$$

From (3.12b) and (3.12c) the mean energy ε_a of accreting electrons for away from the particle is after putting $v_s = -Z\alpha/X$, given by

$$\varepsilon_a^-(E_a/k_BT) = 2X - Z\alpha \quad \text{for } Z<0 \tag{4.10a}$$

and

$$\varepsilon_a^+(E_a/k_BT) = (2 + Z\alpha/X)/(1 + Z\alpha/X) \quad \text{for } Z > 0, \tag{4.10b}$$

From (2.42a) and (2.73d)

$$\varepsilon_{\text{ph}}(Z-1) = -Z\alpha + 2I(\xi)/\Phi_0(\xi) \quad \text{for } Z<0 \tag{4.11a}$$

and

$$\varepsilon_{\text{ph}}(Z-1) = [Z\alpha\, \Phi_0(\xi - Z\alpha) + 2I(\xi - Z\alpha)]/[\psi(\xi, Z\alpha)], \quad \text{for } Z > 0, \tag{4.11b}$$

where $I(\Omega) = \int_0^\infty \eta \ln[1 + \exp(\Omega - \eta)]d\eta$.

In steady state, the energy of photoelectrons emitted per unit time per unit volume should be equal to the energy of accreting electrons per unit volume per unit time. Thus,

$$\begin{aligned} &\left(\sum_{-\infty}^{0} n_{Z-1} n_{\text{ph}}(Z-1)\varepsilon_{\text{ph}}(Z-1) + \sum_{1}^{\infty} n_{Z-1} n_{\text{ph}}(Z-1)\varepsilon_{\text{ph}}(Z-1)\right) \\ &= \left(n_e \sum_{-\infty}^{0} n_Z \left(\frac{n_c(Z)}{n_e}\right)\varepsilon_c(Z) + n_e \sum_{1}^{\infty} n_Z \left(\frac{n_c(Z)}{n_e}\right)\varepsilon_c(Z)\right), \end{aligned} \tag{4.12}$$

where the left-hand side of above equation refers to the total energy of emitted photoelectrons from the surface of dust and the right hand side of the above equation refers to the total energy of accreted electrons on the surface (per unit time per unit volume) of dust. Hence, substituting for $n_{\text{ph}}(Z-1)$ from Sect. 2.3.2, for $n_{\text{ec}}(Z)$ from Sect. 3.1, $\varepsilon_a(Z)$ from (4.10a), (4.10b), and $\varepsilon_{\text{ph}}(Z-1)$ from (4.11a), (4.11b) and n_Z from (4.9a), (4.9b) one obtains.

$$\begin{aligned}&\left(YX^{-3/2}\right)\left(\sum_{-\infty}^{0}\left(YX^{-1/2}\right)^{Z-1}\exp[-Z(Z-1)\alpha/2X]\left(-Z\alpha+\frac{I(\xi)}{\phi_0(\xi)}\right)\right.\\&\quad\left.+\sum_{1}^{\infty}\left(YX^{-1/2}\right)^{Z-1}\left(\prod_{1}^{Z-1}\frac{\Psi(\xi,Z\alpha)}{((1+(Z-1)\alpha/X)\phi_0(\xi))}\right)\frac{\Psi(\xi,Z\alpha)}{\phi_0(\xi)}\left(-Z\alpha+\frac{I(\xi,Z\alpha)}{\Psi(\xi,Z\alpha)}\right)\right)\\&=\left(\sum_{-\infty}^{0}\left(YX^{-1/2}\right)^{Z-1}\exp[-Z(Z+1)\alpha/2X](2-Z\alpha/X)\exp(-Z\alpha/X)\right.\\&\quad\left.+\sum_{1}^{\infty}\left(YX^{-1/2}\right)^{Z-1}(1+Z\alpha/X)((2X+Z\alpha)/(X+Z\alpha))\left(\prod_{1}^{Z}\frac{\Psi(\xi,Z\alpha)}{((1+Z\alpha/X)\phi_0(\xi))}\right)\right)\end{aligned}\tag{4.13}$$

The charge neutrality for the dust cloud can be expressed as:

$$\frac{n_e}{n_0}=\sum_{-\infty}^{\infty}Z\frac{n_Z}{n_0}=\sum_{-\infty}^{0}Z\frac{n_Z}{n_0}+\sum_{1}^{\infty}Z\frac{n_Z}{n_0}.$$

By substitution from (4.8a) and (4.8b)

$$\begin{aligned}\frac{n_e}{n_0}&=\sum_{-\infty}^{0}Z\left(YX^{-1/2}\right)^{Z}\exp[-Z(Z+1)\alpha/2X]\\&\quad+\sum_{1}^{\infty}Z\left(YX^{-1/2}\right)^{Z}\left(\prod_{1}^{Z}\frac{\Psi(\xi,Z\alpha)}{((1+Z\alpha/X)\phi(\xi))}\right).\end{aligned}\tag{4.14}$$

Total density of the dust particle n_d is given by

$$\frac{n_d}{n_0}=\sum_{-\infty}^{\infty}\frac{n_Z}{n_0}=\sum_{-\infty}^{0}\frac{n_Z}{n_0}+\sum_{1}^{\infty}\frac{n_Z}{n_0}$$

or

$$\begin{aligned}\frac{n_d}{n_0}&=\sum_{-\infty}^{0}\left(YX^{-1/2}\right)^{Z}\exp[-Z(Z+1)\alpha/2X]\\&\quad+\sum_{1}^{\infty}\left(YX^{-1/2}\right)^{Z}\left(\prod_{1}^{Z}\frac{\Psi(\xi,Z\alpha)}{((1+Z\alpha/X)\phi(\xi))}\right)\end{aligned}\tag{4.15}$$

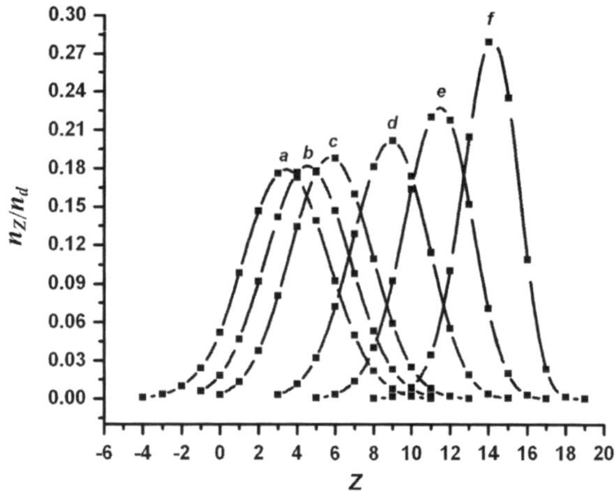

Fig. 4.4 Statistical charge distribution of the dust particles, i.e., variation of (n_Z/n) with Z for $\xi = 141.85$ and $\alpha = 6.39$; the *curves a, b, c, d, e* and *f* refer to $\varsigma = 50.53$, 86.60, 151.75, 555.00, 2005.58, and 9101.45, respectively. Points with (n_Z/n) corresponding to integral values of Z are the only significant ones on the *curves* (after Sodha et al. [9], curtsey authors and publishers IOP)

From a simultaneous solution of (4.13) and (4.5a), (4.5b) and the expressions given earlier n_e, T_e and n_Z can be evaluated.

Figure 4.4 shows the statistical charge distribution [*i.e.* (n_Z/n)] of the dust particles; the curves have been drawn for different values of parameter $\varsigma(=(n_Z/n_d)(8kT_e/m_e\pi)^{1/2})$. Values of (n_Z/n_d) corresponding to integral values of Z are the only significant points on the curves. The curves indicate that the distribution shifts toward lower charging and more broadening with decreasing ς. It is also interesting to mention that for small values of ς [viz. large n_d], a small fraction of the dust particles acquire negative charge. Oppositely charged particles should attract each other, leading to the formation of bigger particles.

There is another method based on the simultaneous solution of rate equations for n_Z and electron density/temperature; the method is detailed in later chapters for the case when the ionization/deionization phenomenon in the host gas are also considered.

4.3 Uniform Charge Theory

When the accuracy of the input data is not good enough, there is little point to the determination of charge distribution over the particles; one may instead assume that all the particles carry the same charge and proceed to evaluate this uniform charge and the electron density and temperature. As an example of this approach, one may consider the electron-dust dynamics in a dust cloud in near space [5].

Only two mechanisms viz. photoelectric emission from and electron accretion on the surface of the dust particles are assumed to determine the charge on the particles and the electron density/temperature. The temperature of the dust particles is obtained by equating the absorbed solar flux to the emitted thermal radiation. If one considers dust of a material with high work function (say 7.8 eV), only solar radiation with wavelength less than 1600 Å will be effective in photoelectric emission. It may be seen from the solar spectrum in near space that in the region $\lambda < 1600$ Å, Lyman Alpha radiation with a wavelength 1215.7 Å has 80 % of photons and 60 % of the energy in this region ($\lambda < 1600$ Å) of interest. Hence, it is a good approximation to assume that the photoelectric emission is caused only by Lyman Alpha radiation.

The electric neutrality, charge balance on the particles and energy balance of electrons require

$$n_e = Zn_d, \tag{4.16a}$$

$$n_{\text{ph}}(Z,T) = n_{\text{ec}}(Z,T_e) \tag{4.16b}$$

and

$$E_{\text{ph}}(Z,T) = E_c(Z,T_e) \tag{4.16c}$$

where n_e and n_d are the densities of electrons and dust particles, $n_{\text{ph}}(Z,T)$ is the rate of photoelectron emission per unit area from the particles and given by (2.72c).

n_p is the rate of photoelectron emission is the product of number of incident photons per unit area per unit time and the photoelectric efficiency,

$$-\upsilon_s = (Z+1)\alpha,$$

$$\alpha = e^2/ak_BT,$$

$E_{\text{ph}}(Z,T)$ is given by (2.73d).

$$n_{\text{ec}} = n_e\pi a^2(8k_BT_e/m_e\pi)^{1/2}(1+Z\alpha_e), \tag{3.11c}$$

$$E_{\text{ec}} = [(2+Z\alpha_e)/(1+Z\alpha_e)]k_BT_e \tag{3.12c}$$

and $\alpha_e = e^2/ak_BT_e$.

The set of three equations (4.16a), (4.16b), (4.16c) can be solved numerically to given n_e, T_e and Z. In what follows some numerical results, obtained from the above considerations for the following set of standard parameters have been presented.

Wavelength of radiation = 1215.7 Å

Work function of dust material = 7.8 eV.

Radius of dust particles a = 100, 175 and 250 Å

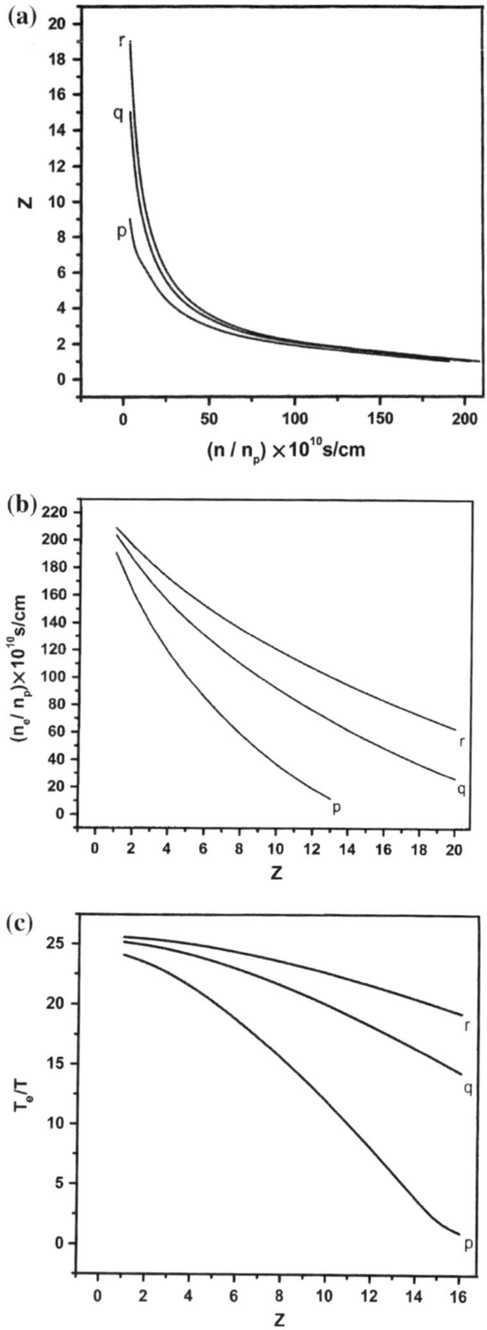

Fig. 4.5 **a** Dependence of particle charge Z on n/n_p for dirty ice particles, irradiated by Lyman-α radiation; the letters p, q and r refer to $a = 100$, 175 and 250 Å (after Sodha et al. [5], curtsey authors and publishers APS). **b** Dependence of electron density n_e/n_p on Z for dirty ice particles, irradiated by Lyman-α radiation; the letters p, q and r refer to $a = 100$ Å, 175 Å and 250 Å (after Sodha et al. [5], curtsey authors and publishers APS). **c** Dependence of electron temperature T_e/T on Z for dirty ice particles irradiated by Lyman-α radiation; the letters p, q and r refer to $a = 100$, 175 and 250 Å (after Sodha et al. [5], curtsey authors and publishers APS)

$$n_p = 1.5 \times 10^{10}/\text{cm}^2\text{s},$$

$$T = 200\,\text{K}.$$

The dependence of the charge on the particles Ze on n_d/n_p is illustrated in Fig. 4.5a for varying a; the dependence of n_e/n_p and T_e/T on Z for the three radii has been illustrated in Fig. 4.5b, c.

4.4 Dust Cloud with Cylindrical Dust Particle

The kinetics of a dust cloud in thermal equilibrium and when illuminated has been developed on similar considerations as in the proceeding sections by Sodha et al. [8].

4.5 Solid State Complex Plasma (After Sodha and Guha [6])

In this section, a summary of section XII of the review by Sodha and Guha [6] on the subject is presented; references may be obtained from the review.

The infrared, E.S.R., optical and NMR spectrum, light scattering measurements, X-ray scattering experiments, electron diffraction studies, and photo conductivity measurements establish the presence of spherical alkali metal particles in additively colored (excess of alkali metal) alkali halide crystals; the size of the particles has been estimated as few hundred atoms of the alkali metal.

The excess amount of alkali metal in an additively colored alkali halide crystal is used in the creation of F Centers (one atom per center) and alkali metal particles. Hence, the number of alkali metal particles n_d is given by

$$n_d = (N_k - N_F)/p, \tag{4.17}$$

where N_k is the number of excess alkali metal atoms per unit volume, N_F is the number of F centers per unit volume and p is the number of alkali metal atoms in a particle.

The number of F centers per unit volume is given by

$$N_F = N_k C \cdot \exp(-\Delta H_c/k_B T) \tag{4.18}$$

where ΔH_c is the energy required to from an F center from an excess alkali metal atom and C is a constant characteristic of the crystal. The radius of the alkali metal particles a can be estimated from the relation

$$p = (4/3)\pi a^3 \rho \Lambda/A, \tag{4.19}$$

where ρ is the density and A is the atomic weight of the alkali metal and Λ is Avogadro's number.

In alkali halide crystals the work function ϕ is replaced by $\phi-\chi$, where χ is the electron affinity of the crystal; hence (4.3) is valid with

$$n_s = \left(2/h^3\right)\left(2\pi m_e k_B T/h^2\right)^{3/2} \exp\left[-(\phi - \chi)/k_B T\right].$$

For experiments of interest $\alpha \ll 6$, and $(n_d/\alpha n_s) \gg 1$.

Therefore

$$n_e = n_s.$$

With the available expression for electron mobility μ_e and n_e, thus obtained the electronic conductivity $\sigma_e = e n_e \mu_e$ may be computed: for additively colored *NaCl* crystal, the evaluated electron conductivity is about 10^{-1} to 10^{-3} times the observed value. The photoconductivity of an illuminated additively colored crystal can also be evaluated likewise.

References

1. S.K. Mishra, M.S. Sodha, S. Misra, Phys. Plasma **19**, 073705 (2012)
2. S. Misra, S.K. Mishra, private communication (2013)
3. K.E. Shuler, J. Weber, A microwave investigation of the ionization of hydrogen and acetylene-oxygen flames. J. Chem. Phys. **22,** 491 (1954)
4. F.T. Smith, On the ionization of solid particles. J. Chem. Phys. **28**, 746 (1958)
5. M.S. Sodha, A. Dixit, S. Srivastava, Phys. Rev. **E 79,** 046407 (2009a); Erratum **E 80,** 069906E (2010)
6. M.S. Sodha, S. Guha, *Physics of Colloidal Plasmas in Advances in Plasma Physics*, vol. 4, ed. by A. Simon, W.B Thompson (Interscience Publishers, New York, 1971), p. 221
7. M.S. Sodha, S.K. Mishra, Phys. Plasma **18**, 044502 (2011)
8. M.S. Sodha, S.K. Mishra, Shikha Misra, Phys. Plasmas **16**, 123701 (2009b)
9. M.S. Sodha, A. Dixit, S. Srivastava, S.K. Mishra, M.P. Verma, L. Bhasin, Plasama Sources Sci. Tech. **19**, 015006 (2010)
10. T.M. Sugden, B.A. Thrush, Nature **168**, 703 (1951)

Chapter 5
Kinetics of Complex Plasmas with Uniform Size Dust

5.1 Introduction

The early studies on the charging of dust particles, suspended in a plasma are characterized by the consideration of the charge balance of the particles taking into account the electron emission from and electron/ion accretion on the dust. These investigations ignored the effect of dust on the plasma parameters viz. on the density and temperature of the constituent electrons, ions, and neutral atoms; thus, these studies were ideally relevant to a single test particle. This chapter presents kinetics of complex plasmas, taking number and energy balance of electrons and ions (in addition to charge balance on particles) and maintenance mechanism of the plasma (without dust).

The next significant advancement in the theory was the inclusion of the number balance of the constituents viz. neutral atoms, electrons, and ions along with the charge balance on the particles. This was the first step in the recognition of the character of openness of complex plasmas, which is an important [26, 27, 29] feature of the kinetics of complex plasma. This feature is related to the continuous flow of electrons and ions toward the surface of the particles and the continuous emission of electrons from the surface. The accreting electrons and ions give their charge to the particles and accreting ions are released as neutral atoms thereafter; thus electron/ion accretion on the surface of the particles provides in effect a mechanism for electron–ion recombination and a sink for plasma electrons and ions. It was also emphasized by Tsytovich and Morfill [27] that the existence of the steady state in complex plasmas requires an agency for ionization of neutral atoms in addition to electron emission from and electron/ion accretion on the dust particles. Diffusion of electrons has also to be considered when the inhomogenity is significant.

Hence, it is necessary [13, 28] to consider the processes for electron/ion generation/annihilation, in addition to electron emission from and electron/ion accretion on the surface of the particles. Since these processes are characterized by significant exchange of energy between the various constituents of the complex plasmas, the rates of these processes depend significantly on the temperatures of the species, which should figure in the kinetics. Thus, the investigations in the

M. S. Sodha, *Kinetics of Complex Plasmas*, Springer Series on Atomic, Optical, and Plasma Physics 81, DOI: 10.1007/978-81-322-1820-3_5,

kinetics of complex plasmas should be based on a system of equations, corresponding to the charge balance on the dust particles and the number and energy balance of the constituents' viz. dust, ions, electrons, and neutral atoms.

The electrical neutrality of the complex plasmas follows from the charge balance on the dust particles and the number balance of electrons and ions and hence it need not to be considered separately. Such an analysis for an illuminated complex plasma having a suspension of dust of uniform size was made by Sodha et al. [21]; appropriate expression Sodha [18] for the photoelectric emission from a charged spherical particle was employed. In a series of papers, this approach was followed up for the investigation of complex plasma kinetics, corresponding to a number of interesting situations. The extension of this approach to complex plasmas, with size distribution of the dust has been discussed in Chap. 7.

The quantum effects in the emission/accretion of electrons, discussed in Chaps. 2 and 3 have in general been ignored in studies on complex plasma kinetics except in three papers [11, 19, 20]. Sodha et. al [21] have also pointed out a serious limitation of the analyses viz. that the mean free path of the electrons should be less than the dimensions of the complex plasma. In this chapter, the kinetics of complex plasmas under various conditions has been discussed.

5.2 Complex Plasma in Thermal Equilibrium

Sugden and Thrush [25] and Shuler and Weber [15] measured the electron density in rich hydrocarbon flames and found the electron density to be much larger than that which could be explained on the basis of the application of Saha's equation to the ionization of the gaseous species. It was also concluded that the observations could be explained on the basis of thermionic emission of electrons from the hot solid carbon particles in the flame; the existence of carbon particles in rich flames has indeed been established by experiments. However, this approach ignores the accretion of ions on the dust (solid carbon) particles and hence is not applicable, when the ion density is appreciable.

Smith [17] extended the analysis [16] of the statistical mechanics of an electron dust cloud, in thermal equilibrium (as in Chap. 4) to the general case to include many species of singly ionized gases and many species of dust with characteristic radius and work function. In this section, the treatment by Smith [17] has been followed.

The single ionization and electron attachment of the jth gas may be represented as

$$n_{sj} = n_{ij} n_e \big/ n_{0j} \tag{5.1a}$$

and

$$L_j = n_e n_{0j} / n_{ij}^- \tag{5.1b}$$

where $n_e, [n_{ij}^-], [n_{0j}]$ and $[n_{ij}]$ represent the number density of electrons, singly charged negative ions, neutral atoms and singly charged positive ions of the jth gas, and n_{sj}/L_j are the equilibrium constants for the ionization/electron attachment reactions of the *j*th gas. The equilibrium constant for single ionization of atoms is given by *Saha's* equation as

$$n_{sj} = \left(n_e n_{ij}/n_{0j}\right) = 2\left(2\pi m_e k_B T/h^2\right)^{3/2} \exp\left(-eV_j/k_B T\right), \tag{5.1c}$$

where V_j is the ionization potential of the *j*th gas.

From (4.6a) and for $\alpha_r > 6$ one obtains

$$N_r\overline{Z_r} = (N_r/\alpha_r)\ln(n_{rs}/n_e) = (N_r/\alpha_r)[\ln(n_{rs}) - \ln(n_e)] \tag{5.1d}$$

where N_r is number density of the *r*th species of particles, $\alpha_r = \left(e^2/a_r k_B T\right)$, a_r is the radius of the *r*th species of particles, n_{rs} is as given by (4.3) and $\overline{Z_r}$ is the mean charge on the dust particles of the *r*th species. Equation (5.2a) has been derived on the basis of (4.3), which does not take into account the phenomenon of ion accretion on dust.

The charge balance requires

$$\sum n_{ij} + \sum N_r\bar{Z}_r = n_e + \sum n_{ij}^- . \tag{5.1e}$$

Substituting for n_{ij}, n_{ij}^- and $N_r\bar{Z}_r$ from (5.1a), (5.1b) and (5.1d), one obtains

$$A - B\ln n_e + (C/n_e) = Dn_e, \tag{5.2a}$$

where

$$A = \sum_r (N_r/\alpha_r)\ln(n_{rs}), \tag{5.2b}$$

$$B = \sum_r (N_r/\alpha_r), \tag{5.2c}$$

$$C = \sum_j n_{sj}\, n_{0j}, \tag{5.2d}$$

and

$$D = 1 + \sum_j [n_{0j}]/L_j. \tag{5.2e}$$

Further, the conservation of the *j*th species of gas requires

$$n_j = n_{ij}^- + n_{0j} + n_{ij}$$

or

$$n_{0j} = n_j/\left[1 + \left(n_{sj}/n_e\right) + \left(n_e/L_j\right)\right]. \tag{5.2f}$$

Smith [17] obtained a different equation for n_e which is valid, only if n_{rs} is the same for all species of dust. Once n_e is known from numerical solutions of (5.2a), $\overline{Z_r}$, n_{ij} and n_{ij}^- may be obtained from (5.1a) and (5.1b); the charge distribution is given by (4.4c). If the gases do not play a significant role, i.e., when the ionization of and electron attachment to atoms are neglected $(C = 0; D = 1)$, (5.2a) reduces to

$$A - B \ln n_e = n_e. \tag{5.2g}$$

The preceding analysis is based on (4.3), which is not valid in the presence of ions. However a simple analysis of the case (neglecting electron attachment) when ions are taken into account is as follows. From (5.1a) and (5.1e), one obtains

$$n_{ij} = n_e n_{sj} / \left(n_e + n_{sj}\right). \tag{5.1f}$$

The charge balance on a particle of rth variety of gas can be represented by

$$N_{\text{rec}} = N_{r\text{th}} + \sum N_{rijc},$$

where $N_{r\text{th}}$, N_{rec} and N_{rijc} are the rates of thermionic electron emission and accretion of electrons and ions of the gas of jth variety corresponding to the particle.

Substituting for the rates of emission and accretion from Chaps. 2 and 3 in the above equation one obtains

$$\pi a_r^2 n_e \upsilon_e (1 + \upsilon_r) = 4\pi a_r^2 N_{\text{tr}} (1 + \upsilon_r) \exp(-\upsilon_r) + \pi a_r^2 \exp(-\upsilon_r) \sum n_{ij} \upsilon_{ij} \text{ for } \upsilon_r > 0 \tag{5.3a}$$

and

$$\pi a_r^2 n_e \upsilon_e \exp(\upsilon_r) = 4\pi a_r^2 N_{\text{tr}} + \pi a_r^2 (1 - \upsilon_r) \sum n_{ij} \upsilon_{ij} \text{ for } \upsilon_r < 0, \tag{5.3b}$$

where
$\upsilon_r = Z_r e^2 / a_r k_B T$,
$Z_r e$ is the charge on particles of rth variety,
$\upsilon_e = (8 k_B T / m_e \pi)^{1/2}$,
$\upsilon_{ij} = \left(8 k_B T / m_j \pi\right)^{1/2}$
and $N_{\text{tr}} = 2\left(2\pi m_e k_B T / h^2\right)^{3/2} \exp(-\phi_r / k_B T)$.
The charge neutrality may be expressed as:

$$\sum N_r Z_r + \sum n_{ij} = n_e$$

or

$$\sum N_r \left(a_r k_B T / e^2\right) \upsilon_r + \sum n_{ij} = n_e \tag{5.4}$$

The unknown parameters viz. $n_{ij}, \upsilon_r(Z_r)$ n_e can be determined from simultaneous solution of (5.4) and sets of (5.1f) and (5.3a). To a higher order of sophistication (4.2) has to be replaced by a system of coupled equations as has been done in the following analysis [22], which considers a singly ionized gas with a single species of spherical dust particles of uniform radius. The analysis can be extended to multiple gases with multiple species of dust. The basic equations of the kinetics of a singly ionized gas and a single species of dust of uniform radius are as follows [22].

5.2.1 Charging

The charge distribution on the particles is given by [9, 10]

$$\begin{aligned}(\mathrm{d}n_Z/\mathrm{d}t) = {} & n_{Z+1}n_{\mathrm{ec}}(Z+1) + n_{Z-1}[n_{\mathrm{th}}(Z-1) + n_{\mathrm{ic}}(Z-1)] \\ & - n_Z[n_{\mathrm{th}}(Z) + n_{\mathrm{ic}}(Z) + n_{\mathrm{ec}}(Z)],\end{aligned} \tag{5.5}$$

where n_z is the fraction of the particles with charge Ze and the expressions for $n_{\mathrm{th}}(Z-1)$, $n_{\mathrm{ec}}(Z)$ and $n_{\mathrm{ic}}(Z)$ are as derived in Chaps. 2 and 3.

The first term on the RHS refers to the production of the particles of charge Ze by electron accretion on particles of charge $(Z+1)e$, the second term refers to the production of particles of charge Ze by thermionic electron emission from and ion accretion on the particle of charge $(Z-1)e$. The last term refers to loss the particles with charge Ze by thermionic emission and electron/ion accretion on the surface.

5.2.2 Conservation of the Sum of Number Densities of Neutral Atoms and Ions

Since a neutral atom produces an ion by ionization and an ion gives rise to an atom by recombination with a neutral atom

$$n_0 + n_i = n_{00} + n_{i0} = N(\text{constant}), \tag{5.6}$$

where n_0 and n_i is the number densities of neutral atoms and ions, respectively; additional subscripts 0 refer to the initial values in a dust free plasmas.

5.2.3 Charge Neutrality

This can be expressed as

$$n_d \sum_{Z_1}^{Z_2} Z n_Z = (n_e - n_i), \tag{5.7}$$

where the consideration is restricted to the particles with charges between Z_1e and Z_2e ($Z_1 \le \overline{Z} \le Z_2$), where $\overline{Z}$ is the mean value of the charge on the particle; it is usual to have $(Z - Z_1)$ and $(Z - Z_2)$ equal to $4 \times 0.5Z^{1/2}$, since the standard deviation for the approximately normal distribution is $0.5\,\overline{Z}^{1/2}$ [3] as obtained from numerical simulation. To determine the mean charge, (5.5) may be replaced by

$$(\mathrm{d}Z/\mathrm{d}t) = n_{\mathrm{th}}(Z) + n_{\mathrm{ic}}(Z) - n_{\mathrm{ec}}(Z). \tag{5.5a}$$

The procedure is to evaluate the mean charge on the particle Z by simultaneous solutions of (5.1c, 5.5a, 5.6 and 5.7) as $t \to \infty$ or $(\mathrm{d}/\mathrm{d}t) \to 0$ (steady state). Using the values of Z, $Z_2 = \overline{Z} + 4 \times 0.5\overline{Z}^{1/2}$ and $Z_1 = \overline{Z} - 4 \times 0.5\overline{Z}^{1/2}$, get fixed and $n\,(Z)$ is obtained from the steady state solution ($t \to \infty$) of the system of (5.5) and (5.1c, 5.6 and 5.7) for $Z_1 \le \overline{Z} \le Z_2$; one has to make the justifiable assumption that n_z is zero for $Z < Z_1$ and $Z > Z_2$.

For a numerical appreciation of the results, the following set of standard parameters may be chosen.

$$N = 10^9\mathrm{cm}^{-3},\ T_{e0} = T_{i0} = T = 1500\,\mathrm{K},\ a = 0.1\ \mu\mathrm{m},\ n_d = 10^3\ \mathrm{cm}^{-3},$$
$$\varphi = 4.0\ \mathrm{eV},\ V_i = 5.2\ \mathrm{eV}\ \text{and}\ m_0 \approx m_i = 30\ \mathrm{amu}.$$

The dependence of the charge distribution on the size of the particle a, the temperature T and the work function of dust (keeping other parameters the same) has been illustrated in Fig. 5.1a, b, and c, respectively.

5.3 Complex Plasma in Absence of Electron Emission from Dust Particles

Consider a gaseous plasma to be characterized by an electron density $n_{\mathrm{e}0}$, electron temperature $T_{\mathrm{e}0}$, ion density $n_{\mathrm{i}0}$, ion temperature $T_{\mathrm{i}0}$, neutral atom/molecule density n_{00} and temperature T_{00}. To a good approximation, the ionization may be modeled as that maintained by the production of $\beta_i n_{00}$ electron ion pairs, per unit volume per unit time and recombination of electrons and ions at the rate of $\alpha_{\mathrm{re}}(T_{\mathrm{e}0})n_{\mathrm{e}0}n_{\mathrm{i}0}$ per unit volume per unit time; β_i and α_{re} are known as the ionization frequency and recombination coefficient. Thus in steady state one has

$$\beta_i n_{00} = \alpha_{\mathrm{re}}(T_{\mathrm{e}0})n_{\mathrm{e}0}n_{\mathrm{i}0} \tag{5.5a}$$

If E_e and E_i denote the mean energy of the electron and ion, produced in the ionization of a neutral atom the energy balance of the electron/ions in the steady state requires

$$E_e = (3/2)k_B T_{\mathrm{e}0}\ \text{and}\ E_i = (3/2)k_B T_{\mathrm{i}0}, \tag{5.5b}$$

Because an electron with mean energy E_e and an ion with E_i recombine to form a neutral atom, the energy of an atom formed by electron ion recombination is given by

$$E_{\mathrm{recomb}} = [(3/2)k_B(T_e + T_i) + V_i], \tag{5.5c}$$

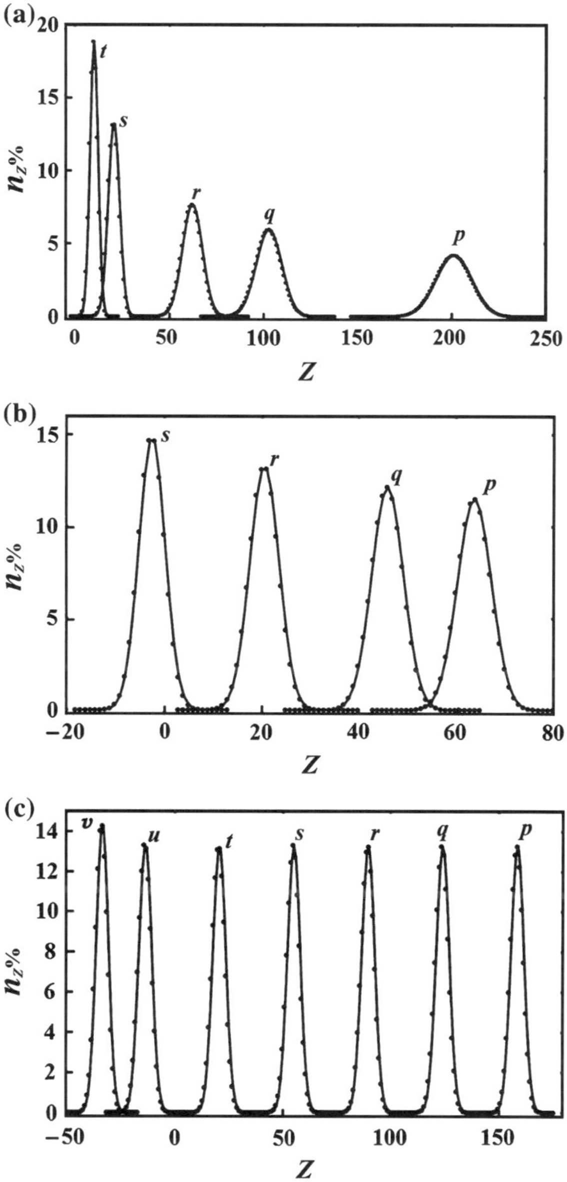

Fig. 5.1 Charge distribution on dust for the standard set of the parameters, given in the text; it is only the values of n_z for integral values of Z, which are meaningful **a** p, q, r, s and t refer to $a = 1.0, 0.5, 0.3, 0.1$ and 0.05 μm, **b** p, q, r and s refer to $T = 2000, 1800, 1500$ and 1200 K, **c** p, q, r, s, t and u refer to $\varphi = 2, 2.5, 3, 3.5, 4.0, 4.5$ and 5.0 eV (after Sodha et al. [22], curtsey authors and publishers IOP)

where V_i is the ionization potential of the atom. The recombination coefficient can usually be expressed as:

$$\alpha_{\mathrm{re}}(T_e) = \alpha_{\mathrm{re}0}(T_e/T_{e0})^{\kappa}.$$

The basic dust parameters, which are relevant to the studies on kinetics of complex plasmas include the rate and mean energy of electrons emitted from the surface of the dust particles which has been discussed in Chap. 2 in the context of thermionic, photoelectric, secondary, normal field and light-induced field emission from spherical and cylindrical particles. An equally important aspect is the rate and mean energy of the accreting electrons and ions, on the surface of the dust particles; this has been discussed in Chap. 3.

In this section, the electron emission from the dust grain has been neglected; a cold dark plasma corresponds to this condition. If dust is introduced in the plasma the basic equations of kinetics are as follows.

5.3.1 Number Balance

$$(\mathrm{d}n_e/\mathrm{d}t) = [\beta_i n_0 - \alpha_{\mathrm{re}}(T_e)n_e n_i] - n_d n_{\mathrm{ec}}(Z) \quad \text{(Electrons)} \tag{5.6a}$$

$$(\mathrm{d}n_i/\mathrm{d}t) = [\beta_i n_0 - \alpha_{\mathrm{re}}(T_e)n_e n_i] - n_d n_{\mathrm{ic}}(Z) \quad \text{(Ions)} \tag{5.6b}$$

$$(\mathrm{d}n_0/\mathrm{d}t) = -(\mathrm{d}n_i/\mathrm{d}t) = [\alpha_{\mathrm{re}}(T_e)n_e n_i - \beta_i n_0] + n_d n_{\mathrm{ic}}(Z) \quad \text{(Neutrals)} \tag{5.6c}$$

Equation (5.6c) is a consequence of the conservation of $(n_0 + n_i)$.

5.3.2 Energy Balance

$$\begin{aligned}\frac{\mathrm{d}}{\mathrm{d}t}\left(\frac{3}{2}n_e k_B T_e\right) = &\left[\beta_i n_0 E_e - \alpha_{\mathrm{re}}(T_e)n_e n_i\left(\frac{3}{2}k_B T_e\right)\right] - n_d n_{\mathrm{ec}}(Z)E_{\mathrm{ec}}(Z)\\ &- \left[n_e \nu_{\mathrm{en}}\delta_{\mathrm{en}}\left(\frac{3}{2}k_B(T_e - T_0)\right) + n_e \nu_{\mathrm{ei}}\delta_{\mathrm{ei}}\left(\frac{3}{2}k_B(T_e - T_i)\right)\right]\\ &+ Q_{\mathrm{ohmic}} \quad \text{(Electrons)}\end{aligned} \tag{5.7a}$$

$$\begin{aligned}\frac{\mathrm{d}}{\mathrm{d}t}\left(\frac{3}{2}n_i k_B T_i\right) = &\left[\beta_i n_0 E_i - \alpha_{\mathrm{re}}(T_e)n_e n_i\left(\frac{3}{2}k_B T_i\right)\right] - n_d n_{\mathrm{ic}}(Z)E_{\mathrm{ic}}(Z)\\ &+ \left[n_e \nu_{\mathrm{ei}}\delta_{\mathrm{ei}}\left(\frac{3}{2}k_B(T_e - T_i)\right) - n_i \nu_{\mathrm{in}}\delta_{\mathrm{im}}\left(\frac{3}{2}k_B(T_i - T_0)\right)\right] \quad \text{(Ions)}\end{aligned} \tag{5.7b}$$

$$\begin{aligned}\frac{\mathrm{d}}{\mathrm{d}t}\left(\frac{3}{2}n_0 k_B T_0\right) = &\left[\alpha_{\mathrm{re}}(T_e)n_e n_i E_{\mathrm{recomb}} - \beta_i n_0\left(\frac{3}{2}k_B T_0\right)\right]\\ &+ \left[n_e \nu_{\mathrm{en}}\delta_{\mathrm{en}}\left(\frac{3}{2}k_B(T_e - T_0)\right) + n_i \nu_{\mathrm{in}}\delta_{\mathrm{im}}\left(\frac{3}{2}k_B(T_i - T_0)\right)\right]\\ &- Q_e \quad \text{(Neutrals)}\end{aligned} \tag{5.7c}$$

5.3.3 Dust Particle Balance

$$(\mathrm{d}Z/\mathrm{d}t) = n_{\mathrm{ic}}(Z) - n_{\mathrm{ec}}(Z) \quad \text{(Charge)} \tag{5.8a}$$

$$\frac{\mathrm{d}}{\mathrm{d}t}\left[m_d C_p T_d\right] = n_{\mathrm{ec}}(Z)E^s_{\mathrm{ec}}(Z) + n_{\mathrm{ic}}(Z)\left[E^s_{\mathrm{ic}}(Z) - \left(\frac{3}{2}k_B T_0\right)\right] - Q_d \quad \text{(Energy)} \tag{5.8b}$$

where n_e, n_i, n_0 and n_d are the number densities of electrons, ions, neutral atoms, and dust particles, T_e, T_i, T_0 and T_d are the temperatures of electrons, ions, neutral atoms, and dust particles, *Ze* is the charge on the particles, β_i is the ionization frequency, $\alpha_{\mathrm{re}}(T_e)$ is the electron–ion recombination coefficient, n_{ec} and n_{ic} are the rates of accretion of electrons and ions on dust particles, E_{ec} and E_{ic} are the mean energies (far from the dust surface) of the accreting electrons and ions on the dust particle, v and δ are the frequency and fraction, characterizing energy exchange in plasma particle collisions, suffixes en, ei, and im refer to collisions between electron–neutral, electron–ion, and ion–neutral collisions, Q_0 is rate of *Ohmic* loss per unit volume, Q_e is the rate of heat loss to the environment per unit volume by the gaseous component (predominantly neutral atoms) and Q_d is the rate of heat loss to the environment per unit volume by the dust particles. From (5.6a), (5.6b), and (5.8a), one obtains $(\mathrm{d}/\mathrm{d}t)(Zn_d + n_i - n_e) = 0$. At $t = 0, Z = 0$ and $n_e = n_i$, and hence one has

$$(Zn_d + n_i - n_e) = 0. \tag{5.9}$$

Thus it is seen that charge neutrality is inherent in the number and charge balance and need not be separately taken into account.

In the steady state $(\mathrm{d}/\mathrm{d}t) = 0$ and the set of (5.6), (5.7) and (5.8a, b) reduces to a set of algebraic equations, which may be solved to get plasma parameters using the appropriate expressions for $n_{\mathrm{ec}}, n_{\mathrm{ic}}, E_{\mathrm{ec}}$ and E_{ic} as in Chap. 3. However, it is convenient to retain the differential form of these equations and solve them by using relevant mathematical software program (*Mathematica* or *Mat-Lab*) using suitable boundary conditions viz. the value of the parameters before the addition of dust and $Z = 0$ at $t = 0$. The steady state $(\mathrm{d}/\mathrm{d}t = 0)$ solution does not depend on the initial conditions, because for $\mathrm{d}/\mathrm{d}t = 0$ the equations reduces to algebraic equations, independent of the boundary conditions. In this section, the typical results of the analysis have not been considered because the more general case of illuminated complex plasma has been discussed later in the text.

5.4 Illuminated Complex Plasmas (After Sodha et al. [23])

5.4.1 Early Investigations

A number of investigations (e.g., [4, 5, 7] on the kinetics of illuminated complex plasma were conducted on the basis of charge balance on the dust particles, i.e. equating electron emission from and ion/electron accretion on the surface of the dust particles. Apart from ignoring the number and energy balance of the constituents of the complex plasmas, these investigations were based on an intuitive but erroneous expression for the rate of photoelectric emission (n_{ph}) from a positively charged spherical particle viz.

$$n_{\text{ph}} = n_p \exp\left(-eV_s/k_B T_p\right) \quad (\text{for } V \geq 0),$$

where $n_p = \pi a^2 \chi J$, a is the radius of the particle, J is the photon flux incident on the particle, $\chi(v)$ is the photo-electric efficiency and T_p is an unspecified temperature of the photoelectrons. The inadequacy of this relation has been discussed at length in Chap. 1.

Sodha et al. (2009b) investigated the kinetics of a dust electron cloud in near space, where the photoelectric emission from the surface of the dust ($\varphi = 7.8$ eV) was caused by the dominant *Lyman-α* radiation (121.57 nm); with photon energy exceeding the work function of the dust particles. The number and energy balance of the electrons was taken into account. The analysis made use of appropriate expressions for the rate of photoelectron emission and the corresponding mean energy of the photoelectrons.

5.4.2 Collisions in Gaseous Plasmas

As seen in Sect. 5.3, the carrier gas (or plasma) plays an important role in complex plasma kinetics and hence it is useful to recapitulate the processes in plasmas/gases, which have a bearing on the kinetics.

5.4.2.1 Ion–Neutral Atom Interaction

The frequency v'_{in} of the collisions of an ion with neutral atoms (not resulting in charge exchange) is given by Banks [1]

$$v'_{\text{in}} = \beta'_{\text{in}} n_o, \tag{5.10a}$$

where β'_{in} is a constant. The values of β'_{in}, corresponding to some neutral-ion pairs has been given by Gurevich [8] in Table 9 of his book.

The frequency of collision of an ion with a neutral atom, resulting in charge exchange is given by Gurevich [8]

$$\nu''_{\text{in}} = \beta''_{\text{in}}(T_i + T_o)^{1/2} n_o \tag{5.10b}$$

where β''_{in} is given in Table 10 of the book by Gurevich [8] for some neutral-ion pairs.

The net ion–neutral collision frequency ν_{in} is given by

$$\nu_{\text{in}} = \left(\nu'_{\text{in}} + \nu''_{\text{in}}\right) \tag{5.10c}$$

This has been tabulated as a function of height, for the ionosphere in Table 11 of the book by Gurevich [8]. The energy exchange in an ion–neutral atom collision is $\delta_{\text{in}}(3k_B/2)(T_i - T_o)$ where $\delta_{\text{in}} = 2m_i/(m_o + m_i) \approx 1$.

5.4.2.2 Electron Collisions

The frequency ν_{en} of the collision of an electron with speed υ with neutral atoms is given by

$$\nu_{\text{en}} = n_0 \upsilon Q(\upsilon), \tag{5.11}$$

where $Q(\upsilon)$ is the collision cross section; for constant mean free path $Q(\upsilon)$ is independent of υ and

$$\nu_{\text{en}} = \nu_{\text{en0}}(T_e/T_{\text{e0}})^{1/2} \text{ with } \nu_{\text{en0}} = n_0 Q(3k_B T_{\text{e0}}/m_e)^{1/2}.$$

The rate of energy loss to the neutral atoms per unit volume is given by $n_e \nu_{\text{en}} \delta_{\text{en}}(3k_B/2)(T_e - T_0)$ where

$$\delta_{\text{en}} = 2m_e/(m_0 + m_e) \approx (2m_e/m_0)$$

The frequency ν_{ei} of the electron collision with ions having charge $Z_i e$ [6, 14] is given by

$$\nu_{\text{ei}} = n_i \upsilon Q_{\text{ei}}(\upsilon) \tag{5.12}$$

where $Q_{\text{ei}}(\upsilon) = \left(2\pi Z_i^2 e^4/m_e \upsilon^4\right) \ln \Lambda$, $\ln \Lambda = \ln\left[1 + \left(D^2 m_e \upsilon^4/e^4\right)\right] \approx 10$ and D is the *Debye* length.

The rate of energy gain by ions from the electrons per unit volume can be given by $n_e \nu_{\text{ei}} \delta_{\text{ei}}(3k_B/2)(T_e - T_i)$ where $\delta_{\text{ei}} = 2m_e/(m_i + m_e) \approx (2m_e/m_i)$.

Following Rose and Clark [14], Sodha [18] obtained the following expression for elastic collisions of electrons with charged particles viz.

$$\nu'_{\text{ed}} = n_d \upsilon Q'_{\text{ed}}(\upsilon), \tag{5.13}$$

where

$$Q_{\mathrm{ed}}(\upsilon) = \left(4\pi Z^2 e^4/m_e^2\upsilon^4\right)\ln[\sin(\chi_1/2)/\sin(\chi_2/2)],\ \cot(\chi_1/2) = \left(m_e^2 b\upsilon^2/Ze^2\right),$$

$$\cot(\chi_2/2) = \left(m_e^2 D\upsilon^2/Ze^2\right) \text{ and } b^2 = a^2\left[1 + \left(2Ze^2/am_e\upsilon^2\right)\right].$$

It is common to use a typical value of the ln term viz. 10.

5.4.3 Specific Problem and Approach (After Sodha et al. [23])

In this section, the kinetics of a complex plasma, characterized by the suspension of the dust of Cesium (Cs) coated Bronze, LaB_6 and CeO_2 with the work function 1.5, 2.0, and 3.0 eV respectively in near space plasma has been studied; the photoelectric ionization of the neutral atoms and the photoelectric emission from the surface of the dust particles by the incident solar radiation are the dominant mechanisms for the generation of electrons.

In view of the nonavailability of the spectral dependence of the absorption coefficient and emissivity of the surface of the dust particles, the energy balance of the dust particles has not been considered but a plausible value viz. 250 K for the temperature of the dust particles has been assumed instead. The attenuation of solar radiation in the complex plasma has also been neglected. The spectral distribution of the solar irradiance can be approximated [1] by a black body (sun) at a temperature of 5800 K for the wavelengths, higher than 122.5 nm the lower wavelength behavior can be approximated [2] by the flux of 3.25×10^{11} photons, corresponding to *Lyman-*α radiation of the wavelength 121.57 nm. For low work function dust (low threshold frequency) the photoelectric efficiency, becomes negligible; hence, the photoelectric emission from solar radiation of wavelengths below 121.57 nm has not been considered.

5.4.4 Rate of Emission and Mean Energy of Photoelectrons

The number of photons incident on a unit area, normal to the direction of incidence per unit time having frequencies between ν and $(\nu + d\nu)$ is given by Fortov et al. [4]

$$\mathrm{d}n_{\mathrm{inc}} = (r_s/r_d)^2 \Lambda(\nu)\mathrm{d}\nu, \tag{5.14}$$

where $\Lambda(\nu) = \left(4\pi\nu^2/c^2\right)f_{\mathrm{tr}}(\nu)[\exp(h\nu/k_B T_s) - 1]^{-1}\mathrm{d}\nu$; $f_{\mathrm{tr}}(\nu)$ is the transmission function of the experimental chamber (it is unity for a dust cloud in near space), $r_s(\approx 6.96 \times 10^{10}\,\mathrm{cm})$ is the radius of the radiating surface of the sun,

$r_d(\approx 1.5 \times 10^{13}\,\text{cm})$ is the mean distance between the sun and the ensemble of the dust particles.

After photoemission of an electron, the charge Ze on the particle changes to $(Z+1)e$. Hence using the results in Chap. 2, the rate of photoemission of electrons from the particles corresponding to $\mathrm{d}n_{\text{inc}}$ is given by

$$n_{\text{ph}}(Z) = \int_{\nu_0}^{\nu_m} \mathrm{d}n_{\text{ph}}(Z) = \pi a^2 (r_s/r_d)^2 \int_{\nu_0}^{\nu_m} \chi(\nu)\Lambda(\nu)\left[\frac{\Psi(\xi, \overline{Z+1}\alpha_d)}{\Phi(\xi)}\right]\mathrm{d}\nu, \tag{5.15}$$

where $h\nu_0 = \varphi + \left[(Z+1)e^2/a\right]$ is the minimum frequency for photoemission and ν_m is the upper limit of the incident solar radiation spectrum.

The spectral dependence of the photoelectric efficiency $\chi(\nu)$ is given by Spitzer [24] as

$$\chi(\nu) = 0 \quad \text{for } \nu < \nu_{00} \tag{5.16a}$$

and

$$\chi(\nu) = b\frac{1}{\nu^4}\left(1 - \frac{\nu_{00}}{\nu}\right)^2, \quad \text{for } \nu \geq \nu_{00}, \tag{5.16b}$$

where $\nu_{00} = (\varphi/h)$ is the threshold frequency and b is a constant for a given surface. In terms of maximum value χ_m of $\chi(\nu)$, (5.16b) can be expressed as

$$\frac{\chi(\nu)}{\chi_m} = \frac{729}{16}\left(\frac{\nu_{00}}{\nu}\right)^4\left(1 - \frac{\nu_{00}}{\nu}\right)^2. \tag{5.16c}$$

The mean energy of the photoelectrons, emitted by a particle with positive charge Ze is given by

$$\begin{aligned} n_{\text{ph}}(Z)\bar{E}_{\text{ph}}(Z) &= \int_{\nu_0}^{\nu_m} n_{\text{ph}}(Z)\left[\mathrm{d}n_{\text{ph}}(Z)/\mathrm{d}\nu\right]\mathrm{d}\nu \\ &= \pi a^2 (r_s/r_d)^2 \int_{\nu_0}^{\nu_m} E_{\text{ph}}(Z,\nu)\chi(\nu)\Lambda(\nu)\left[\frac{\Psi(\xi, \overline{Z+1}\alpha_d)}{\Phi(\xi)}\right]\mathrm{d}\nu \end{aligned} \tag{5.17}$$

where $-\nu_s = (Z+1)e^2/ak_BT_d$, $E_{\text{ph}}(Z,\nu) = k_BT_d\varepsilon_{\text{ph}}$, ε_{ph} is given by (2.73d) for $Z > 0$, $\varepsilon_{\text{ph}} = \nu_s + \langle\varepsilon''\rangle$ and is given for $Z < 0$ by (2.42a).

5.4.5 Analytical Model

A. Conservation of the sum of densities of neutral atoms and ions

$$n_0 + n_i = n_{00} + n_{i0} = N \tag{5.6}$$

B. Charge neutrality

$$Zn_d + n_i = n_e \tag{5.18}$$

C. Charging of the dust particles

$$(\mathrm{d}Z/\mathrm{d}t) = n_{\mathrm{ic}} + n_{\mathrm{ph}} - n_{\mathrm{ec}}, \tag{5.19}$$

D. Number balance of electron

$$(\mathrm{d}n_e/\mathrm{d}t) = \beta_i n_0 + n_{\mathrm{ph}} n_d - \alpha_r n_e n_i - n_{\mathrm{ec}} n_d, \tag{5.20}$$

where β_i is the coefficient of ionization and $\alpha_r(T_e) = 5.0 \times 10^{-7}(300/T_e)^{1.2}\,\mathrm{cm}^3/\mathrm{s}$ is a typical [2, 8] coefficient of electron/ion recombination. In the absence of dust one has $\beta_i n_{00} = \alpha_r n_{e0} n_{i0} = \alpha_r n_{e0}^2$.

E. Number balance of Ions

$$(\mathrm{d}n_i/\mathrm{d}t) = \beta_i n_0 - \alpha_r n_e n_i - n_{\mathrm{ic}} n_d. \tag{5.21}$$

F. Energy balance for electrons

$$\begin{aligned}\frac{\mathrm{d}}{\mathrm{d}t}\left(\frac{3}{2}k_B n_e T_e\right) &= \beta_i n_0 \varepsilon_e + n_{\mathrm{ph}} n_d \varepsilon_{\mathrm{ph}} - \alpha_r n_e n_i (3k_B T_e/2) - n_d n_{\mathrm{ec}} \varepsilon_{\mathrm{ec}} \\ &\quad - \nu_{\mathrm{em}} \delta_{\mathrm{em}} (3k_B/2)(T_e - T_0) n_e - \nu_{\mathrm{ei}} \delta_{\mathrm{ei}} (3k_B/2)(T_e - T_i) n_e\end{aligned} \tag{5.22}$$

where $\varepsilon_e \approx (3/2)k_B T_{e0}$ is the energy of an electron produced by neutral ionization, expression for $\varepsilon_{\mathrm{ec}}$ has been given in Chap. 3 and the significance of ν_{em}, δ_{em}, ν_{ei} and δ_{ei} has been explained in the Sect. 5.4.2.

G. Energy balance for ions

$$\begin{aligned}\frac{\mathrm{d}}{\mathrm{d}t}\left(\frac{3}{2}k_B n_i T_i\right) &= \beta_i n_0 \varepsilon_i + \nu_{\mathrm{ei}} \delta_{\mathrm{ei}} (3k_B/2)(T_e - T_i) n_e - n_{\mathrm{ic}} n_d \varepsilon_{\mathrm{ic}} \\ &\quad - \alpha_r n_e n_i (3k_B T_i/2) - \nu_{\mathrm{im}} \delta_{\mathrm{im}} (3k_B/2)(T_i - T_0) n_i,\end{aligned} \tag{5.23}$$

where $\varepsilon_i \approx (3/2)k_B T_{i0}$ is the energy of an ion produced by neutral ionization, expression for $\varepsilon_{\mathrm{ic}}$ has been given in Chap. 3 and the significance of ν_{im} and δ_{im} has been explained in the Sect. 5.4.2.

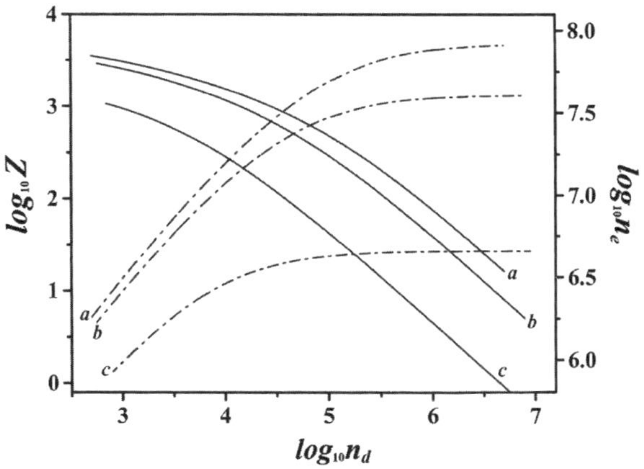

Fig. 5.2 Dependence of charge of dust particles *Ze* and electron density n_e, on the number density of the particles n_d, for the standard set of parameters. The labels *a, b* and *c* on the *curves* refer to the dust of Cs coated Bronze, LaB_6 and CeO_2 respectively; *solid* and *broken curves* correspond to *left* and *right hand scales* respectively (after Sodha et al. [23], curtsey authors and publishers AIP)

H. Energy balance for atoms

$$\begin{aligned}\frac{\mathrm{d}}{\mathrm{d}t}\left(\frac{3}{2}kn_0T_0\right) &= \alpha_r n_e n_i\left[(3k/2)(T_e+T_i)+I_p\right]+n_{\mathrm{ic}}n_d(3k_BT_d/2)\\ &\quad+\nu_{\mathrm{em}}\delta_{\mathrm{em}}(3k_B/2)(T_e-T_0)n_e+\nu_{\mathrm{im}}\delta_{\mathrm{im}}(3k_B/2)(T_i-T_0)n_i\\ &\quad-\beta_i n_0(3k_BT_0/2)-E_{\mathrm{diss}}\end{aligned} \tag{5.24}$$

where E_{diss} is the rate of energy loss per unit volume to the environment.

I. Energy balance for dust particles

$$\begin{aligned}\frac{\mathrm{d}}{\mathrm{d}t}\left(m_dC_pT_d\right) &= \pi a^2\alpha_a S-4\pi a^2 \quad \in \sigma\left(T_d^4-T_{\mathrm{env}}^4\right)\\ &\quad-4\pi a^2 n_0(8k_BT_0/\pi m_0)^{1/2}(T_d-T_0).\end{aligned} \tag{5.25}$$

where α_a is the absorption coefficient of the material of the dust particles for solar radiation, $\in$ is the emissivity of the material of the dust particles and σ is the Stefan-Boltzmann constant. Electron collisions with gaseous species can be taken into account as in Sect. 5.4.2.

The first term on the right-hand side of (5.25) is the power of incident radiation absorbed by a dust particle and the second term is the power loss by dust particle by radiation; the last term is the rate of heat transfer from the dust particle to the low pressure neutral gas.

The system of (5.6) and (5.19) to (5.24) can be solved by *Mathematica Software* with the initial conditions, conforming to the plasma, without dust. As $t \to \infty$, $(\mathrm{d}/\mathrm{d}t) \to 0$, the solution conforms to the steady state, independent of the initial conditions. Since the charge neutrality, expressed by (5.18) can be derived from (5.19), (5.20) and (5.21), it has been omitted in the analysis (as redundant).

Sodha et al. [23] have made a parametric analysis of the charge on the particles and other plasma parameters by choosing the following set of standard parameters and study the effect of changing a single parameter keeping others unchanged.

$$n_{e0} = n_{i0} \approx 3 \times 10^5/\mathrm{cm}^3,\ n_{00} = 10^3 n_{e0},\ T_{e0} = 800\ \mathrm{K},\ T_{i0} = T_{00} = 670\ \mathrm{K},$$
$$T_d = 250\ \mathrm{K},\ \nu_{\mathrm{ei0}} = 480\ \mathrm{s}^{-1},\ \nu_{\mathrm{em0}} = 910\ \mathrm{s}^{-1},\ \nu_{\mathrm{im0}} = 100\ \mathrm{s}^{-1},\ \chi_m = 0.05,$$
$$a = 5.0\ \mu,\ V_i = 9.3\ \mathrm{eV},\ m_i 0 = 30\ \mathrm{amu},\ S = 1.368 \times 10^6\ \mathrm{ergs/cm}^2,$$
$$C_p = 4.47 \times 10^6\ \mathrm{ergs/gK},\ \rho = 7.87\ \mathrm{g/cm}^3\ \text{and}\ m_d = (4/3)^3 \rho.$$

The computations correspond to the dust of Cs coated Bronze, LaB_6 and CeO_2 (as an illustration) with the work functions 1.5, 2.0, and 3.0 eV, respectively. The dependence of the charge on the particles Ze and electron density n_e for the dust of the three materials on the number density of the dust particles is shown in Fig. 5.2.

In this analysis, photoelectric emission is the only mechanism for the emission of electrons from the surface of the dust particles, which has been considered. Other mechanisms viz. thermionic emission, electric field emission [20], light-induced field emission [19], and secondary electron emission [12] have also been considered in other analyses.

References

1. P.M. Banks, Proc. I.E.E.E. **57**, 258 (1959)
2. S.J. Bauer, *Physics of Planetary Ionospheres* (Springer, New York, 1973)
3. C. Cui, G. Goree, IEEE Trans. Plasma Sci. **22**, 151 (1994)
4. V.E. Fortov, A.P. Nefedov, O.S. Vaulina, A.M. Lipaev, V.I. Molotokov, A.A. Samarian, V.P. Nikitski, A.I. Ivanov, S.F. Savin, A.V. Kalinikov, A. Ya Solov'ev, P.V. Vinogradov, J. Exp. Theor. Phys. **87**, 1087 (1998)
5. V.E. Fortov, A.P. Nefedov, O.S. Vaulina, O.F. Petrov, I.E. Dranzhevski, A.M. Lipaev, Yu P Semenov, New J. Phys. **5**, 102 (2003)
6. V.L Ginzburg, A.V Gurevich, Sov. Phys. Usp. **3**, 115 (1960)
7. J. Goree, Plasma Sources Sci. Technol. **03**, 400 (1994)
8. A.V. Gurevich, *Some Nonlinear Phenomena in the Ionosphere* (Springer, New York, 1978)
9. T. Matasoukos, M. Russel, J. Appl. Phys. **77**, 4285 (1995)
10. T. Matasoukos, M. Russel, M. Smith, J. Vac. Sc.Tech. A **14**, 624 (1996)
11. S. Misra, S.K. Mishra, M.S. Sodha, Phys. Plasmas **19**, 043702 (2012)
12. S. Misra, S.K. Mishra, M.S. Sodha, Phys. Plasmas **20**, 013702 (2013)
13. K.N. Ostrikov, M.Y. Yu, L. Stenflo, Phys. Rev. E **61**, 4314 (2000)
14. D.J. Rose, M. Clark, *Plasmas and Controlled Fusion* (John Wiley, New York, 1961)
15. K.E. Shuler, J. Weber, J. Chem. Phys. **22**, 491 (1954)
16. F.T. Smith, J. Chem. Phys. **28**, 746 (1958)
17. F.T. Smith, in *Proceedings of Third Conference on Carbon*, University of Buffalo, Buffalo, New York 1957, Pergamm Prev., New York, p. 301 (1959)
18. M.S. Sodha, Brit. J. Appl. Phys. **14**, 172 (1963)
19. M.S. Sodha, A. Dixit, Appl. Phys. Lett. **95**, 101502 (2009)

20. M.S. Sodha, A. Dixit, G. Prakash, J. Plasma Phys. **76**, 159 (2010)
21. M.S. Sodha, A.Dixit, S. Srivastava, Phys. Rev. E **79**, 046407 (2009), *Erratum* E **80**, 06990 (2009)
22. M.S. Sodha, S.K. Mishra, S. Misra, Phys. Scripta **83**, 015502 (2011)
23. M.S. Sodha, S. Misra, S.K. Mishra Phys. Plasmas **16**, 123705 (2009a), *Erratum* **17**, 049902 (2010)
24. L. Spitzer, Astrophys. J. **107**, 6 (1948)
25. T.M. Sugden, B.A. Thrush, Nature **168**, 703 (1951)
26. V.N. Tsytovich, Sov. Phys. Uspekhi **40**, 53 (1997)
27. V.N. Tsytovich, G.E. Morfill, Plasma Physics Rep. **28**, 171 (2002)
28. V.N. Tsytovich, G.E. Morfill, S.V. Vladimirov, H.M. Thomas *Elementary physics of complex plasmas*, (Springer, Berlin, 2008)
29. S.V. Vladimirov, Phys. Plasmas **1**, 2762 (1994)

Chapter 6
Kinetics of Flowing Complex Plasma

6.1 Introduction

In the last chapter, the kinetics of a uniform, isotropic, and stationary complex plasma with dust particles of uniform size has been discussed. However frequently, the plasma exhibits a drift relative to the dust particles; it is a common feature in *rf discharge* laboratory experiments and the charging of satellites, spacecraft, and meteors/meteorites in the ionospheric and near earth space plasmas [2–5, 12, 15, 24]. In such situations, the symmetrical nature of the charging phenomena breaks down, and the angular momentum of the accreting ions is no longer conserved. This asymmetric behavior modifies the accretion cross-section and velocity distribution function of electrons and ions, and in turn affects the particle charge and other plasma characteristics and transport properties. In an elegant review, Whipple [24] analyzed the charging of moving bodies (satellite and spacecraft) in space by applying the balance between the various charging currents on the bodies; an expression for the ion flux toward a moving spherical body, originally derived by Kanal [8] was employed. By reviewing the status of the subject, Melandso [11] pointed out that the mesothermal flow (speed larger than thermal speed of ions but smaller than electron thermal speed) of plasmas with respect to dust is a typical situation in industrial plasmas and laboratory plasma experiments; in particular, the characteristic features of the crystal structure formation and their heating in a flowing plasma were discussed. Discussing an interesting aspect of flowing plasma *viz.* meteor showering, Sorasio et al. [20] analyzed the dynamics and charging of a meteoroid entering in the Earth's atmosphere by solving the simultaneous equations for conservation of charge, mass, momentum, and energy. The plasma flow occurring in the fabrication of devices and the interactions of flowing plasmas with near-wall impurities and/or dust significantly affects the efficiency and lifetime of such devices [21, 23]. Considering the relevance of the complex plasmas characterized by relative motion of dust and plasma to natural phenomena, laboratory experiments, nano/micro device fabrication, fusion and industrial applications, the kinetics of such plasmas has been discussed in this chapter, and the charge distribution on the dust particles has been obtained.

M. S. Sodha, *Kinetics of Complex Plasmas*, Springer Series on Atomic, Optical, and Plasma Physics 81, DOI: 10.1007/978-81-322-1820-3_6,

6.2 Modification in Electron/Ion Accretion Current to Particles

The electrons/ions in an isotropic plasma in equilibrium are characterized by the *Maxwellian* distribution function of energy; however, in the presence of a relative flow of a plasma constituents, the distribution function gets modified to shifting *Maxwellian*, which leads to a significant modification of the accretion current. Following Chap. 2, the shifting *Maxwellian* distribution [13] can be expressed as:

$$f(c) = n\left(\pi \upsilon_T^2\right)^{-3/2} \exp\left[-(\mathbf{c}-\mathbf{C})^2 / \upsilon_T^2\right] \tag{6.1}$$

where C is the velocity of electron/ions relative to the dust particle and $\upsilon_T^2 = (2k_BT/m)$ refers to the thermal speed.

Utilizing the alternate method based on the *OML* approach, described in Sect. 2.3, the cross-section for electron/ion accretion on a dust particle surface (having charge Ze) may be expressed as [17]:

$$\sigma(c) = \pi a^2\left[1 - \left(2V_s/mc^2\right)\right], \tag{6.2}$$

where $V_s = \left(\pm Ze^2/a\right)$ and the $\pm$ signs correspond to ions/electrons, respectively.

Hence, the rate of accretion (n_c) on the surface of the dust particle corresponding to the shifted *Maxwellian* distribution [$f(c)$] of speeds is given by Mishra et al. [13].

$$\begin{aligned} n_c(V_s) &= \int c\cdot\sigma(c) f(c) \mathrm{d}^3 c \\ &= \pi a^2 n\left(\pi\upsilon_T^2\right)^{-3/2} \int c\left(1-\frac{2V_s}{mc^2}\right)\exp\left[-\frac{(\mathbf{c}-\mathbf{C})^2}{\upsilon_T^2}\right]\mathrm{d}^3 c \\ &= \pi a^2 n\left(\pi\upsilon_T^2\right)^{-3/2} \int_{0,V_s}^{\infty}\int_{-1}^{1} c\left(1-\frac{2V_s}{mc^2}\right)\exp\left[-\frac{(c^2+C^2-2cC\cos\theta)}{\upsilon_T^2}\right](2\pi c^2)\mathrm{d}c\mathrm{d}(\cos\theta) \\ &= \frac{(\pi a^2 n_e)\upsilon_T}{\pi^{1/2}p}\int_{x_m}^{\infty}\left(x^2-\upsilon_s\right)\left[\exp\left[-(x-p)^2\right]-\exp\left[-(x+p)^2\right]\right]\mathrm{d}x \end{aligned} \tag{6.3}$$

where $\upsilon_s = \left(2V_s/m\upsilon_T^2\right)$, $p = (C/\upsilon_T)$, $x = (c/\upsilon_T)$ and $x_m^2 = \upsilon_s$ for $\upsilon_s > 0$ and $x_m^2 = 0$ for $\upsilon_s \le 0$, respectively. Using the accretion conditions viz. $x_m^2 = \upsilon_s$, 0 corresponding to $\upsilon_s > 0$ and $\upsilon_s \le 0$, respectively (6.3) gives *for* $\upsilon_s \ge 0$.

$$\begin{aligned} n_c(\upsilon_s) = \left(\pi a^2 n\right)\left(\upsilon_T/4\pi^{1/2}p\right)&\left[2\left[p_m\exp\left(-p_n^2\right)-p_n\exp\left(-p_m^2\right)\right]\right. \\ &\left.+\pi^{1/2}(1-2p_mp_n)[\mathrm{erf}(p_m)-\mathrm{erf}(p_n)]\right] \end{aligned} \tag{6.4a}$$

and *for* $\upsilon_s < 0$,

$$n_c(\upsilon_s) = (\pi a^2 n)(v_T/2\pi^{1/2}p)[2p\exp[-p^2] + (1 + 2p^2 + 2\upsilon_s)\pi^{1/2}\mathrm{erf}[p]] \quad \text{for } \upsilon_s < 0, \tag{6.4b}$$

where $p_n = (\upsilon_s^{1/2} - p)$ and $p_m = (\upsilon_s^{1/2} + p)$.

Further, the mean energy associated with the accreting electrons/ions can be written as [13]:

$$\begin{aligned} E_c(V_s) &= \int (m/2)(\boldsymbol{c} - \boldsymbol{C})^2[c \cdot \sigma(c)f(c)]\mathrm{d}^3c \\ &= \pi a^2 n(\pi\upsilon_T^2)^{-3/2}\int (m/2)(\boldsymbol{c} - \boldsymbol{C})^2 c\left(1 - \frac{2V_s}{mc^2}\right)\exp\left[-\frac{(\boldsymbol{c} - \boldsymbol{C})^2}{\upsilon_T^2}\right]\mathrm{d}^3c. \end{aligned}$$

Following the algebraic treatment and simplifications, similar to that in derivation of the accretion current, one obtains *for* $\upsilon_s \geq 0$

$$\begin{aligned} n_c(\upsilon_s)E_c(\upsilon_s) = k_BT(\pi a^2 n)\left(\upsilon_T/8\pi^{1/2}p\right)&[6(p_m\exp[-p_n^2] - p_n\exp[-p_m^2]) \\ &+4\upsilon_s^{1/2}(\exp[-p_n^2] - \exp[-p_m^2]) \\ &+\pi^{1/2}(5 - 6p_m p_n)(\mathrm{erf}[p_m] - \mathrm{erf}[p_n])] \end{aligned} \tag{6.5a}$$

and *for* $\upsilon_s < 0$

$$\begin{aligned} n_c(\upsilon_s)E_c(\upsilon_s) = k_BT(\pi a^2 n)\left(\upsilon_T/4\pi^{1/2}p\right)&[p(6 - 4\upsilon_s)\exp(-p^2) \\ &+ \pi^{1/2}(5 + 6\upsilon_s + 6p^2)\mathrm{erf}[p]] \end{aligned} \tag{6.5b}$$

Further, the mean energy associated with the accreting electrons/ions can be expressed in the dimensionless form as $\varepsilon_c = (E_c/k_BT)$.

6.3 Kinetics

6.3.1 Charge Distribution

As discussed in the Chap. 5, the charge on a dust particle in a complex plasma undergoes fluctuations around the mean charge due to the random nature of the dust charging processes (emission/accretion) [1, 6], (Sodha et al. [18]); this charge fluctuation leads to a charge distribution on the dust particles in the complex plasma. Matasoukas et al. [10, 11] have analyzed the charge distribution considering a *Markov process* [22] in which the probability density is governed by a master difference equation for the population balance ensuring the discrete nature of charge on the particles. Using this master difference equation, Sodha et al. [18] developed the statistical mechanics of the charge distribution on uniform size dust particles in dark, thermal, and illuminated complex plasmas on the basis of the number and energy balance of the plasma constituents; the study shows good agreement with the fluctuation theory [1], (Sodha et al. [18]) for large values of charge. Mishra et al. [13] have extended the theory of the charge distribution to study the effect of relative motion of dust and plasma system.

6.3.2 Master Equation for the Population Balance

In a complex plasma, the dust particles undergo random charging processes viz. emission of the electrons from and accretion of electrons/ions on the surface of the dust particles, the inherent charging processes can be represented as:

$$P_{Z-1} \rightleftharpoons P_Z + e,\ P_{Z+1} + e \rightleftharpoons P_Z \text{ and } P_{Z-1} + i \rightleftharpoons P_Z.$$

where P_Z refers to particles with charge Ze; e and i represent electrons and ions, respectively.

As discussed in Sect. 5.2, the population balance equation, describing the generation and depletion of dust particles with charge Ze can be written as [19]:

$$\begin{aligned}(\mathrm{d}n_Z/\mathrm{d}t) = {} & n_{Z+1}n_{\mathrm{ec}}(Z+1) + n_{Z-1}[n_{\mathrm{ee}}(Z-1) + n_{\mathrm{ic}}(Z-1)] \\ & - n_Z[n_{\mathrm{ee}}(Z) + n_{\mathrm{ic}}(Z) + n_{\mathrm{ec}}(Z)]\end{aligned} \tag{6.6}$$

where $n_Z(= N_Z/n_d)$ is the charge distribution viz. the fraction of particles carrying a charge Ze, n_{ec}, and n_{ic} refer to electron/ion accretion currents on the surface of dust grain and n_{ee} represents the rate of electron emission from the dust particles.

6.4 Other Kinetic Equations

6.4.1 Conservation of Neutral Plus Ionic Species

As described in Chap. 5, for a plasma in the steady state

$$n_{\mathrm{o}} + n_i = n_{00} + n_{i0} = N(\text{constant}), \tag{6.7}$$

where n_{o} and n_i are the number density of neutral atoms and ions, respectively, and the additional subscript "0" corresponds to number densities in dust-free plasmas.

6.4.2 Charge Neutrality

$$n_d \sum_{Z_1}^{Z_2} Zn_Z = (n_e - n_i). \tag{6.8}$$

To keep the algebra/mathematics tractable, one restricts the consideration to the values of Z lying between Z_1 and Z_2 $\left(Z_1 \leq \overline{Z} \leq Z_2\right)$; n_d is the number density of dust particles, n_e is the electron density and $\overline{Z}$ refers to the average charge on dust particles. The values of Z_1 and Z_2 can be chosen arbitrarily such that $n_Z \to 0$ for $Z < Z_1$ and $Z > Z_2$.

6.4.3 Electron and Ion Kinetics

$$(\mathrm{d}n_e/n_e\mathrm{d}t\mathrm{d}t) = (\beta_i n_\mathrm{o} - \alpha_r n_e n_i) - n_d\left(\sum_{(Z_1+1)}^{Z_2} n_Z n_\mathrm{ec}(Z) - \sum_{Z_1}^{(Z_2-1)} n_Z n_\mathrm{ee}(Z)\right) \quad (6.9)$$

and

$$(\mathrm{d}n_i/\mathrm{d}t) = (\beta_i n_o - \alpha_r n_e n_i) - n_d \sum_{Z_1}^{(Z_2-1)} n_Z n_\mathrm{ic}(Z) \quad (6.10)$$

where β_i is the coefficient of ionization, $\alpha_r(T_e) = \alpha_{r0} T_e^{-\eta}\,\mathrm{cm}^3/\mathrm{s}$ is the coefficient of recombination of electrons and ions [6], T_e is the electron temperature and α_{r0} and η are constants.

The first term on the right in both (6.9, 6.10) refer to the net gain in electron and ion density per unit time on account of ionization of neutral atoms and electron–ion recombination processes, respectively. The second term represents the loss in electron/ion density on account of the net accreting electron/ion flux over the dust particle and electron emission.

6.4.4 Energy Balance for Electrons and Ions

$$\begin{aligned}\frac{\mathrm{d}}{\mathrm{d}t}\left(\frac{3}{2}k_B n_e T_e\right) &= [\beta_i n_0 \varepsilon_e - \alpha_r n_e n_i (3k_B T_e/2)] \\ &\quad - n_d\left[\sum_{(Z_1+1)}^{Z_2} n_Z n_\mathrm{ec}(Z)\varepsilon_\mathrm{ec}(Z) - \sum_{Z_1}^{(Z_2-1)} n_Z n_\mathrm{ee}(Z)\varepsilon_\mathrm{ee}(Z)\right] - Q_\mathrm{ec}\end{aligned} \quad (6.11)$$

and

$$\frac{\mathrm{d}}{\mathrm{d}t}\left(\frac{3}{2}k_B n_i T_i\right) = [\beta_i n_0 \varepsilon_i - \alpha_r n_e n_i (3k_B T_i/2)] - n_d \sum_{Z_1}^{(Z_2-1)} n_Z n_\mathrm{ic}(Z)\varepsilon_\mathrm{ic}(Z) - Q_\mathrm{ic}. \quad (6.12)$$

Here (ε_e, ε_i) are the mean energies of electrons and primary ions produced by ionization of neutral atoms, (ε_{ec}, ε_{ic}) are the mean energies (far away from the surface) of electrons and ions accreting on the surface of the dust grain, ε_{ee} refers the mean energy of emitted electrons at a large distance from the surface of the dust grain, T_i is the temperature of the ionic species and the numerical values of $\varepsilon_e/\varepsilon_i$ and β_i can be evaluated in same way as in Sect. 5.3.

The first term on the right-hand side in (6.11) and (6.12) represents the net power gained per unit volume by the electrons and ions through the ionization of the neutral atoms and electron–ion recombination. The next term corresponds to the net power loss due to net electron and ion flux from the dust grain. The rest of the terms in both the equations represents the net power transferred to electrons (Q_{ec}) and ions (Q_{ic}) on account of elastic collisions between electrons, ions, and neutral species, respectively; the characteristic plasma elastic collisions have been discussed in detail in Chap. 5; usually Q_{ec} and Q_{ic} are negligible as compared to the other terms.

For an arbitrarily chosen range of Z values, one can easily obtain the steady state charge distribution and other plasma parameters (*i.e.* n_e, n_i, n_o, T_e, T_i) by simultaneous numerical integration of (6.6–6.12) for a given initial set of parameters as $t \rightarrow \infty$. The energy balance of neutral atoms/dust has been ignored on account of their large thermal capacity and efficient energy exchange between them; it is a good approximation to assume their temperature to be the same viz. T (hence no energy exchange). The neutral atom/dust energy balance can be written as in Sect. 5.3. To illustrate the effect of plasma flow on dust charging and the numerical appreciation of the analysis a specific situation has been discussed in Sect. 6.5.

6.5 Specific Situations (After Mishra et al. [13])

In this section, the kinetics of the uniformly dispersed mono-sized dust grains in a gaseous complex plasma characterized by mesothermal flow has been analyzed and the charge distribution on the dust particles has been evaluated in *mesothermal flow* regime, which is a situation when the thermal velocity of electrons is much higher than the plasma flow speed ($\upsilon_e \parallel C$) across the dust particle while ions move with random velocity comparable to that of the plasma flow speed ($\upsilon_i \approx C$). In the case of dark flowing plasmas, i.e., in the absence of electron emission from the dust particle, electron/ion accretion on the dust particles is only dust charging mechanism; the numerical results correspond to the following set of parameters.

$n_{e0} = n_{i0} \approx 10^6/\text{cm}^3$, $n_{00} = 10^{10}/\text{cm}^3$, $T_{e0} = 10^3\,\text{K}$, $T_{i0} = 400\,\text{K}$, $T_d = T = 300\,\text{K}$, $m_n \approx m_i = 20\,\text{amu}$, $a = 0.1\,\mu$, $\eta = 0.7$, $\alpha_{r0} = 10^{-7}\text{cm}^3/\text{s}$, $n_d = 100\,\text{cm}^{-3}$ and $p_0 = (2/3)^{1/2}$, $p = (2/3)^{1/2}(C/\upsilon_{Ti})$.

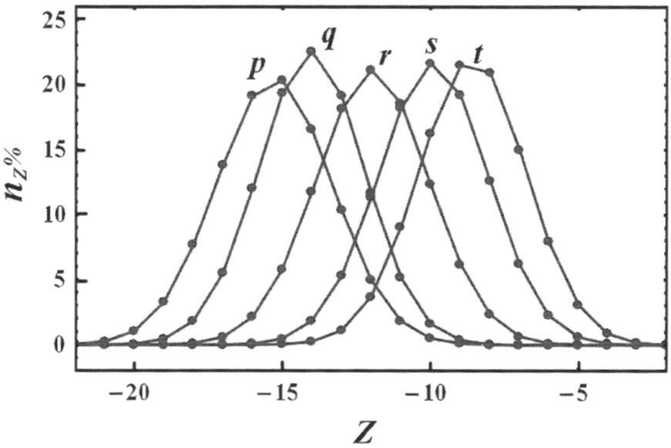

Fig. 6.1 Charge distribution on dust in dark complex plasma for the standard set of the parameters, given in the text; the labels p, q, r, s, and t refer $p_0 = 0$, 5, 10, 15, and 20, respectively. Only integral values of Z on the *curves* are significant (after Mishra et al. [13], curtsey authors and publishers AIP)

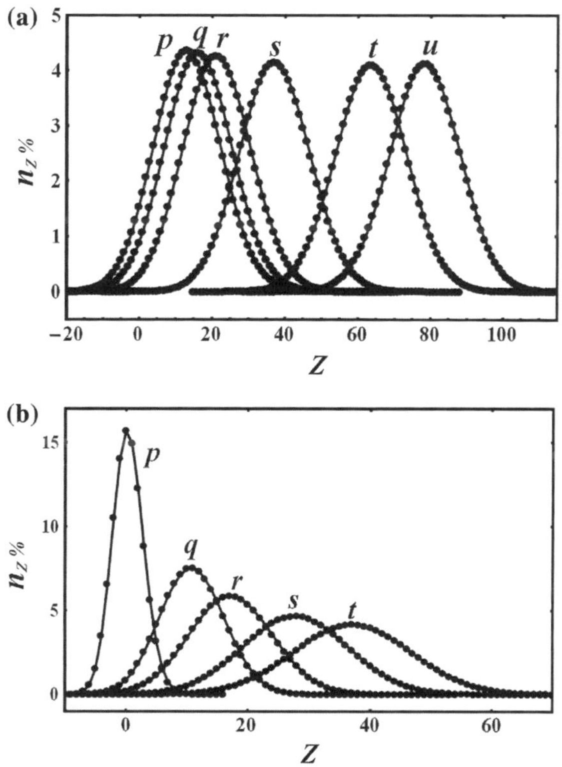

Fig. 6.2 Charge distribution on dust in an irradiated complex plasma for the standard set of the parameters, given in the text; the labels **a** p, q, r, s, t, and u refer to $p_0 = 0$, 2, 3, 5, 8, and 10, respectively for $a = 0.1\ \mu$, **b** p, q, r, s, and t refer to $a = 10$, 30, 50, 80, and 100 nm, respectively. For $p_0 = 5$ only integral values of Z on the *curves* are significant (after Mishra et al. [13], curtsey authors and publishers AIP). The figure corresponds to $n_p = 10^{14}/\text{cm}^2$ s, where n_p is the rate of photoemission of electrons from an uncharged plane surface per unit area (see Chap. 2)

The dependence of the charge distribution on the dust particles on the plasma flow parameter (p) has been displayed in Fig. 6.1. It is noticed that the charge distribution shifts toward larger values of negative charge with decreasing relative speed of the plasma flow (p) and tends to the stationary charge distribution (*i.e.* $C \to 0$) corresponding to an isotropic plasma. This behavior can be attributed to the larger accretion of ions over dust surface on account of the availability of high energy ions in a flowing plasma for large values of p.

Next we consider a complex plasma irradiated by a monochromatic source of radiation and hence the photoelectric emission of the electrons from the dust surface has been taken as an additional charging mechanism along with electron/ion accretion. The photoemission rate has been modified by including *Mie Scattering* factor as a function of dust size and wavelength of incident radiation; the coefficient of *Mie Scattering* (m_f) has been evaluated for monochromatic incident radiation of wavelength $\lambda = 224.3$ nm (corresponding to He–Ag$^+$ *laser*) and suspended dust grains of *Na* (with work function $\varphi = 2.3$ eV) as a function of the size parameter ($2\pi a/\lambda$) from Web available computational program viz. *Mie Scattering Calculator* [16].

Charge distribution in the case of the illuminated flowing plasma has been displayed as a function of plasma flow parameter (p) and grain size (a) in Fig. 6.2. Large accretion of ionic species over dust surface on account of the presence of high energy ions in the flowing plasma in addition to photoemission leads the charge distribution on the dust grains widening and shifting toward more positive values; the effect is more pronounced with increasing particle radius.

References

1. C. Cui, J. Goree, IEEE Trans. Plasma Sci. **22**, 151 (1994)
2. V.E. Fortov, A.G. Khrapak, S.A. Khrapak, V.I. Molotkov, O.F. Petrov, Phys. Usp. **47**, 447 (2004)
3. V.E. Fortov, A.V. Ivlev, S.A. Khrapak, A.G. Khrapak, G.E. Morfill, Phys. Rep. **421**, 01 (2005)
4. C.K. Goertz, Rev. Geophys. **27**, 271 (1989)
5. J. Goree, Plasma Sources Sci. Technol. **03**, 400 (1994)
6. A.V. Gurevich, *Nonlinear Phenomena in the Ionosphere* (Springer, New York, 1978)
7. Ch. Hollenstein, Plasma Phys. Control. Fusion **42**, R93 (2000)
8. M. Kanal, Theory of Current Collection of Moving Spherical Probes. Science Report, No. **JS-5**, Space Physics Research Laboratory, University of Michigan, Ann Arbor, 1962
9. T. Matasoukas, M. Russel, J. Appl. Phys. **77**, 4292 (1995)
10. T. Matsoukas, M. Russel, M. Smith, J. Vac. Sci. Technol. **A14**, 624 (1996)
11. F. Melandso, Phys. Rev. E **55**, 7495 (1997)
12. D.A. Mendis, Plasma Sources Sci. Technol. **11**, A219 (2002)
13. S.K. Mishra, S. Misra, M.S. Sodha, Phys. Plasmas **18**, 103708 (2011)
14. G.E. Morfill, E. Grun, T.V. Johnson, Planet. Space Sci. **28**, 1087 (1980)
15. T.G. Northrop, Phys. Scr. **45**, 475 (1992)
16. S. Prahl, Mie scattering calculator, http://omlc.ogi.edu/calc/mie_calc.html
17. M.S. Sodha, S. Guha, in *Physics of Colloidal Plasma*, ed. by A. Simon, W.B. Thomas. Advantage of Plasma Physics, vol 4 (Interscience, New York, 1971), p. 219
18. M.S. Sodha, S.K. Mishra, S. Misra, S. Srivastava, Phys. Plasmas, **17**, 073705 (2010)
19. M.S. Sodha, S.K. Mishra, S. Misra, Phys. Scr. **83**, 015502 (2011)
20. G. Sorasio, D.A. Mendis, M. Rosenberg, Planet. Space Sci. **49**, 1257 (2001)

21. V.N. Tsytovich, G.E. Morfill, S.V. Vladimirov, H.M. Thomas, *Elementary Physics of Complex Plasmas* (Springer, Berlin, 2008)
22. N.G. Van Kampen, *Stochastic Processes in Physics and Chemistry* (North Holland, New York, 1990)
23. S.V. Vladimirov, K. Ostrikov, A.A. Samarian, *Physics and Applications of Complex Plasmas* (Imperial College Press, London, 2005)
24. E.C. Whipple, Rep. Prog. Phys. **44**, 1197 (1981)

Chapter 7
Kinetics of the Complex Plasmas Having Dust with a Size Distribution

7.1 Introduction

The fact that the dust particles in complex plasmas, occurring in nature, laboratory and engineering are not of the same size, but have a size distribution that has been known for a long time, but relatively few papers on the kinetics of the complex plasmas having a size distribution have been published. In this Chapter an approach, based on number/energy balance of constituents, charge balance of the particles, ionization/deionization mechanisms and Mie scattering have been given.

In early work [2–4, 10, 11], an intuitive rather than the appropriate [16, 17] expression for the photoelectric electron emission from a positively charge spherical particle was employed and the dependence of the absorption coefficient of light by a particle (Q_a) on the size of the particle (as analyzed by Mie [13] was ignored. Further, the energy balance of the constituents of the complex plasma was also not considered.

Recently Sodha et al. [18] have analyzed the kinetics of a complex plasma, with a size distribution of dust, taking into account the charge balance on the particles and the number and energy balance of the constituents; the dependence of Q_a on size was neglected in the analysis which is justified for particles with radius larger than $1.5 = (10/2\pi)$ times the wavelength of the incident radiation, causing the photoelectric emission. The analysis was of course applicable to dark plasmas, with or without thermionic emission from the dust particles. Considerable simplification in the analysis was made by using the fact that in the steady state, the electric potential of the dust particles is independent of the particle size. This analysis was extended to the case of a complex plasma, having dust of different materials (with a size distribution) by Sodha et al. [19]. In a later paper [20] the dependence of Q_a on size was also taken into account. The charge distribution, thus obtained is the distribution of the mean charge and does not take into account the discrete nature of the charge, whereby the charge on the particle can change only by a discrete amount viz the electronic charge. The charge distribution/ fluctuation of charge on particles of a specific radius will be considered in a later chapter.

M. S. Sodha, *Kinetics of Complex Plasmas*, Springer Series on Atomic, Optical, and Plasma Physics 81, DOI: 10.1007/978-81-322-1820-3_7,

7.2 Size Distribution

The size distribution of dust in a complex plasma depends on a number of processes, associated with the formation, coalescing/breakup and annihilation of the particles. The size distribution of the dust in the interplanetary [5] and interstellar [8] space is given by the *MRN* [12] power law viz.

$$f(a)\mathrm{d}a = Aa^{-s}\mathrm{d}a, \tag{7.1}$$

where $f(a)\mathrm{d}a$ is the number of particles per unit volume having radii between a and $(a + \mathrm{d}a)$, $0.9 < s < 4.5$ for the interplanetary space, $s \approx 3.5$ for the interstellar space and A is a normalizing constant given by

$$n_d = A\int_{a_1}^{a_2} f(a)\mathrm{d}a, \tag{7.2}$$

where a_1 and a_2 denote the extreme limits of the radii in the region characterized by the size distribution function $f(a)$ and n_d is the number density of the dust particles.

Raadu [14], in an investigation on electrostatic waves in a complex plasma, with size distribution of dust assumed that $f(a)$ varies exponentially with mass for large size and according to a power law for smaller particles; both the features are inherent in the *Kappa* distribution [15] of the size of dust particles, given by

$$f(a)\mathrm{d}a = Aa^{-s}\exp\left(-\alpha^3 a^3\right)\mathrm{d}a, \tag{7.3}$$

where A, α, and s are constants.

The two useful parameters, which are determined by the size distribution viz. mean and root mean square values of the radius of the dust grains, are given by

$$\bar{a} = \left(\int_{a_1}^{a_2} af(a)\mathrm{d}a\right) \Big/ \left(\int_{a_1}^{a_2} f(a)\mathrm{d}a\right) = a_m \tag{7.4a}$$

and

$$\overline{a^2} = \left(\int_{a_1}^{a_2} a^2 f(a)\mathrm{d}a\right) \Big/ \left(\int_{a_1}^{a_2} f(a)\mathrm{d}a\right) = a_{\mathrm{rms}}^2. \tag{7.4b}$$

The parameters a_m and a_{rms} can be evaluated for the power law distribution by straightforward integration. The value of a_m and a_{rms} for the *Kappa* distribution have been tabulated by Sodha et al. [18].

7.3 Uniform Electric Potential on All Dust Particles of Same Material

The charging of dust particles is represented by

$$\begin{aligned}(\mathrm{d}Z/\mathrm{d}t) &= n_{\mathrm{ee}} + n_{\mathrm{ic}} - n_{\mathrm{ec}} \\ &= \pi a^2[f_{\mathrm{ee}}(V,\ T) + n_i f_{\mathrm{ic}}(V,\ T_i) - n_e f_{\mathrm{ec}}(V,\ T_e)],\end{aligned} \tag{7.5}$$

where n_{ee}, n_{ec}, and n_{ic} have usual meanings, f_{ec} and f_{ic} are independent of the particle size a and f_{ee} is also independent of a, when (i) photoelectric emission is not present or when (ii) the dependence *Mie* absorption coefficient (Q_a) on a is neglected viz. when large particles are considered (a_m, $a_{\mathrm{rms}} > 1.5\lambda$). The fact that $f_{\mathrm{ee}}, f_{\mathrm{ic}}$, and f_{ec} are independent of a, can be verified from the expressions of n_{ee}, n_{ec}, and n_{ic} in Chaps. 2 and 3.

In the steady state (d/dt) → 0, the term in the parentheses is zero and V is determined by an algebraic equation in which a does not occur. Hence, V is independent of a or the electric potential on all the particles is the same in the steady state. In what follows we have assumed that V is independent of a, even under transient conditions; since we are only looking at the steady state solution, this approximation is not relevant.

7.4 Kinetics with Uniform Electric Potential on Dust Particles (After Sodha et al. [20])

7.4.1 Basic Equations

Putting $V = (Ze/a)$ in (7.5) , multiplying both sides by $f(a)\mathrm{d}a$ and integrating in the limits $a_1 \le a \le a_2$ and making use of (7.2) and (7.4a) one obtains

$$(\mathrm{d}V/\mathrm{d}t) = e\pi a_m[f_{\mathrm{ee}}(V,\ T) + n_i f_{\mathrm{ic}}(V,\ T_i) - n_e f_{\mathrm{ec}}(V,\ T_e)]. \tag{7.6a}$$

The number balance of electrons is given by

$$(\mathrm{d}n_e/\mathrm{d}t) = \beta_i n_0 - \alpha_r n_e n_i - n_d(n_e f_{\mathrm{ec}} - f_{\mathrm{ee}}) \int_{a_1}^{a_2} \pi a^2 f(a)\mathrm{d}a.$$

Using (7.2) and (7.4b), the above equation can be expressed as

$$(\mathrm{d}n_e/\mathrm{d}t) = \beta_i n_0 - \alpha_r n_e n_i - \pi n_d a_{\mathrm{rms}}^2(n_e f_{\mathrm{ec}} - f_{\mathrm{ee}}), \tag{7.7a}$$

where β_i is the coefficient of ionization, $\alpha_r(T_e) = \alpha_{r0} T_e^{-\kappa}$ cm^3/s, is the coefficient of electron–ion recombination and α_{r0} and κ are constants [9].

Similarly, the number balance of ions can be expressed as:

$$(dn_i/\mathrm{d}t) = \beta_i n_0 - \alpha_r n_e n_i - \pi n_d a_{\mathrm{rms}}^2 n_i f_{ic}. \tag{7.8a}$$

Since the charge neutrality can be derived from (7.5), (7.7a), and (7.8a), it is not considered separately. The energy balance for electrons and ions can be expressed as:

$$\frac{\mathrm{d}}{\mathrm{d}t}\left(\frac{3}{2}k_B n_e T_e\right) = \beta_i n_0 E_e - \alpha_r n_e n_i (3k_B T_e/2) - \pi n_{\mathrm{d}} a_{\mathrm{rms}}^2 (n_e f_{\mathrm{ec}} \varepsilon_{\mathrm{ec}} - f_{\mathrm{ee}} \varepsilon_{\mathrm{ee}}) \tag{7.9a}$$

and

$$\frac{\mathrm{d}}{\mathrm{d}t}\left(\frac{3}{2}k_B n_i T_i\right) = \beta_i n_0 E_i - \alpha_r n_e n_i (3k_B T_i/2) - \pi n_{\mathrm{d}} a_{\mathrm{rms}}^2 n_i f_{\mathrm{ic}} \varepsilon_{\mathrm{ic}}. \tag{7.10a}$$

Further

$$n_0 + n_i = n_{00} + n_{i0} = N_t\,. \tag{7.11}$$

Expressions for f_{ee}, f_{ic}, f_{ec}, E_e, E_i, $\varepsilon_{\mathrm{ec}}$, and $\varepsilon_{\mathrm{ic}}$ are given in Chaps. 2 and 3. For the sake of simplicity, the temperatures of dust and neutral species may be taken to be T (known).

The system of equations (7.6a–7.10a) and (7.11) may be solved, using *Mathematica* software to obtain V, n_e, n_i, n_0, T_e, and T_i as functions of t. The initial conditions at $t = 0$ may be the ones, corresponding to $Z = 0$ and the plasma in the absence of dust. As discussed earlier, the steady state values ($t \to \infty$) of V, n_e, n_i, n_0, T_e, and T_i are independent of the initial conditions.

It is seen from equations (7.6a–7.10a) and (7.11), that the steady state kinetics of a complex plasma with dust having a size distribution is very similar to that of the complex plasma with the same number density of the dust particles of uniform radius equal to the root mean square radius of the particle size distribution. The only difference is that in the case of complex plasma with a size distribution the charge on the particle is proportional to the radius. Thus, the charge distribution on the particles with a size distribution can be expressed as:

$$F(Z)\mathrm{d}Z = A(V/e)^{1-s} Z^{-s}\mathrm{d}Z \quad (\textit{MRN power law}) \tag{7.12a}$$

and

$$F(Z)\mathrm{d}Z = A(V/e)^{1-s} Z^{-s}\exp\left[-(e\alpha/V)^3 Z^3\right]\mathrm{d}Z \quad (\textit{Kappa distribution}). \tag{7.12b}$$

V is of course the same for all the particles and, the charge Ze on a particle of radius a, as referred to here before is just the mean charge. The charge distribution on such particles is a normal distribution with the standard deviation $\sigma_Z \approx 0.5\ Z^{1/2}$ [6] when $Z > 30$.

7.4.2 Kinetics of the Complex Plasmas with a Mixture of Dust of Different Materials: Uniform Electric Potential Theory (After Sodha et al. [19])

Usually a mixture of dust of different materials is present in a complex plasma. For different applications, the electron density may be increased or decreased by addition of dust of low or high work function. In such cases, one has to write separate equations for the electric potential on the *j*th species of dust and consider the effect of all species of dust on the number and energy balance of the electrons/ions; hence the (7.6a–7.10a) are replaced by

$$\left(\mathrm{d}V_j/\mathrm{d}t\right) = e\pi a_{m,j}\left[f_{\mathrm{ee,j}}\left(V_j,\ T\right) + n_i f_{\mathrm{ic},j}\left(V_j,\ T_i\right) - n_e f_{\mathrm{ec,j}}\left(V_j,\ T_e\right)\right], \tag{7.6b}$$

$$(\mathrm{d}n_e/\mathrm{d}t) = \beta_i n_0 - \alpha_r n_e n_i - \sum_j \pi a^2_{\mathrm{rms},\,j} n_j \left(n_e f_{\mathrm{ec},\,j} - f_{\mathrm{ee,\,j}}\right), \tag{7.7b}$$

$$(\mathrm{d}n_i/\mathrm{d}t) = \beta_i n_0 - \alpha_r n_e n_i - \sum_j \pi a^2_{\mathrm{rms},\,j} n_j n_i f_{ic,\,j}, \tag{7.8b}$$

$$\frac{\mathrm{d}}{\mathrm{d}t}\left(\frac{3}{2}kn_eT_e\right) = \beta_i n_0 E_e - \alpha_r n_e n_i (3kT_e/2) - \sum_j \pi n_j a^2_{\mathrm{rms,\,j}}\left(n_e f_{\mathrm{ec,\,j}}\varepsilon_{\mathrm{ec,\,j}} - f_{\mathrm{ee,\,j}}\varepsilon_{\mathrm{ee,\,j}}\right) \tag{7.9b}$$

and

$$\frac{\mathrm{d}}{\mathrm{d}t}\left(\frac{3}{2}kn_iT_i\right) = \beta_i n_0 E_i - \alpha_r n_e n_i (3kT_i/2) - \sum_j \pi n_j a^2_{\mathrm{rms,\,j}} n_i f_{\mathrm{ic},\,j}\varepsilon_{\mathrm{ic},\,j}, \tag{7.10b}$$

Equation (7.11) is of course valid in this case.

Thus one has $(p + 5)$ equations $(j = 1, 2, \ldots, p)$ to determine V_j and n_e, n_i, n_0, T_e, and T_i when there are p species of dust. The fluctuation of charge on an individual particle is as before viz. $0.5Z_j^{1/2}$; where Z_j refers to the average charge on the particle of radius a_j. The set of equations can as before be solved by using the *Mathematica* software.

7.5 Kinetics of the Complex Plasma in Thermal Equilibrium

In the exploration of the kinetics of such systems in thermal equilibrium, two approximations have been made in some papers, viz.

(1) Applicability of *Saha's* equation to the electron emission/accretion process in the case of dust particles; this is only justified in the absence of ions (ideally for a dust-electron cloud system).

(2) Using the number balance of electrons/ions with uncertain values of recombination/ionization coefficients (α_r/β_i) which are in general not consistent with *Saha's* equation for the ionization of gases.

In this section following Sodha et al. [19] a formulation of the kinetics, which is free of these handicaps is presented. As before only single ionization of atoms has been allowed. Consider a number of species of gases (p) and number of species (j) of dust particles with a size distribution.

From *Saha's equation*

$$\left(n_e n_{ip}/n_{0\mathrm{p}}\right) = k_p, \tag{7.12}$$

where $k_p = 2(2\pi m_e k_B T/h^2)^{3/2}\exp(-eI_p/k_B T)$ and I_p is the ionization potential of the pth gas.

The conservation of the sum of the ions and neutral atom densities of the pth gas requires

$$n_{0\mathrm{p}} + n_{\mathrm{ip}} = n_p \tag{7.11}$$

From (7.11 and 7.12), one obtains,

$$n_{\mathrm{ip}} = x_p n_p/\left(1 + x_p\right), \tag{7.13}$$

where

$$x_p = \left(k_p/n_e\right)$$

Further the charging equation is [18]

$$(\mathrm{d}V_j/\mathrm{d}t) = e\pi a_{\mathrm{mj}}\left[f_{\mathrm{eej}}(V_j,\ T) - n_e f_{\mathrm{ecj}}(V_j,\ T) + \sum n_{\mathrm{ip}} f_{\mathrm{icpj}}(V_j,\ T)\right] \tag{7.13b}$$

where V_j, a_{mj}, and a_{rmsj} are the surface potential, mean radius and root mean radius of the jth species of the dust; $f_{\mathrm{ee},j} = n_{\mathrm{ee},j}/\pi a_j^2, f_{\mathrm{ec},j} = n_{\mathrm{ec},j}/\pi a_j^2$, and $f_{\mathrm{icp},j} = n_{\mathrm{icp},j}/\pi a_j^2$ are the rates of electron emission, electron accretion and accretion of ion of pth gas specie on a particle of jth species of dust with a radius a_j.

The equation for the charge neutrality takes the form

$$\sum (V_j/e) n_j a_{m,j} + \sum n_{\mathrm{ip}} = n_e \tag{7.14}$$

where V_j is the number density of the dust particles of the jth species. Thc system of (7.11, 7.12, 7.13b, and 7.14) is adequate to determine V_j and n_e, $n_{i,j}$, $n_{0,j}$.

7.6 Inclusion of *Mie* Scattering by Dust in Complex Plasma Kinetics (After Sodha et al. [20])

In most of the investigations on the kinetics of the complex plasma, particularly the ones with a size distribution of dust, the number of photons, incident on the spherical particle of radius a per unit time is taken as $\Lambda\pi a^2$, where Λ is the number of photons incident per unit area on an infinite plane surface. However, Mie[32] has long back pointed out that in general the number of photons, incident on a sphere is $\pi a^2 \Lambda . Q_a(2\pi a/\lambda, \mu)$, where Q_a is a function of $(2\pi a/\lambda)$ and the complex refractive index μ of the material of the sphere and λ is the wavelength of incident radiation. Sodha et al. [19] have investigated the kinetics of a complex plasma having dust with size distribution governed by the *MRN* power law [12], taking into account the variation of Q_a with $(2\pi a/\lambda)$, as given by *Mie* theory [13]. In this section, it is proposed to present an outline of the approach to the kinetics and discussion of the results, thus obtained for typical values of the relevant parameters. Since Q_a is a function of a, the uniform potential theory is not applicable, when photoelectric effect is the only or dominant source of the electron emission from the dust. Hence, the rate of photoelectric emission (n_{ph}) as derived in Chap. 2, n_{ph} should be multiplied by Q_a; the expression for E_{ph} or $\varepsilon_{\mathrm{ph}}$ remains unchanged. The uniform potential theory is applicable, when the dust particles have a large radius so that $Q_a \approx 1$.

It is convenient to proceed further in terms of the dimensionless radius $\rho = (a/a_0)$ and divide the region of the radii $\rho_1 < \rho < \rho_2$ in $N(=100)$ regions; the jth region is described by $(\rho_j - \delta\rho) < \rho < (\rho_j + \delta\rho)$ where $2\delta\rho = 0.1$. The number of particles in this region having the mean radius ρ_j is given by

$$n_j = A_0 n_d \rho_j^{-s}(2\delta\rho) = B n_d \rho_j^{-s} \tag{7.15a}$$

where

$$B = 1/\sum \rho_j^{-s} \tag{7.15b}$$

The conservation of number density of neutral atoms and ions requires

$$n_0 + n_i = n_{00} + n_{i0} = N. \tag{7.16}$$

The rate of charging of the dust particles in the jth region $(\rho_j - \delta\rho) < \rho < (\rho_j + \delta\rho)$ is given by

$$\left(\mathrm{d}Z_j/\mathrm{d}t\right) = n_{\mathrm{ic}}\left(\rho_j,\ Z_j\right) - n_{\mathrm{ec}}\left(\rho_j,\ Z_j\right) + n_{\mathrm{ph}}\left(\rho_j,\ Z_j\right). \tag{7.17}$$

The number and energy balance of the electrons/ions may be expressed as:

$$(\mathrm{d}n_e/\mathrm{d}t) = \beta_i n_0 - \alpha_r n_e n_i - \sum_1^N n_j\left[n_{\mathrm{ec}}\left(\rho_j,\ Z_j\right) - n_{\mathrm{ph}}\left(\rho_j,\ Z_j\right)\right], \tag{7.18}$$

$$(\mathrm{d}n_i/\mathrm{d}t) = \beta_i n_0 - \alpha_r n_e n_i - \sum_1^N n_j n_{\mathrm{ic}}\left(\rho_j,\ Z_j\right), \tag{7.19}$$

$$\begin{aligned}\frac{\mathrm{d}}{\mathrm{d}t}\left(\frac{3}{2}k_B n_e T_e\right) = \beta_i n_0 E_e - \alpha_r n_e n_i (3k_B T_e/2) - \sum_1^N n_j\left[n_{\mathrm{ec}}\left(\rho_j,\ Z_j\right)\varepsilon_{\mathrm{ec}}\left(\rho_j,\ Z_j\right)\right.\\ \left. - n_{\mathrm{ee}}\left(\rho_j,\ Z_j\right)\varepsilon_{\mathrm{ph}}\left(\rho_j,\ Z_j\right)\right]\end{aligned} \tag{7.20}$$

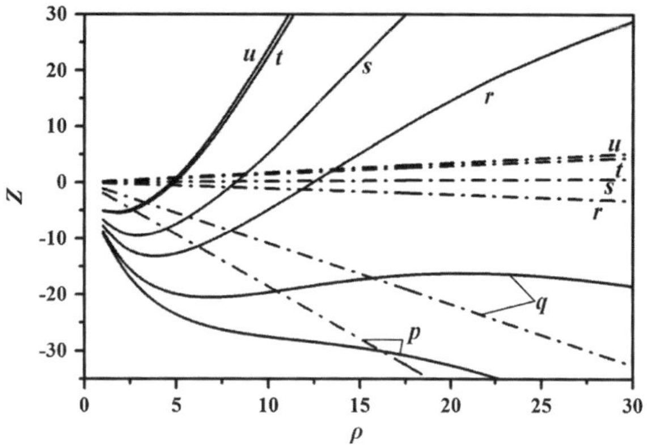

Fig. 7.1 Dependence of the charge Z on the dust particles, on the particle radius $\rho(=a/a_0)$; p, q, r, s, t, and u refer to $n_d = 50$, 100, 500, 1000, 5,000, and 10,000 cm^{-3}, respectively, the plasma parameters are per text, following (7.19) (after Sodha et al. [20], curtsey authors and publishers AIP)

and

$$\frac{\mathrm{d}}{\mathrm{d}t}\left(\frac{3}{2}k_B n_i T_i\right) = \beta_i n_0 E_i - \alpha_r n_e n_i (3k_B T_i/2) - \sum_1^N n_j n_{\mathrm{ic}}(\rho_j, Z_j)\varepsilon_{\mathrm{ec}}(\rho_j, Z_j), \quad (7.21)$$

The system of (7.16) and (7.17–7.21) can be solved (with a set of initial conditions, conforming to the absence of dust) to obtain the steady state ($t \to \infty$) values of Z_j and n_e, n_i, n_0, T_e, and T_i. As pointed out before the steady state values are independent of the initial condition.

For a numerical appreciation of the results, Sodha et al. [20] considered a complex plasma comprising of high work function pure ice particles suspended in a plasma, corresponding to near earth environment. Since ice has a high work function (8.7 eV), only the EUV part of the solar radiation is effective in causing the photoelectric emission of electrons from the surface of the particles. The only significant EUV radiation from the sun in the outer region of the earth is (3×10^{11} $\mathrm{cm}^{-2}\mathrm{s}^{-1}$) photons of *Lyman Alpha radiation* ($\lambda = 121.57$ nm) [1]. The computed [7] dependence of Q_a on $(2\pi a/\lambda)$ can to a very good approximation be expressed [20] as:

$$Q_a = 1.128 - \left(1.31 \times 10^{-3}\right)(2\pi a/\lambda) - 1.125\exp\left[-0.135(2\pi a/\lambda)^{1.277}\right]. \quad (7.22)$$

For a parametric analysis of the results, the following set of parameters may be chosen and the effect of changing one of these, keeping others the same may be studied.

$$\begin{aligned}
&n_{e0} = n_{i0} \approx 10^5/\mathrm{cm}^3,\ n_{00} = 10^3 n_{e0},\ n_p = 10^{12}\mathrm{s}^{-1}\mathrm{cm}^{-2},\ T_{e0} = 1000\,\mathrm{K},\\
&T_{i0} = 400\,\mathrm{K},\ T = 250\,\mathrm{K},\ 0.5 \le \rho \le 100.5,\ \xi = 73.61(\text{corresponding to}\\
&Lyman - \alpha\,\text{radiation and}\ \varphi = 8.7\,\mathrm{eV}),\ a_0 = 0.01\,\mu\mathrm{m},\ n_d = 10^3\mathrm{cm}^{-3},\\
&m_i = 30\,\mathrm{amu},\ m_0 \approx m_i,\ s = 2, \kappa = 1.2\ \text{and}\ \alpha_{r0} = 5 \times 10^{-7}\mathrm{cm}^3/\mathrm{s}.
\end{aligned}$$

Some of the results are presented in Fig. 7.1; it is seen that (i) the charge on the particles increases with increasing number density of the particles and (ii) since Q_a

increases with increasing radius of the particle, the photoelectric emission from the smaller particles gets reduced in a much larger proportion compared to the case of large particles, for which $Q_a \approx 1$. This leads to the interesting conclusion, evident from the figure that in some conditions the large particles are positively charged while small particles are negatively charged. This is in contrast to the case (Sect. 7.3), when *Mie* scattering is neglected; then all the particles carry the charge of same sign, which is proportional to radius. The other dependences have been illustrated and discussed by Sodha et al. [20].

References

1. S.J. Bauer, *Physics of Planetary Ionosphere* (Springer, New York, 1973)
2. L. Bringol-Barge, T.W. Hyde, Adv. Space Res. **29**, 1277 (2002)
3. L. Bringol-Barge, T.W. Hyde, Adv. Space Res. **29**, 1283 (2002)
4. L. Bringol-Barge, T.W. Hyde, Adv. Space Res. **29**, 1289 (2002)
5. J.A. Burns, M.R. Showalter, G.M. Morfill, *The Ethereal Rings of Jupiter and Saturn in Planetary Rings*, ed. by R. Greenberg, A. Braphic (The University of Arizona Press, Tuscon, 1984), p. 200
6. C. Cui, J. Goree, IEEE Trans. Plasma Sci. **22**, 151 (1994)
7. J. Dorschner, Astr. Nach. **292**, 71 (1908)
8. B.T. Draine, B. Sutin, Astrophys. J. **320**, 803 (1987)
9. A.V. Gurevich, *Nonlinear Phenomena in the Ionosphere* (Springer, New York, 1978)
10. O. Havnes, T.K. Aanesan, F. Melandso, J. Geophys. Res. **95**, 6581 (1990)
11. H. Houpis, E.C. Whipple, J. Geophys. Res. **92**, 12057 (1987)
12. J.S. Mathis, W. Rumpl, K.H. Nordsieek, Astrophys. J. **217**, 425 (1977)
13. G. Mie, Ann. Phys. Leipzig **25**, 377 (1908)
14. M.A. Raadu, IEEE Trans. Plasma Sci. **29**, 182 (2001)
15. M. Shafiq, Doctoral thesis on *Test charge response of a dusty plasma with grain size distribution and charging dynamics*, Space and Plasma Physics, School of Electrical Engineering, Royal Institute of Technology, Stockholm, 2006
16. M.S. Sodha, Brit. J. Appl. Phys. **14**, 172 (1963)
17. M.S. Sodha, S. Guha, *Physics of Colloidal Plasma*, *Adv. Plasma Phys.*, ed. by A. Simon, W.B. Thomas. Interscience, vol 4 (Wiley, New york, 1971), p. 219
18. M.S. Sodha, S. Mishra, S.K. Mishra, Phys. Plasmas **17**, 113705 (2010)
19. M.S. Sodha, S.K. Mishra, S. Misra, IEEE Trans. Plasma Sci. **39**, 1141 (2011)
20. M.S. Sodha, S.K. Mishra, S. Misra, J. Appl. Phys. **109**, 01303 (2011)

Chapter 8
Theory of Electrical Conduction

8.1 Phenomenological Theory (After Sodha [2])

Soon after the discovery of the electron, the flow of the electrical current (synonymous with the transport of charge) in the ionized gases was ascribed to the movement of electrons and ions, along and counter (respectively) to the direction of the applied electric field. On account of the heavier mass of ions, the role of ions in the electrical conduction is not significant and hence not taken into account in this chapter. A theory of electrical conduction, based on the average behavior of electrons/ions is usually referred to as the phenomenological theory. This theory brings out the physics of the electrical conduction, but it ignores the energy/velocity distribution of electrons/ions.

8.1.1 Motion of Electrons

The motion of an electron in the presence of a constant electric field $\boldsymbol{E}$ is in the first instance expressed by

$$(\mathrm{d}\upsilon/\mathrm{d}t) = -(e\boldsymbol{E}/m_e), \tag{8.1}$$

where υ is the drift (or average) velocity of the electrons in the direction of the electric field (the average of random velocities is of course zero because for any value of random velocity υ_{random} the value $(-\upsilon_{\text{random}})$ is equally probable) and $-e$ and m_e are the electronic mass and charge, respectively.

On integration the above equation leads to an absurd conclusion that the drift velocity υ (shown later to be proportional to the electrical current density) keeps increasing with time and does not attain a steady state value. This is contrary to experience and one has to look for a damping term (or process), which would ensure a steady state. Collision of electrons (as may be seen in the following) represents a damping process, which leads to the steady state. One may define a collision as an event, which randomizes a fraction (say α_o) of the momentum of an electron.

M. S. Sodha, *Kinetics of Complex Plasmas*, Springer Series on Atomic, Optical, and Plasma Physics 81, DOI: 10.1007/978-81-322-1820-3_8,

If an electron has an average momentum $m_e\upsilon$ before collision, the momentum lost by the electron in a collision is $\alpha_o m_e \upsilon$; it still retains an average momentum $(1-\alpha_o)m_e\upsilon$. It should be emphasized that the space average of randomized momentum is zero. Hence, the force on an electron on account of collisions, i.e., the rate of change of momentum is equal to the change in the momentum per collision $\alpha_o m_e \upsilon$, multiplied by the number of collisions per unit time ν_0. Thus the damping force $\boldsymbol{F}_\nu$ on an electron is given by

$$\boldsymbol{F}_\nu = -\alpha_o m_e \upsilon \nu_0 = -m_e \upsilon \nu_e \tag{8.2}$$

where $\nu_e (= \alpha_o \nu_0)$ refers to the electron momentum transfer collision frequency or simply the electron collision frequency; the parameters α_o and ν_0 do not occur separately but only as $\alpha_o \nu_0$ and hence the use of $\nu_e (= \alpha_o \nu_0)$ is physically more meaningful. The negative sign of $\boldsymbol{F}_\nu$ indicates that it is a damping force with corresponding negative rate of change (loss) of momentum. Hence, the equation of motion of an electron in a plasma can be rewritten as:

$$(\mathrm{d}\upsilon/\mathrm{d}t) = -(e\boldsymbol{E}/m_e) - \upsilon\nu_e. \tag{8.3}$$

The above equation is equally applicable to the ordinary and complex plasmas, where electron collisions with other plasma constituents occur naturally. The physics of the electron collisions in complex plasmas and the expressions for corresponding collision frequencies are given in Sect. 5.4.2.2. Without loss of generality with $\upsilon = 0$ at $t = 0$ when an external electric field is switched on, one obtains from (8.3)

$$\upsilon = -(e\boldsymbol{E}/m_e\nu_e)[1 - \exp(-\nu_e t)]. \tag{8.4a}$$

It can be seen from (8.4a, b) that in the steady state $t \gg \tau (= 1/\nu_e)$ the drift velocity attains a value

$$\upsilon_o = -(e\boldsymbol{E}/m_e\nu_e). \tag{8.4b}$$

Another parameter of basic interest is the drift mobility, which can be defined as:

$$\mu_e = (|\upsilon/E|) = (e/m_e\nu_e). \tag{8.5}$$

8.1.2 Current Density/Electrical Conductivity/Resistivity

Consider a cuboid (Fig. 8.1) of unit area of cross section and of length υ, with its length along the electric field. In unit time, all the electrons entering from left end (A) will travel a distance υ and cross the unit area (shaded) at the right end (B); thus the charge flowing across the unit area per unit time, i.e., the current density is given by the number of such electrons, i.e., the number of electrons in the cuboid; thus

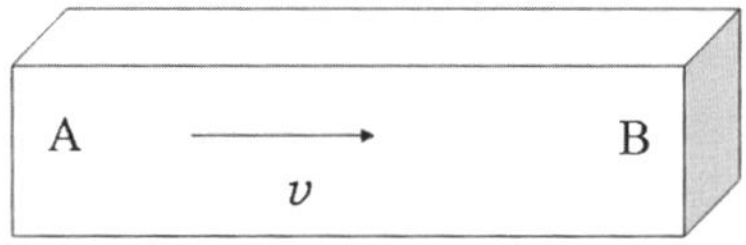

Fig. 8.1 Cuboid of unit area of cross section and length equal to drift velocity

$$\boldsymbol{J} = -n_e e \boldsymbol{\upsilon}, \tag{8.6}$$

where n_e is the electron density of the plasma.

Substituting for $\boldsymbol{\upsilon}$ from (8.4b) one gets

$$\boldsymbol{J} = \sigma \boldsymbol{E}, \tag{8.7}$$

where $\sigma = (1/\rho) = \left(n_e e^2/m_e v_e\right) = n_e e \mu_e$ is the electrical conductivity and ρ represents the resistivity.

8.1.3 Einstein Relation

Consider an isolated conductor, located in an electric field, which pushes the electrons in the opposite direction, setting up an electron density gradient. In the steady state, the net current density in the conductor on account of electron diffusion (caused by the density gradient) and the electric field should be zero. Thus one has

$$J_x = eD_e(\partial n_e/\partial x) + n_e e \mu_e E = 0,$$

or,

$$D_e(\partial n_e/\partial x) = -n_e \mu_e E, \tag{8.8}$$

where D_e is the coefficient of diffusion.

For partially ionized plasmas the density of electrons/ions, usually characterized by the *Maxwellian Boltzmann* distribution can be, expressed as:

$$n_e(x) = n_{eo} \exp[e\varphi(x)/k_B T_e], \tag{8.9}$$

where $\varphi(x)$ is the electric potential corresponding to electric field $\boldsymbol{E} = -(\partial\varphi/\partial x)\hat{x}$.

Substituting for n_e from (8.8), (8.9) reduces to

$$(D_e/\mu_e) = (k_B T_e/e). \tag{8.10}$$

This relation is known as *Einstein* relation and is valid when the charge carriers obey classical (*Boltzmann*) statistics.

8.1.4 Electrical Conductivity in Presence of an Alternating Electric Field

In the presence of an alternating electric field $\boldsymbol{E}[= \boldsymbol{E}_0 \exp(i\omega t)]$, the equation of motion gets modified as

$$(\mathrm{d}\boldsymbol{\upsilon}/\mathrm{d}t) + \boldsymbol{\upsilon}\nu_e = -(e\boldsymbol{E}_0/m_e)\exp(i\omega t). \tag{8.11}$$

The steady state solution of (8.11) for the drift velocity of electron can be written as:

$$\begin{aligned}\boldsymbol{\upsilon} = \mathrm{Re}\left(\frac{-e\boldsymbol{E}_0}{m_e(\nu_e + i\omega)}\exp(i\omega t)\right) &= \mathrm{Re}[(\boldsymbol{\upsilon}_r - i\boldsymbol{\upsilon}_i)\exp(i\omega t)] \\ &= [\boldsymbol{\upsilon}_r \cos(\omega t) + \boldsymbol{\upsilon}_i \sin(\omega t)]\end{aligned} \tag{8.12}$$

where $\boldsymbol{\upsilon}_r = -\left(\frac{e\boldsymbol{E}_0 \nu_e'}{m_e\omega(1+\nu_e'^2)}\right)$, $\boldsymbol{\upsilon}_i = -\left(\frac{e\boldsymbol{E}_0}{m_e\omega(1+\nu_e'^2)}\right)$, and $\nu_e' = (\nu_e/\omega)$.

The current density in this case can then be expressed as:

$$\begin{aligned}\boldsymbol{J} = \sigma\boldsymbol{E} = -n_e e\boldsymbol{\upsilon} &= \mathrm{Re}[(\sigma_r - i\sigma_i)\boldsymbol{E}_0 \exp(i\omega t)] \\ &= [\sigma_r \cos(\omega t) + \sigma_i \sin(\omega t)]\boldsymbol{E}_0,\end{aligned} \tag{8.13}$$

where $\sigma_r = \left(\frac{n_e e^2 \nu_e'}{m_e\omega(1+\nu_e'^2)}\right)$ and $\sigma_i = \left(\frac{n_e e^2}{m_e\omega(1+\nu_e'^2)}\right)$.

It is interesting to notice that the drift velocity and hence the current density have a component in phase with the electric field while other one is out of phase by $(\pi/2)$.

8.1.5 Electrical Conductivity in Presence of Magnetic Field

When an electric field (*say* $\boldsymbol{E} = \hat{x}E_x + \hat{y}E_y$) and a magnetic field $(\boldsymbol{B} = \hat{z}B_o)$ are simultaneously operative, the equation of motion of an electron is

$$(\mathrm{d}\boldsymbol{\upsilon}/\mathrm{d}t) + \nu_e\boldsymbol{\upsilon} = -(e/m_e)[\boldsymbol{E} + (1/c)\boldsymbol{\upsilon} \times \boldsymbol{B}]. \tag{8.14}$$

In terms of velocity components, this equation can be written as:

$$(\mathrm{d}\upsilon_x/\mathrm{d}t) + \nu_e\upsilon_x = -(e/m_e)E_x - \omega_c\upsilon_y \tag{8.15a}$$

and

$$(\mathrm{d}\upsilon_y/\mathrm{d}t) + \nu_e\upsilon_y = -(e/m_e)E_y + \omega_c\upsilon_x, \tag{8.15b}$$

where $\omega_c = eB/mc$.

The z-velocity component is zero in the absence of any driving force in the z direction. In the steady state (i.e., $\mathrm{d}\upsilon/\mathrm{d}t = 0$) and with a little algebra (8.15a, b) lead to

$$\upsilon_x = -\left(\frac{e\left(v'_{\mathrm{ec}}E_x - E_y\right)}{m_e\omega_c\left(1 + v'^2_{\mathrm{ec}}\right)}\right) \quad \text{and} \quad \upsilon_y = -\left(\frac{e\left(E_x + v'_{\mathrm{ec}}E_y\right)}{m_e\omega_c\left(1 + v'^2_{\mathrm{ec}}\right)}\right), \tag{8.16}$$

where $v'_{\mathrm{ec}} = (v_e/\omega_c)$.

Hence, the two components of the current density are given by

$$J_x = -en_e\upsilon_x = \left(\sigma_{||}E_x - \sigma_{\perp}E_y\right) \tag{8.17a}$$

and

$$J_y = -en_e\upsilon_x = \left(\sigma_{\perp}E_x + \sigma_{||}E_y\right), \tag{8.17b}$$

where

$$\sigma_{||} = \left(\frac{n_e e^2 v'_{\mathrm{ec}}}{m_e\omega_c\left(1 + v'^2_{\mathrm{ec}}\right)}\right) \text{ and } \sigma_{\perp} = \left(\frac{n_e e^2}{m_e\omega_c\left(1 + v'^2_{\mathrm{ec}}\right)}\right) \tag{8.18}$$

It is remarkable that the expressions for $\sigma_{||}$ and $\sigma_{\perp}$ are identical to those for σ_r and σ_i, when the cyclotron frequency ω_c is replaced by the frequency of the applied electric field ω. It is noticed that (8.17a and b) provide two relations between four parameters E_x, E_y, J_x and J_y and hence to uniquely define a relationship between two parameters, an additional auxiliary condition has to be specified. The popular auxiliary condition is $J_y = 0$ which can easily be realized if E_y is measured by a potentiometer; this yields an important parameter viz. *Hall coefficient*. Using (8.17a), (8.17b) the *Hall coefficient* (R_H) may be expressed as:

$$R_H = \left[E_y/B_oJ_x\right]_{J_y=0} = (-1/en_e c). \tag{8.19a}$$

The nature (positive or negative charge) of the carrier and the number density of the carriers can thus be obtained from the knowledge of the *Hall coefficient*. In a way, similar to that followed in the derivation of *Hall coefficient* it can be shown that

$$(J_x/E_x)_{J_y=0} = \left(e^2 n_e/m_e v_e\right). \tag{8.19b}$$

This relation leads to the fact that in this particular situation the conductivity (i.e., σ) is independent of the applied magnetic field B_o. Another auxiliary condition, which can be easily realized, is that by electric shorting in the y direction, i.e., $E_y = 0$; relations similar to (8.18–8.19a, b) can be derived, as in the case $J_y = 0$.

In the presence of an alternating electric field say $\boldsymbol{E}\ \left[= \left(\hat{x}E_x + \hat{y}E_y\right)\exp(i\omega t)\right]$, an approach similar to that in Sect. 8.1.5 leads to

$$\upsilon_x = -(e/m_e)\left(\frac{(\nu_e + i\omega)E_x - \omega_c E_y}{(\nu_e + i\omega)^2 + \omega_c^2}\right)$$

and

$$\upsilon_y = -(e/m_e)\left(\frac{\omega_c E_x + (\nu_e + i\omega)E_y}{(\nu_e + i\omega)^2 + \omega_c^2}\right), \tag{8.20}$$

Hence, in the oscillatory case the current density is given by $\boldsymbol{J}\exp(i\omega_o t)$ $\left[= \left(\hat{x}J_x + \hat{y}J_y\right)\exp(i\omega_o t)\right]$ where J_x and J_y can be expressed by (8.17a, b) and

$$\sigma_{||} = \left(n_e e^2/m_e\right)\left(\frac{(\nu_e + i\omega)E_x - \omega_c E_y}{(\nu_e + i\omega)^2 + \omega_c^2}\right)$$

and

$$\sigma_{\perp} = \left(n_e e^2/m_e\right)\left(\frac{\omega_c E_x + (\nu_e + i\omega)E_y}{(\nu_e + i\omega)^2 + \omega_c^2}\right). \tag{8.21}$$

It is interesting to notice that the expressions for $\sigma_{||}$ and $\sigma_{\perp}$ in this case are identical to those in (8.19a, b) and can be obtained by substituting $(\nu_e + i\omega)$ for ν_e in (8.19a, b).

8.1.6 Nonlinear Effects: Hot Electrons

As seen in the earlier sections, the current density in a plasma is a linear function of the electric field and the transport coefficients are independent of the electric field. However, it is observed that at high electric fields this is no longer valid and the relation between the current density and the electric field becomes nonlinear. In this section, a model to explain the nonlinearity has been discussed, which brings out the essential physics of the nonlinear collision phenomena.

Consider, for example, a slightly ionized gas with dominant elastic electron-neutral atom collisions with a constant mean free path, which corresponds to

$$\nu_e \propto T_e^{1/2} \text{ or } \nu_e = \nu_{e0}(T_e/T)^{1/2}, \tag{8.22}$$

where T_e and T are the temperatures of the electrons and the gas.

When an electric field is applied, the electrons gain energy from the electric field and in the steady state transfer this energy to the neutral gas through collisions; such an energy transfer between the electrons and the gas takes place only when the electrons are at a higher temperature than that of the gas. When the temperature difference between the electrons and the gas is significant such electrons are known as hot electrons. The energy transfer per collision is

$$\Delta\varepsilon = \delta\left(\frac{3}{2}k_B T_e - \frac{3}{2}k_B T\right), \tag{8.23a}$$

where

$$\delta \approx 2m_e/M \tag{8.23b}$$

and M is the mass of the neutral atoms.

The energy balance of electrons may be expressed as:

$$J.\, E = n_e v_e \Delta\varepsilon \tag{8.24}$$

Using (8.22) and (8.23a, 8.23b), (8.24) leads to

$$\frac{(\mu_o E)^2}{3k_B T/M} = \frac{T_e}{T}\cdot\left(\frac{T_e}{T} - 1\right) = \alpha E^2, \tag{8.25}$$

where $\mu_o = e/mv_o$ is the electron mobility in the absence of the electric field $(T_e = T)$ and $\alpha = \mu_o^2/(3k_B T/M)$. The parameter $(3k_B T/M)$ is equal to the mean square speed of the neutral atoms. Hence, T_e/T is appreciably different from unity when the ratio of the drift velocity $\mu_o E$ to the r.m.s speed of neutral atoms is significant.

The current density J, corresponding to an electric field E can be obtained by evaluating T_e from (8.25) and $(ii)J$ from (8.7) and (8.22). Since (8.25) is quadratic and T_e/T is positive, this can be done for any value of αE^2. However, two cases are of much interest.

1. $\alpha E^2 \gg 1$, $(T_e/T) \approx 1$ and from (8.25), $T_e/T = 1 + \alpha E^2$; hence

$$J = \frac{n_e e^2}{m_e v_e}E = \frac{e^2 n_e}{m_e v_{eo}}E(v_e/v_{eo})^{-1} = \sigma_0 E(T_e/T)^{-1/2} = \sigma_o E\left(1 - \alpha E^2/2\right). \tag{8.26a}$$

where $\sigma_o = e^2 n_e/m_e v_{eo}$ is the electrical conductivity at low electric fields.

2. Similarly when $\alpha E^2 \gg 1, (T_e/T) \gg 1$ and $(T_e/T) = (\alpha E^2)^{1/2}$ from (8.25); hence,

$$J = \frac{e^2 n_e}{m_e v_{eo}}E.(v_{eo}/v_e) = \sigma_o E(T_e/T)^{-1/2} = \sigma_o \alpha^{-1/4} E^{1/2}. \tag{8.26b}$$

Equations (8.26a) and (8.26b) indicate the departure from linearity.

It is also interesting to consider the case when the electric field is very high, leading to large electron temperatures, so that the dominant collisions are the inelastic ones; for simplicity, it may be assumed that the energy transfer from an electron to the neutral atom in such a collision is a constant amount $(\Delta\varepsilon_{in})$. In this case the energy balance of the electrons may be written as.

$$\sigma E^2 = n_{e.} \frac{e^2}{m_e v_e} E^2 = n_e v_e(\Delta\varepsilon_{\text{in}}) \text{ or } (\Delta\varepsilon_{\text{in}}/\text{m}_\text{e}) = (e/m_e v_e)^2 \text{E}^2$$
$$J = en_e(e/m_e v_e)\, E = en_e(\Delta\varepsilon_{\text{in}}/m)^{1/2}. \tag{8.26c}$$

Thus in this extreme case, there is a drastic departure from linearity and the current density is independent of the electric field.

This theory can be extended to include alternate electric or/and magnetic fields.

8.2 Kinetic Theory (After Mishra and Sodha [1])

In contrast to the phenomenological theory, the transport phenomena can be investigated on the basis of the kinetic theory, which takes into account the energy/ speed dependence of the electron collision frequency. The kinetic approach is based on Boltzmann's transfer equation, which explores the change in the electron velocity distribution function on application of an electric field or any other stimulus like temperature and density gradients.

8.2.1 *Boltzmann's Transfer Equation*

This equation is primarily concerned with the electron velocity distribution function $f(x, y, z, \upsilon_x, \upsilon_y, \upsilon_z, , t)\, \mathrm{d}x\, \mathrm{d}y\, \mathrm{d}z\, \mathrm{d}\upsilon_x\, \mathrm{d}\upsilon_y\, \mathrm{d}\upsilon_z$, which represents the number of electrons in the phase space element $\mathrm{d}^3 r \mathrm{d}^3\upsilon = \mathrm{d}x\, \mathrm{d}y\, \mathrm{d}z\, \mathrm{d}\upsilon_x\, \mathrm{d}\upsilon_y\, \mathrm{d}\upsilon_z$ at time t; in equilibrium the distribution function is isotropic and given by classical statistics as

$$f_o = A \exp(-m\upsilon^2/2k_B T_e); \tag{8.27a}$$

the normalizing constant A is in terms of electrons density given by

$$n_e = \int_{-\infty}^{\infty}\int_{-\infty}^{\infty}\int_{-\infty}^{\infty} A \exp(-m\upsilon^2/2k_B T_e) \mathrm{d}\upsilon_x \mathrm{d}\upsilon_y \mathrm{d}\upsilon_z$$
$$= \int_{o}^{\infty} 4\pi A\upsilon^2 \exp(-mv^2/2\, k_B\, Te) \mathrm{d}\upsilon, \tag{8.27b}$$

where by $A = n_e(m_e/2\pi k_B T_e)^{3/2}$.

The triple integral has been replaced by a single one as in Chap. 2. Equation (8.27a) is valid for non degenerate plasma, which is in general the case for gaseous plasmas.

In the presence of an electric field or electron temperature/density gradient the distribution function has an anisotropic component, which gives rise to the transport of electronic charge and energy and causes electrical and thermal currents.

The distribution function f may vary with time on account of the velocity and acceleration of electrons; such a variation is known as drift variation $(\Delta f)_d$. Remembering that at time $t+\mathrm{d}t$, the electrons having the coordinates $x-\upsilon_z\mathrm{d}t$, $y-\upsilon_y\mathrm{d}t$ and $z-\upsilon_z\mathrm{d}t$ and velocity components $\upsilon_x-a_x\mathrm{d}t$, $\upsilon_{y-}a_y\mathrm{d}t$ and $\upsilon_z-a_z\mathrm{d}t$ at time t will be characterized by the space and velocity coordinates $x, y, z, \upsilon_x, \upsilon_y$ and υ_z at time $t+\mathrm{d}t$. In the interval between t and $t+\mathrm{d}t$, the electrons characterized by $x, y, z, \upsilon_x, \upsilon_{y,}\upsilon_z$ at time t have moved out of the phase space $\mathrm{d}x, \mathrm{d}y, \mathrm{d}z, \mathrm{d}\upsilon_x, \mathrm{d}\upsilon_y, \mathrm{d}\upsilon_z$. Hence,

$$\begin{aligned}(\Delta f)_d &= f\left(x-\upsilon_x\mathrm{d}t, y-\upsilon_y\mathrm{d}t, z-\upsilon_z\mathrm{d}t, \upsilon_x-a_x\mathrm{d}t, \upsilon_y-a_y\mathrm{d}t, \upsilon_z-a_z\mathrm{d}t, t\right)\\ &\quad - f\left(x, y, z, \upsilon_x, \upsilon_y, \upsilon_z, t+\mathrm{d}t\right)\\ &= -\left\{\frac{\partial f}{\partial t}+\upsilon_x\frac{\partial f}{\partial x}+\upsilon_y\frac{}{\partial y}+\upsilon_z\frac{\partial f}{\partial z}+a_x\frac{\partial f}{\partial \upsilon_x}+a_y\frac{}{\partial \upsilon_y}+a_z\frac{\partial f}{\partial \upsilon_z}\right\}\mathrm{d}t.\end{aligned}$$

or

$$\left(\frac{\partial f}{\partial t}\right)_d = -\left\{\frac{\partial f}{\partial t}+\boldsymbol{\upsilon}\cdot\nabla f+\boldsymbol{a}\cdot\nabla_{\upsilon} f\right\}$$

To ensure a steady state

$$\left(\frac{\partial f}{\partial t}\right)_d+\left(\frac{\partial f}{\partial t}\right)_c = 0,$$

where $\left(\frac{\partial f}{\partial t}\right)_e$ denotes the rate of change of f due to electron collisions.

Hence

$$\frac{\partial f}{\partial t}+\boldsymbol{\upsilon}\cdot\nabla f+\boldsymbol{a}\cdot\nabla_{\upsilon} f = \left(\frac{\partial f}{\partial t}\right)_c. \tag{8.28}$$

This equation is known as Boltzmann's transfer equation. In the presence of electric and magnetic field and linear gradient of electron density and temperature, the electron velocity distribution function can to a first approximation be written as

$$f(\boldsymbol{\upsilon}, \boldsymbol{r}) = f_o(\upsilon)+\sum(\upsilon_x/\upsilon)F_x(\upsilon). \tag{8.29}$$

For isotropic collisions and validity of (8.29), it can be shown [3] that

$$\left(\frac{\partial f}{\partial t}\right)_c = -\nu_e(\upsilon)\cdot(f-f_o) + \text{Another termwhich takes into account the energy exchange in collisions} \tag{8.30}$$

This another term can be readily written down only for elastic collisions; however, it is of little interest to complex plasmas since in complex plasmas the collisions are primarily inelastic.

8.2.2 Electrical Current/Electrical Conductivity

Consider a uniform plasma with an electric field $\underline{E} = \underline{i}\mathrm{E}_x$ in the x direction. Thus $\nabla\cdot f = 0, a_y = a_z = 0$ and $a_x = -e\mathrm{E}_x/m_e$; then from (8.28), (8.29), and (8.30) one obtains.

$$\frac{eE_x}{m_e}\cdot\frac{\partial}{\partial \upsilon_x}\{f_o(\upsilon) + (\upsilon_x/\upsilon)F_x(\upsilon)\} = \nu_e\frac{\upsilon_x}{\upsilon_x}F_x(\upsilon)$$

Since $f_o(\upsilon) \gg (\upsilon_x/\upsilon)F_x(\upsilon)$

$$\frac{eE_x}{m_e}\cdot\frac{\partial\upsilon}{\partial\upsilon_x}\cdot\frac{\mathrm{d}f_o}{\mathrm{d}\upsilon} = \frac{eE_x}{m_e}\cdot\frac{\mathrm{d}fo}{\mathrm{d}\upsilon_x}\cdot\frac{\upsilon_x}{\upsilon} = \nu_e\cdot\frac{\upsilon_x}{\upsilon}F_x(\upsilon) \tag{8.31}$$

or

$$F_x(\upsilon) = \frac{eE_x}{m_e\nu_e}\frac{\partial f_o}{\partial\upsilon}$$

The electrical current density J_x is given by

$$J_x = -e\int_{-\infty}^{+\infty}\int_{-\infty}^{+\infty}\int_{-\infty}^{+\infty}\upsilon_x[f_o(\upsilon) + (\upsilon_x/\upsilon)F_x/\upsilon]\mathrm{d}\upsilon_x\mathrm{d}\upsilon_y\mathrm{d}\upsilon_z$$

Putting for υ_x^2 its average value $\upsilon^2/3$,

$$J_x = -e\int_o^{\infty}\frac{4\pi\upsilon^2}{3}\cdot\frac{eE_x}{m_e\nu_e}\frac{df_0}{d\upsilon}\mathrm{d}\upsilon.$$

Integrating by parts

$$J_x = \frac{e^2}{3m_e} E_x \int_o^\infty 4\pi \upsilon^2 f_o \frac{\mathrm{d}}{\mathrm{d}\upsilon}\left(\frac{\upsilon^3}{v_e}\right) \mathrm{d}\upsilon = \left(e^2/3m_e\right) E_x \int_o^\infty \frac{\mathrm{d}}{\mathrm{d}\upsilon}\left(\upsilon^3/v_e\right) \mathrm{d}n_e$$

where $\mathrm{d}n_e$ is the number of electrons per unit volume with speed between υ and $\upsilon + \mathrm{d}\upsilon$.

Hence,

$$J_x = \frac{e^2 n_e}{3m_e} \left\langle \frac{1}{\upsilon^2} \frac{\mathrm{d}}{\mathrm{d}\upsilon} \left(\upsilon^3/v_e\right) \right\rangle E_x = \sigma_{||} E_x, \tag{8.32a}$$

where $< >$ denotes average over the velocity distribution of electrons and σ_e is the electrical conductivity given by

$$\sigma = \frac{e^2 n_e}{3m_e} \left\langle \frac{1}{\upsilon^2} \frac{\mathrm{d}}{\mathrm{d}} \left(\upsilon^3/v_e\right) \right\rangle. \tag{8.32b}$$

When the electron collisions frequency v is independent of the electron velocity (as in the phenomenological theory)

$$\sigma = e^2 n_e / m_e v_e.$$

8.2.3 Other Transport Parameters

In a way, similar to that used in the Sect. 8.2.2 one can obtain [1, 3] expressions for the other transport coefficients as follows:

$$D_e = \left(k_B T_e/3m_e\right) \left\langle \frac{1}{\upsilon^2} \frac{\mathrm{d}}{\mathrm{d}\upsilon} \left(\frac{\upsilon^3}{v_e}\right) \right\rangle, \tag{8.33}$$

$$\sigma_r = \frac{e^2 n_e}{3m_e} \left\langle \frac{1}{\upsilon^2} \frac{\mathrm{d}}{\mathrm{d}\upsilon} \left[\frac{v_e \upsilon^3}{v_e^2 + \omega^2}\right] \right\rangle \tag{8.34a}$$

and

$$\sigma_i = \frac{e^2 n_e}{3m_e} \left\langle \frac{1}{\upsilon^2} \frac{\mathrm{d}}{\mathrm{d}\upsilon} \left[\frac{\omega_o \upsilon^3}{v_e^2 + \omega^2}\right] \right\rangle; \tag{8.34b}$$

the expressions for $\sigma_{||}$ and $\sigma_\perp$ are identical to (8.32a) and (8.32b), with ω, replaced by ω_e, the cyclotron frequency of electrons.

The parameters $D_e, \sigma_r, \sigma_{\mathrm{i}}, \sigma_{||}$, and $\sigma_\perp$ have been defined and corresponding expressions based on the phenomenological theory given in Sect. 1.8 of this chapter.

In the presence of an electric field $\boldsymbol{E} = \left(\boldsymbol{i}E_x + \boldsymbol{j}E_y\right)\exp(i\omega t)$ and a magnetic field $\boldsymbol{B} = \boldsymbol{k}B(\boldsymbol{\omega}_c = \boldsymbol{k}\omega_c)$ the current density $\boldsymbol{J}$ is given by

$$J_x = \left(\sigma_{||}E_x - \sigma_{\perp}E_y\right)\exp(i\omega t) \tag{8.35a}$$

and

$$J_y = \left(\sigma_{\perp}E_x - \sigma_{||}E_y\right)\exp(i\omega t) \tag{8.35b}$$

where

$$\sigma_{||} = \frac{e^2 n_e}{3m_e}\left\langle \frac{1}{\upsilon^2}\frac{\mathrm{d}}{\mathrm{d}\upsilon}\left[\frac{(\nu_e + i\omega_o)\upsilon^3}{(\nu_e + i\omega_o)^2 + \omega_c^2}\right]\right\rangle \tag{8.36a}$$

and

$$\sigma_{\perp} = \frac{e^2 n_e}{3m_e}\left\langle \frac{1}{\upsilon^2}\frac{\mathrm{d}}{\mathrm{d}\upsilon}\left[\frac{\omega_c\upsilon^3}{(\nu_e + i\omega_o)^2 + \omega_c^2}\right]\right\rangle \tag{8.36b}$$

8.2.4 Ohmic Power Loss

The Ohmic power loss per unit volume is given by [3]

(1) dc electric field

$$W = \sigma E_x^2 \tag{8.37}$$

(2) de electric field with perpendicular magnetic field

$$W = J_x E_x + J_y E_y = \sigma_{||}\left(E_x^2 + E_y^2\right) \tag{8.38}$$

where $E_y = K\upsilon B$ (MHD generators); $\boldsymbol{\upsilon} = \boldsymbol{k}\upsilon$

(a) Continuous electrodes

$$E_x = 0;\ W = \sigma_{||}(K\upsilon B)^2 \tag{8.39a}$$

(b) Segmented electrode

$$J_x = 0,\ W = \left[\sigma_{||} + \sigma_{\perp}^2/\sigma_{||}\right](K\upsilon B)^2 \tag{8.39b}$$

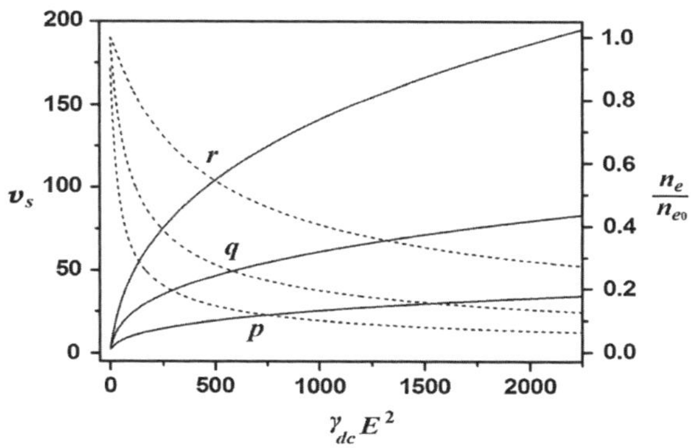

Fig. 8.2 v_s and (n_e/n_{e0}) dependence on $\gamma_{\text{dc}}E^2$ in the presence of a dc field for the standard set of parameters as stated in the text. The labels p, q, and r refer to $n_d = 10000, 1000$ and $100/\text{cm}^3$ respectively while solid and *broken lines* refer to *left*- and *right-hand scales*, respectively (after Sodha [2], curtsey authors and publishers AIP)

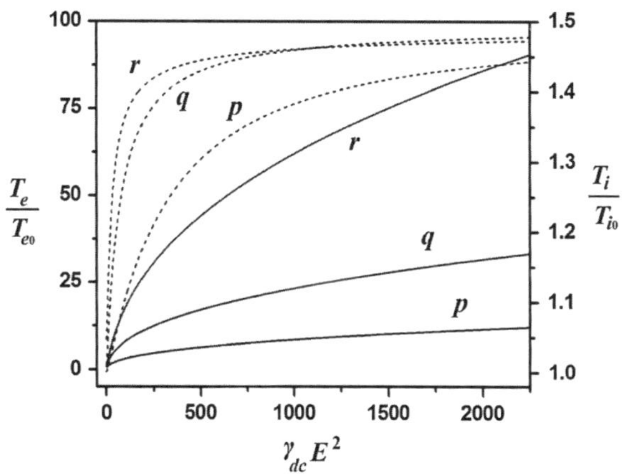

Fig. 8.3 (T_e/T_{e0}) and (T_i/T_{i0}) dependence on $\gamma_{dc}E^2$; the parameters and labels on the *curve* are the same as in Fig. 8.2 (after Mishra and Sodha [1], curtsey authors and publishers AIP)

(3) Alternating electric field with perpendicular magnetic field

$$
\begin{aligned}
W = \left(\frac{e^2 n_e}{24 m_e}\right)\left(A_1 A_1^* + A_2 A_2^*\right)\left[\left\langle \frac{1}{\upsilon^2}\frac{\mathrm{d}}{\mathrm{d}\upsilon}\left(\frac{v_e \upsilon^3}{(\omega_o - \omega_c)^2 + v_e^2}\right)\right\rangle\right] \\
+ \left[\left\langle \frac{1}{\upsilon^2}\frac{\mathrm{d}}{\mathrm{d}\upsilon}\left(\frac{v_e \upsilon^3}{(\omega_o + \omega_c)^2 + v_e^2}\right)\right\rangle\right]
\end{aligned}
\tag{8.40}
$$

where $A_1 = E_{\text{ox}} + iE_{\text{oy}}$ and $A_2 = E_{\text{ox}} - iE_{\text{oy}}$

The Ohmic power loss is an important term in the energy balance of electrons, for significant electric fields.

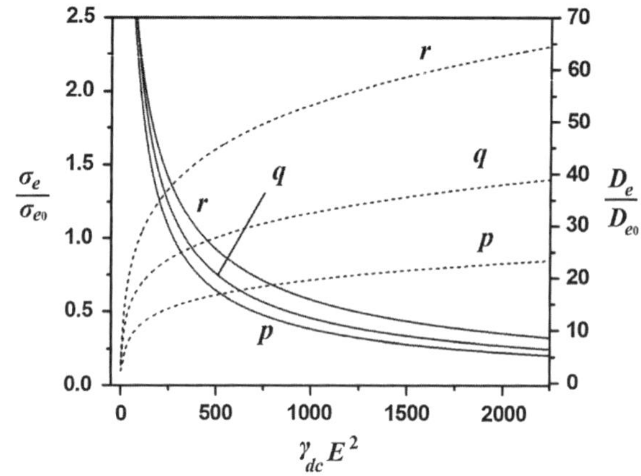

Fig. 8.4 (σ_e/σ_{e0}) and (D_e/D_{e0}) dependence on $\gamma_{\mathrm{dc}}E^2$ with $\sigma_{e0} = 11.11/\Omega\mathrm{m}$ and $D_{e0} = 10^3\ \mathrm{m}^2/\mathrm{s}$; the other parameters and labels on the *curve* are the same as in Fig. 8.2 (after Mishra and Sodha [1], curtsey authors and publishers AIP)

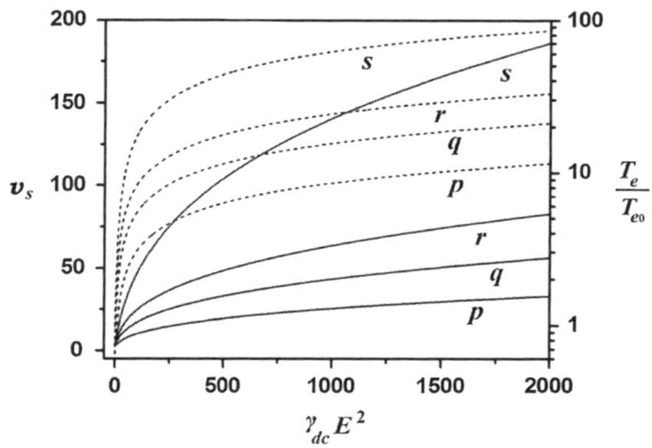

Fig. 8.5 v_s and (T_e/T_{e0}) dependence on $\gamma_{\mathrm{dc}}E^2$ in the presence of a dc field for the standard set of parameters as stated in the text. The labels p, q, r, and s correspond to $a = 1.0, 0.5, 0.3$ and $0.1\,\mu\mathrm{m}$ respectively *while solid* and *broken lines* refer to *left-* and *right-hand scales*, respectively (after Mishra and Sodha [1], curtsey authors and publishers AIP)

8.3 Kinetics of Complex Plasma with a D.C. Electric Field

Consider a plasma with suspended dust having a size distribution. The uniform electric potential on the dust particles V_s and the electron/ion, density/temperature can be obtained by simultaneous solution of (7.6a), (7.7a), (7.8a), (7.9a), (7.10a), and (7.11a). In the presence of a d.c. /alternating electric field with or without a magnetic field, the energy balance of electrons, expressed by Eq. (7.9a) gets modified by the inclusion of the Ohmic loss, given by (8.37) to (8.40) of this Chapter in different cases. In this Chapter the case of a d.c. electric field, with W, given by (8.36a, b) has been considered as an example. Thus in the presence of a d.c. electric field one has to obtain the simultaneous solution of (7.6a), (7.7a), (7.8a), and (7.11), (8.37) for Ohmic loss, (8.32b) for electrical conductivity, and the following (modified energy balance of electrons) equation to obtain $n_o, V_s, n_e, n_i, T_e, T_i$ and the electrical conductivity [from (8.30)]:

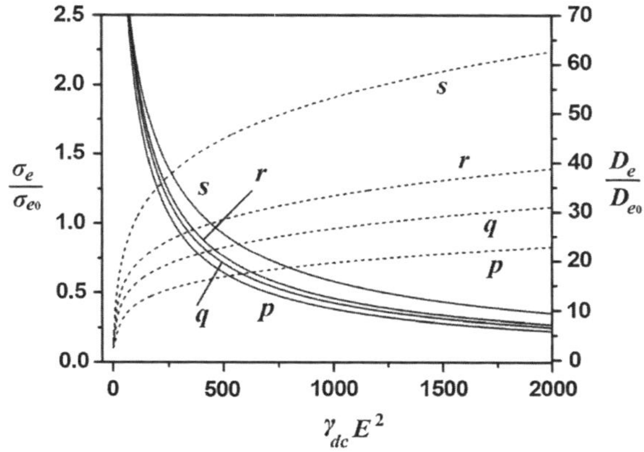

Fig. 8.6 (σ_e/σ_{e0}) and (D_e/D_{e0}) dependence on $\gamma_{dc}E^2$, the parameters and labels on the *curve* are the same as in Fig. 8.5 (after Mishra and Sodha [1], curtsey authors and publishers AIP)

$$\frac{\mathrm{d}}{\mathrm{d}t}\left(\frac{3}{2}k_Bn_eT_e\right) = \beta_in_oE_e - \alpha_rn_en_i(3k_BT_e/2) - n_d\cdot\pi a_{\mathrm{rms}}^2(n_ef_{\mathrm{ec}}\varepsilon_{\mathrm{ec}} - f_{\mathrm{ec}}\varepsilon_{\mathrm{ec}}) + W \tag{8.41}$$

Electron collisions with gaseous species can be taken into account as in Sect. 5.4.2 .

For a numerical appreciation of the effect of the electric field, computations have been made for the following set of standard parameters; the parametric study is conducted by changing one parameter keeping others the same.

Standard parameters:

$n_{\mathrm{eo}} = n_{\mathrm{io}} = 10^{10}/\,\mathrm{cm}^3, n_{\mathrm{oo}} = 10^4n_{\mathrm{eo}}, (v_e/\omega_c) = 0.2, n_d = 10^4\,\mathrm{cm}^{-3}, a = 1\ \mu\mathrm{m}, T_{\mathrm{eo}} = 1000\ \mathrm{K}, T_{\mathrm{io}} = 400\ \mathrm{K}, T = T_d = 300\ \mathrm{K}, \kappa = 1, \alpha_{\mathrm{ro}} = 10^{-7}\,\mathrm{cm}^3/s, \gamma_{\mathrm{dc}} = \left(4\pi e^2/m_e\delta_{\mathrm{en}}k_BT_{\mathrm{eo}}v_{\mathrm{eo}}^2\right)$ $[\sigma_{\mathrm{eo}} = 11.11/\,\Omega\mathrm{m}$ and $D_{\mathrm{eo}} = 10^3\,\mathrm{m}^2/\mathrm{s}$ are thecomputed values for $\gamma_{\mathrm{dc}} = 0]$.

The results of the computations have been illustrated in Figs. 8.2, 8.3, 8.4, 8.5, and 8.6. It is seen that an electric field can significantly affect the kinetics of complex plasmas and the transports parameters.

In a similar manner the effect of the magnetic field and alternating electric field can be investigated (see [1]).

References

1. S.K. Mishra, M.S. Sodha, Phys. Plasmas **20**, 033701 (2013)
2. M.S. Sodha, *IAPT Bull.* **31**, 240 (2007)
3. I.P. Shakarofsky, T.W. Johnston, M.P. Prachynski, *The Particle Dynamics of Plasmas.* (Addison Wesley Publishing Co., Reading, 1966)

Chapter 9
Electromagnetic Wave Propagation in Complex Plasma

9.1 Linear Propagation

9.1.1 Wave Equation

Electromagnetic phenomena in a medium are governed by Maxwell's equations. As derived in all texts on electromagnetic theory, Maxwell's equations lead to the equation for the propagation of an electromagnetic wave in an electrically neutral medium. Nonlinear propagation of electromagnetic waves with uniform and non-uniform irradiance on the wave front have been explored.

$$\nabla^2 \boldsymbol{E} = \left(\in \mu/c^2\right)\frac{\partial^2 \boldsymbol{E}}{\partial t^2} + \left(4\pi\mu/c^2\right)\frac{\partial \boldsymbol{J}}{\partial t}, \tag{9.1a}$$

where

$\boldsymbol{E}$ is the electric vector

$\in$ is the dielectric function of the medium

μ is the magnetic permeability, and

$\boldsymbol{J}$ is the current density

The above equation is known as the wave equation and has been herein expressed in the Gaussian system of units; this system is the most popular one in plasma physics.

For complex plasmas of interest $\in = 1$ and $\mu = 1$; thus for an electromagnetic wave $\boldsymbol{E} = jE_0 \exp(i\omega t)$ propagating in the z direction, (9.1a) reduces to

$$\frac{\mathrm{d}^2 E_0}{\mathrm{d}\xi^2} = -[E_0 - (4\pi i/\omega)J_0], \tag{9.1b}$$

where $\boldsymbol{J} = \boldsymbol{j}J_0 \exp(i\omega t)$ and $\xi = (\omega z/c)$.

M. S. Sodha, *Kinetics of Complex Plasmas*, Springer Series on Atomic, Optical, and Plasma Physics 81, DOI: 10.1007/978-81-322-1820-3_9,

9.1.2 Complex Refractive Index for Linear Propagation in the Absence of a Magnetic Field

Substituting for $\boldsymbol{J}_0$ from (8.13) in (9.1b) one obtains

$$\frac{d^2E_0}{d\xi^2} = -\beta^2 E_0, \tag{9.2}$$

where

$$\beta^2 = (1 - 4\pi\sigma_i/\omega) - i(4\pi\sigma_r/\omega) = (-n + ik)^2$$
$$2n^2 = \left\{(1 - 4\pi\sigma_i/\omega)^2 + (4\pi\sigma_r/\omega)^2\right\}^{1/2} + (1 - 4\pi\sigma_i/\omega)$$
$$2k^2 = \left\{(1 - 4\pi\sigma_i/\omega)^2 + (4\pi\sigma_r/\omega)^2\right\}^{1/2} - (1 - 4\pi\sigma_i/\omega)$$

and σ_r, σ_i are given by (8.13).

It is convenient to express $(4\pi\sigma_i/\omega)$, and $(4\pi\sigma_r/\omega)$, as $4\pi\sigma_i/\omega = \omega_p^2/(v^2 + \omega^2)$ and $4\pi\sigma_r/\omega = \omega_p^2(v/\omega)/(v^2 + \omega^2)$ in the phenomenological theory approximation, where $\omega_p = (4\pi n_e e^2/m_e)^{1/2}$, is known as the plasma frequency; the electron collisions and collision frequency have been discussed in earlier chapters, specifically Sect. 5.4.2.

In the linear case, corresponding to low electric field $J_0 \propto E_0$, and (σ_r, σ_i), are independent of E_0; thus the solution of (9.2) may be written as

$$\begin{aligned} E &= E_0 \exp(iwt) = E_1 \exp[i(\omega t + \beta\xi)] + E_2 \exp[i(\omega t - \beta\xi)] \\ &= E_1 \exp\{i[\omega t - (2\pi n/\lambda)z]\} \exp[(-2\pi k/\lambda)z] \\ &\quad + E_2 \exp\{i[\omega t + (2\pi n/\lambda)z]\} \exp[(2\pi k/\lambda)z], \end{aligned} \tag{9.3}$$

where $\lambda = 2\pi c/\omega$ is the wavelength of the wave in free space and E_1 and E_2 are constants.

The first term on the RHS of (9.3) represents a wave, propagating in the z direction, while the second term refers to a wave propagating in the opposite $(-z)$ direction. In both the cases, n and k represent the refractive index and the attenuation constant.

Sodha et al. [7] have shown that the amplitude and phase of an *EM* wave $E\exp[i(\omega t + \psi)]$, propagating in a nonlinear medium in the *JWKB* approximation are given by

$$E^{-1}[dE/d\xi] = -k - (2\mu_0)^{-1}(d\mu_0/d\xi) \tag{9.4a}$$

or

$$[\beta E^2]^{-1}[d(\beta E^2)/d\xi] = -2k - (\mu_0)^{-1}(d\mu_0/d\xi) \tag{9.4b}$$

and

$$(\mathrm{d}\psi/\mathrm{d}\xi) = -n + 0.5\,(\mathrm{d}\theta/\mathrm{d}\xi), \tag{9.4c}$$

where $\beta E^2(\xi)$ is the dimensionless irradiance of the beam, $\xi = (\omega/c)z$ is the dimensionless distance of propagation along the z axis, $\mu = \mu_0 \exp(-i\theta) = (n - ik)$, $\mu_0 = (n^2 + k^2)^{1/2}$, $\tan\theta(\xi) = [k(\xi)/n(\xi)]$ and $\beta = \left(e^2 m_0/3m_e^2 k_B T\omega^2\right)$ is a constant. It may be remembered that n, k, μ_0, and θ are all functions of βE^2.

Knowing n and k dependence on axial irradiance (βE^2) of the EM beam, the set of (9.13) can be solved numerically to obtain the dependence of the irradiance (βE^2) and phase (ψ) on the dimensionless distance of propagation (ξ).

9.1.3 Electromagnetic Propagation Along the Magnetic Field

An expression for the current density, corresponding to an electric field, represented by $\boldsymbol{E}_0 = \boldsymbol{i}E_{0x} + \boldsymbol{j}E_{0y}$ and a magnetic field $\boldsymbol{B}_0 = \boldsymbol{k}B_0$, has been derived in Chap. 8. After substitution for $\boldsymbol{J}_0$ from (8.21), (9.1b) can be written as

$$\frac{\mathrm{d}^2 E_{0x}}{\mathrm{d}\xi^2} = -E_{0x} + (4\pi i/\omega)J_{0x} = -AE_{0x} - BE_{0y} \tag{9.5a}$$

and

$$\frac{\mathrm{d}^2 E_{0y}}{\mathrm{d}\xi^2} = BE_{0x} - AE_{oy}, \tag{9.5b}$$

where

$A = 1 - (4\pi i/\omega)\sigma_{\|}$,
$B = (4\pi i/\omega)\sigma_{\perp}$

and $\sigma_{\|}/\sigma_{\perp}$ are given by (8.21). Equations (9.5a) and (9.5b) imply that E_{0x} and E_{0y} are coupled. To proceed further, one may multiply (9.5b) by α and add it to (9.5a); thus.

$$\frac{\mathrm{d}^2}{\mathrm{d}\xi^2}\left(E_{0x} + \alpha E_{0y}\right) = -(A - B\alpha)E_{0x} - (A\alpha + B)E_{0y} = -\beta^2\left(E_{0x} + \alpha E_{0y}\right), \tag{9.5c}$$

where $\beta^2 = A - B\alpha$ and $\beta^2\alpha = A\alpha + B$, whence $\alpha^2 = -1$ and $\alpha = \pm i$.

Thus (9.5c) may be expressed as:

$$\frac{\mathrm{d}^2}{\mathrm{d}\xi^2}\left(E_{0x} + iE_{0y}\right) = -\beta_+^2\left(E_{0x} + iE_{0y}\right) = -(n_+ - ik_+)^2\left(E_{0x} + iE_{0y}\right) \tag{9.6a}$$

and

$$\frac{d^2}{d\xi^2}\left(E_{0x} - iE_{0y}\right) = -\beta_-^2\left(E_{0x} - iE_{0y}\right) = -(n_- - ik_-)^2\left(E_{0x} - iE_{0y}\right), \tag{9.6b}$$

where $\beta_+^2 = (-n_+ + ik_+)^2 = A - iB$ and $\beta_-^2 = (-n_- + ik_-)^2 = A + iB$.

From (9.6a) and (9.6b) it is evident that $E_{0x} + iE_{0y}$ and $E_{0x} - iE_{0y}$ propagate as separate waves with refractive indices n_+ and n_- and attenuation coefficients k_+ and k_- (the waves of course get coupled when the amplitudes are large enough to produce appreciable nonlinearities).

It is instructive to explore the nature of the waves $E_x + iE_y$ and $E_x - iE_y$ known as extraordinary and ordinary waves. Thus to explore the nature of $E_x + iE_y$, we put the amplitude of the other wave as zero; thus $E_{\mathrm{ox}} - iE_{\mathrm{oy}} = 0$ or $E_x = e^{i\pi/2}E_{\mathrm{oy}}$, which implies that $|E_{0x}| = |E_{0y}|$ and that E_x is $\pi/2$ ahead in phase with respect to E_y; in other words, the electric vector is rotating clockwise as viewed along $\boldsymbol{B}$. Similarly $E_{\mathrm{ox}} - iE_{\mathrm{oy}}$ represents a wave in which the electric vector rotates counter clockwise as viewed along $\boldsymbol{B}$. The $E_x + iE_y$ mode is called the extraordinary wave and the $E_x - iE_y$ mode is known as the ordinary wave.

9.1.4 Simplified Expressions for Transport Parameters

On account of the mathematically complex expressions, obtained by the use of kinetic theory in Sect. 8.2 it is common to use simpler expressions (at the cost of rigor), based on the phenomenological theory (Sect. 8.1) which ignores the electron velocity distribution. Thus (8.32b), (8.33), (8.34a, b), and (8.36a, b), reduce to

$$\sigma = e^2 n_e / m\nu \tag{9.7a}$$

$$D_e = k_B T_e / m\nu, \tag{9.7b}$$

$$\sigma_r = \left(e^2 n_e / m\right)\left[\nu / \nu^2 + \omega^2\right], \tag{9.7c}$$

$$\sigma_i = \left(e^2 n_e / m\right)\left[\omega / \nu^2 + \omega^2\right], \tag{9.7d}$$

$$\sigma_{\parallel} = \left(e^2 n_e / m\right)(\nu + i\omega) / \left[(\nu + i\omega)^2 + \omega_c^2\right], \tag{9.7e}$$

and

$$\sigma_{\perp} = \left(e^2 n_e / m\right)\omega_c / \left[(\nu + i\omega)^2 + \omega_c^2\right];$$

the symbols have been introduced earlier in the book.

9.2 Physical Basis of Nonlinear Propagation of Electromagnetic Waves

In the presence of a low electric field, associated with an electromagnetic wave, propagating in a plasma, the Ohmic loss and consequent heating of electrons is negligible and hence the temperature and composition of the complex plasma remains almost unaltered. Under these conditions, the refractive index and absorption coefficient may be evaluated by using the expressions, developed in Chap. 8 (for σ_r and σ_i) and Sect. 9.1 (for n and k).

The physics of heating of electrons by an electric field has been developed in Chap. 8 and an expression for the Ohmic loss per unit volume has been given. Basically, the electrons gain power from the electric field; the Ohmic loss or power gained by the electrons from the electric field per unit volume is just the product of the real part of the electrical conductivity and mean square value of the electric field. Electrons also lose power in energy exchange during collisions. Hence, in the steady state, the electrons acquire a higher temperature such that the Ohmic loss equals collisional loss; the electron density and electron temperature, thus get altered, affecting σ_r and σ_i and hence the propagation parameters; this change of the propagation parameters (n and k) with the electric field gives rise to non-linear electromagnetic propagation. In such studies the electron collision frequency is as an important parameter.

The studies on the propagation of intense electromagnetic beams in complex plasmas, based on the number and energy balance of the constituents (considered earlier in the book) and the electromagnetic wave equation should include the following considerations.

(a) Generation and annihilation of ions/electrons in the gaseous component,
(b) Elastic collisions of electrons with ions, neutral atoms, and dust particles,
(c) Elastic collisions of ions with neutral atoms and elastic collision of ions with dust particles may be ignored on account of the slow ion speeds,
(d) Electron and ion accretion on dust particles,
(e) Electron emission from the surface of the dust,
(f) Energy exchange in the processes,
(g) Number and energy balance of the constituents of complex plasmas,
(h) Charge balance on dust particles,
(i) Electrical neutrality follows from (g) and (h),
(j) Electrical conduction in the complex plasma, and
(k) Electromagnetic wave equation.

These phenomena and their impact on the number/energy balance equations have been considered at length in the former Chapters of this book. The distinguishing characteristics of the present approach are (i) inclusion of the Ohmic loss term (Chap. 8) in the energy balance equation of electrons and the evaluation of the real and imaginary parts of the electrical conductivity as a function of the electric field of the wave and (ii) using this dependence of the electrical conductivity on the electric vector of the wave to investigate specific nonlinear phenomenon through the electromagnetic wave equation.

In the rest of the chapter, the nonlinear propagations in dark and illuminated complex plasma have been discussed and typical numerical results have been presented. Some interesting phenomena in PMSE plasmas have also been discussed. Lastly, the self-focusing of electromagnetic beams in complex plasmas has been investigated.

9.3 Nonlinear Complex Plasma Parameters

The basic parameters viz. neutral atom/electron/ion densities and temperatures and the charge or charge distribution on the particles can be determined as indicated in Chaps. 4, 5, and 7 by the simultaneous solution of equations representing conservation of charge, number density, and energy of the constituents; under the action of an electric field, the Ohmic loss has to be accounted for in the energy balance of electrons. Besides electron density, the conduction parameters depend on the electron collision frequency ν_e; the dependence of ν_e on the electron temperature has been elucidated in Sect. 5.4.2. From a knowledge of the electron density and collision frequency, a number of transport parameters may be evaluated as discussed in Chap. 8. The dependence of the transport parameters (n and k) on the electric field leads to a host of nonlinear phenomena like demodulation, cross modulation, harmonic generation, frequency mixing, self-focusing etc. associated with the propagation of high irradiance electromagnetic waves. The physics of these processes is well known (e.g., Ginzburg and Gurevich [2], Gurevich [3] and Sodha et al. [8]) and is outside the scope of the present monograph. It is suffice to state that all these phenomena are analyzed on the basis of the electric field dependence of the refractive index and attenuation factor; this has been adequately discussed before.

For a numerical appreciation of the nonlinear effects, Sodha et al. [9] have analyzed four specific situations viz.

(i) Effect of an applied electric field in a dark complex plasma,
(ii) Effect of an applied electric field in an illuminated complex plasma,
(iii) Effect of a periodic RF field on PMSE structures, and
(iv) Self-focusing of Gaussian electromagnetic beam in a complex plasma.

The approach to such studies consists of

(a) Writing down the equations of the kinetics of complex plasmas as in Chap. 5 viz. (i) conservation of total number of atoms and ions, (ii) charging of dust grains, (iii) number and energy balance of electrons/ions/neutral atoms (the energy balance for electrons has a term, corresponding to Ohmic loss), (iv) expressions for the electron collision frequency as per Sect. 5.4.2 and transport and propagation parameters and (iv) the electromagnetic wave equation.

(b) Simultaneous solution of these equations
It is a reasonable approximation to assume the temperature of the neutral particles and dust to be the same and invariant.

For a numerical appreciation of the nonlinear propagation, it is rational to consider a set of parameters of the complex plasma in absence of an electric field and consider dependence of the effect on a chosen parameter by varying it, keeping the other parameters the same.

$$\begin{aligned}(\omega_p^2/\omega^2) &= 0.8,\ \omega = 10^8\mathrm{s}^{-1},\ n_{e0} = n_{i0} = 2.52 \times 10^6\,\mathrm{cm}^{-3},\ n_{00} = 10^5 n_{e0},\ n_d = 100\,\mathrm{cm}^{-3},\\ a &= 10^2\,\mathrm{nm},\ T_{e0} = 1000\,\mathrm{K},\ T_{i0} = 400\,\mathrm{K},\ T = 300\,\mathrm{K},\ m_0 = 30\,\mathrm{amu},\ m_i = 30\,\mathrm{amu},\\ \kappa &= 1.2,\ \alpha_{r0} = 5 \times 10^{-7}\,\mathrm{cm}^3/\mathrm{s},\ Q_0 = 1.5 \times 10^{-16}\mathrm{cm}^2,\ \nu_{e0} = 6.24 \times 10^3\,\mathrm{s}^{-1},\\ D_0 &= 2.43 \times 10^{10}\,\mathrm{cm}^2/\mathrm{s}.\end{aligned}$$

The set of Fig. 9.1 illustrates the effect of externally applied alternating electric field on different parameters of the complex plasma, as a function of the number density of dust particles and the field strength in the absence of electron emission. Figure 9.1a displays monotonic increase of electron temperature with increasing field strength (βE^2) and consequent *Ohmic* heating, $\beta = \left(e^2 m_0 / 3m_e^2 k_b T \omega^2\right)$. The effective collision frequency (ν_e) of electrons first decreases slightly and then increases with increasing field strength (βE^2), synonymous with increasing electron temperature. This can be understood by considering the term ν_{ec}, the component of effective electron collision frequency (ν_e) corresponding to accretion [$\propto T_e^{1/2} \exp(Z\alpha_e)$], which initially decreases with increasing T_e because the magnitude of the negative charge ($|Z|$) increases with increasing T_e; later the other components of ν_e dominate and the net effect is the increase in ν_e with increasing βE^2 (or T_e). Figure 9.1a also shows the decrease in the electron temperature with increasing n_d; it may be ascribed to larger power loss by electrons with increasing n_d, as a consequence of increasing number of collisions with dust particles and the decrease in the electron temperature.

Figure 9.1b displays the dependence of electron density (n_e) and ion density (n_i) on the electric field strength (βE^2). With increasing electric field or electron temperature the accretion of electrons on dust particles increases, resulting in decreased electron density. The decrease in electron density with increase of (n_d) can also be ascribed to enhanced accretion of electrons on the dust particles. This fact along with the charge neutrality and charge balance (on the particles) explains the observed decrease of ion density with increasing βE^2.

Figure 9.1c shows that the magnitude of mean charge on dust particles increases with increasing field strength (βE^2); this is explained by larger accretion of electrons on the particles with corresponding increase of the electron temperature (T_e). The increase in $|Z|$ with decrease in n_d is explained by corresponding higher rate of electron accretion per particle, due to the availability of more

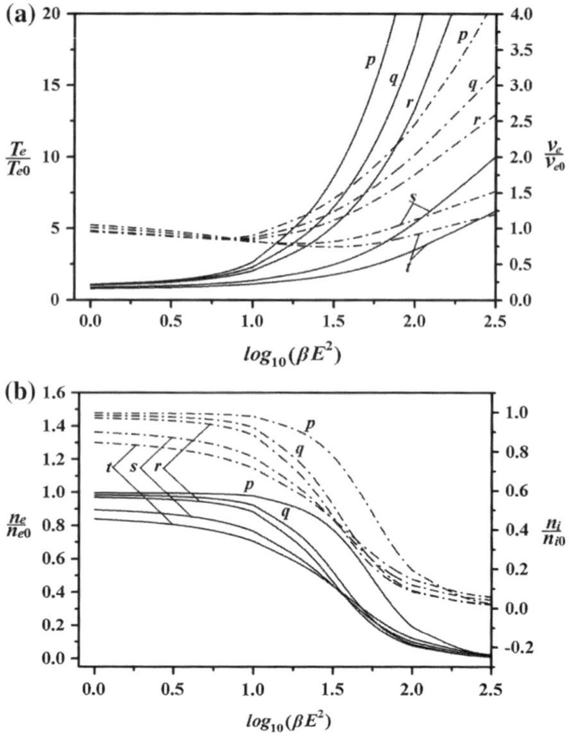

Fig. 9.1 **a** Dependence of the temperature (T_e) and effective collision frequency (ν_e) of electrons on the electric field strength (βE^2) as a function of n_d, for the standard set of parameters as stated in this section. The letters p, q, r, s, and t on the curves refer to $n_d = 10$, 50, 100, 500 and $10^3\,\text{cm}^{-3}$, respectively. The *solid* and *broken lines* correspond to (T_e/T_{e0}) and (ν_e/ν_{e0}), respectively. **b** Dependence of the electron (n_e) and ion (n_i) densities on the electric field strength (βE^2) as a function of n_d. The parameters and labels are the same as given in (**a**). The *solid* and *broken lines* correspond to (n_e/n_{e0}) and (n_i/n_{i0}), respectively. **c** Dependence of the magnitude of mean charge ($|Z|$) on dust particles and the coefficient of diffusion (D_e) of electrons on the electric field strength (βE^2), as a function of n_d. The parameters and labels are the same as given in (**a**). The *solid* and *broken lines* correspond to ($\log_{10}|Z|$) and (D_e/D_0), respectively. **d** Dependence of the refractive index (n) and the absorption coefficient (k) on the electric field strength (βE^2), as a function of n_d. The parameters and labels are the same as given in (**a**). The *solid* and *broken lines* correspond to (n) and (k), respectively. **e** Effect of the size of the dust particle on the magnitude of the mean charge ($|Z|$) on dust particles and the diffusion coefficient with the electric field strength (βE^2), for the standard set of parameters as stated in the text. The letters p, q, r, s, and t on the curves refer to $a = 0.01, 0.05, 0.1, 0.5$ and $1.0\,\mu\text{m}$, respectively. The *solid* and *broken lines* correspond to ($\log_{10}|Z|$) and (D_e/D_0), respectively. (after Sodha et al. [9], curtsey authors and publishers AIP)

electrons per particle, for accretion. The increase in the coefficient of diffusion of electrons ($D_e \propto T_e^{1/2}$), with increasing βE_{00}^2 is the net effect of corresponding enhanced electron temperature and electron collision frequency, enhanced by a

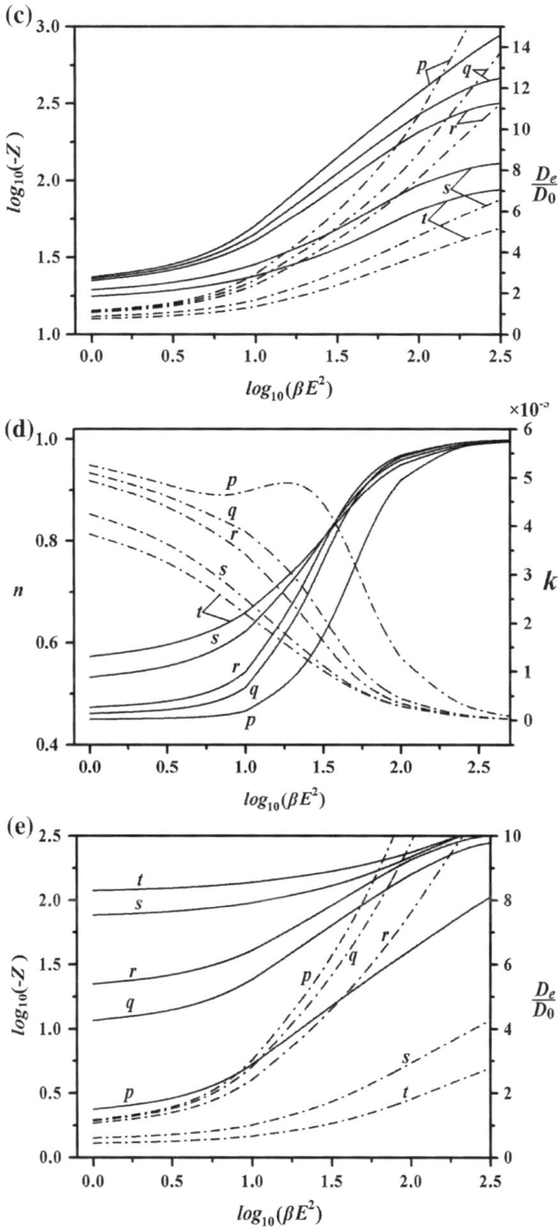

Fig. 9.1 (continued)

lesser extent. The nonlinear dependence of the complex refractive index $(n - ik)$ for the complex plasma with βE^2 and n_d is illustrated in Fig. 9.1d. It is a consequence of the dependence of n_e and v_e on βE^2 (Fig. 9.1a and b).

Figure 9.1e indicates that the charge on the particles increases with increasing radius a of the particles; this is primarily due to enhanced accretion of electrons on the particle, associated with larger size of the dust grains. It is seen (the figure is not given here) that the electron temperature (and hence v_e) increases with decreasing size of the dust grains; this is on account of lesser energy lost by electrons due to their accretion on dust particle surface. The net effect is the increase in coefficient of diffusion as indicated in Fig. 9.1e. The enhancement in the electron temperature for the same *mean square* electric field is more in the case of static field because $\sigma_D > \sigma_r$. Thus it is seen that a radio wave leads to an enhancement in the coefficient of electron diffusion, which can cause disappearance of the PMSEs.

Computation indicates that case of illuminated plasma the dependence of charge Z on the coefficient of diffusion on βE^2 and n_d shows a trend similar to the case of dark plasma. However the values of D_2, are higher on account of larger electron temperature.

9.4 PMSE Structures

Sodha et al. [9, 10] have among others discussed the physics of polar mesospheric summer echoes (PMSE). At an altitude of around 82 km in the polar region, the lower ionosphere has sufficient electron density to cause enhanced radar back scattering (PMSE) this region which is characterized by the presence of ice particles with radii between 3 nm to 20/80 nm. It has been pointed out by Rapp and Lubken [6] that the increase in the coefficient of diffusion by increase of electron temperature, caused by the electric field of a radio wave accounts for the disappearance of PMSEs by the incidence of a r.f. wave and its reappearance on switching off the r.f. transmitter.

The electric field E_0 at the PMSE altitudes can be estimated from the relation (Milikh et al. [4])

$$E_0 = \left(PG/2\pi\varepsilon_0 c z_0^2\right)^{1/2},$$

where P is the power of the transmitter, G is the antenna gain, ε_0 is the permittivity of free space, z_0 is the PMC altitude and all parameters are in SI. Units. The other parameters used in the computation are as follows:

$\omega = 2\pi(2.24 \times 10^8)\,\mathrm{s}^{-1},\ n_{e0} = n_{i0} = 4.5 \times 10^3\,\mathrm{cm}^{-3}$
(at 85 km day time ionospheric altitude)

$n_{00} = 10^{14},\ a = 50\,\mathrm{nm},\ T_{e0} = T_{i0} = T_d = T = 130\,\mathrm{K},\ m_0 = 30\,\mathrm{amu},$

$m_i = 30\,\mathrm{amu},\ \kappa = 1.2,\ G = 10,\ P = 95\,\mathrm{dB},\ w = 3.16 \times 10^9\,\mathrm{watt},\ z_0 = 85\,\mathrm{km}$ and

$\alpha_{r0} = 5 \times 10^{-7}\,\mathrm{cm}^3/\mathrm{s}$. The data corresponds to $\beta E_0^2 \approx 19$.

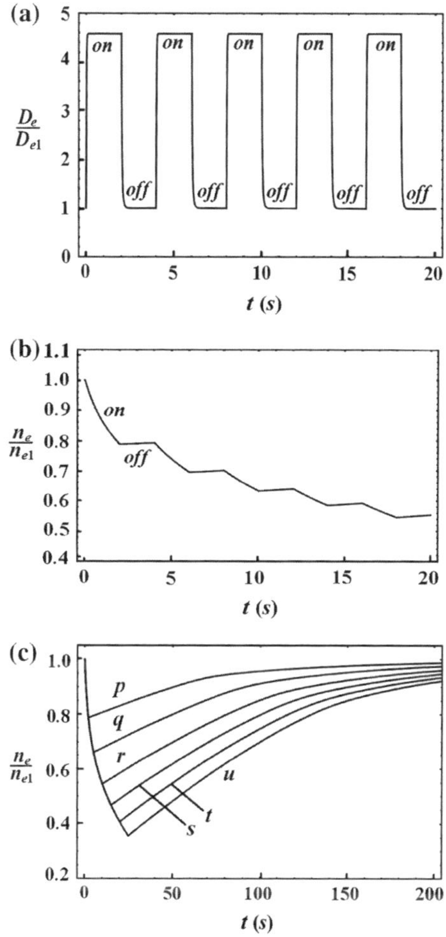

Fig. 9.2 **a** Transient character of diffusivity (D_e/D_{e1}) of electrons in PMSE, for the standard set of parameters as stated in the text with $\beta E^2 = 20$. The pulse duration of the RF field (switching it on and off) is 2 s. **b** Transient character of corresponding (**a**) electron density (n_e/n_{e1}) in PMSE, for the standard set of parameters as stated in the text with $\beta E^2 = 20$. **c** Effect of RF field duration on the electron density (n_e/n_{e1}) in PMSE structures as a function of time, for the standard set of parameters as stated in the text with $\beta E^2 = 20$. The letters p, q, r, s, t and u on the curves correspond to $t_{\text{on}} = 2, 5, 10, 15, 20$ and 25 s, respectively (after Sodha et al. [9], curtsey authors and publishers AIP)

Chilson et al. [1] have demonstrated that the *RF*-heating of electrons reduces the echo power, associated with *PMSE.* This phenomenon has been attributed to the enhancement of electron diffusivity causing reduction of electron density gradients. There is no theory available which correlates the electron density gradients quantitatively to the electron diffusivity or other plasma parameters, but the fact that the gradient gets reduced by high values of diffusivity is obvious and well appreciated.

Figure 9.2a shows the time dependence of the coefficient of electron diffusion, when transmitter is switched (Chilson et al. [1]) ON or OFF, every 2 s. It is seen that the electron diffusivity almost immediately ($t \approx 0.1$ s) acquires a steady-state value after switch ON or OFF. This is on account of the almost instantaneous response of the electron temperature to the changes in the electric field. It is seen that the electron diffusivity gets enhanced by a factor of 4.5 after the field is switched on and comes back to original value after the field is switched off. The corresponding time dependence of the electron density is displayed in Fig. 9.2b. It is seen that the electron density decreases after the field is switched on and it recovers little in the next 2 s, when the field is switched off. This results in steady decrease of electron density with time. The time needed for the electron density to relax is relatively large because the electron density is determined by the charging/discharging process of ice dust grains, where ions play an important part. Figure 9.2c shows the time dependence of the electron density when the electric field is switched off after 2, 5, 10, 15, 20, and 25 s. The nature of the dependence is well explained by what has been stated in the explanation of Fig. 9.2b. In these figures n_{e1} and D_{e1} correspond to the electron density and diffusivity in field-free plasmas with dust particles in *PMSE* structure.

9.5 Self-Focusing of a Gaussian Electromagnetic Beam in a Complex Plasma (After Mishra et al. [5])

9.5.1 Self-Focusing

Consider the propagation of an initially parallel electromagnetic beam with radially decreasing irradiance in a complex plasma, characterized by increasing refractive index with increasing irradiance. In such a complex plasma, the irradiance and hence the refractive index is maximum near the axis of the beam and it decreases with increasing radius. Hence the wave (phase) velocity of the electromagnetic beam decreases with increasing distance from the axis. Hence an initially parallel beam with a plane wave front will be transformed into an increasingly concave wave front as the beam propagates in the plasma or in other words the beam gets increasingly focused as the beam propagates. Such a focusing caused by the dependence of the dielectric function on the irradiance (or electric field) of the wave is known as self-focusing. Since the magnitude of any nonlinear phenomenon depends on the radial distribution of the irradiance, this phenomenon affects the magnitude of all the nonlinear processes; thus self-focusing should be taken into account in the evaluation of the magnitude of all nonlinear phenomena in a complex plasma. The dependence of the refractive index on irradiance EE^* has been discussed in Chap. 8 and Sect. 9.3.

The nonuniform irradiance distribution of the beam causes nonuniform heating of electrons. The electrons are hottest around the axis of the beam, which causes largest electron accretion on the surface of dust particles; this results in the creation of a depleted electron density channel around the axis, even when the diffusion of electrons/ions is taken into account. The depleted electron density channel is responsible for the self-focusing of the beam. Mishra et al. [5] have analyzed the phenomenon of self-focusing of a Gaussian electromagnetic beam in a complex plasma, taking into account

(i) the paraxial theory of self-focusing, which means that all the terms may be expanded to terms in $r^2/r_0^2 f^2$
(ii) inhomogeneous Ohmic heating of electrons/ions due to the inhomogeneous electric field of the beam and radial distribution of electron density/temperature. Ohmic heating of ions has been neglected on account of their large mass,
(iii) evaluation of the net radial electron/ion currents due to the electron/ion density and temperature gradients and space charge field (in steady state, the electron and ion currents are equal in magnitude),
(iv) all parameters except the thermal conductivity and diffusion coefficients have been expanded up to terms in $\left(r^2/r_0^2 f^2\right)$, which is consistent with the paraxial approximation; the diffusion coefficient and the thermal conductivity conform to electron/ion temperatures on the axis. (r_0 is the initial beam width (at $z = 0$) and f is the beam width parameter),
(v) corresponding to a low level of ionization, the density/temperature of the neutral gas atoms has been assumed to be constant,
(vi) generation and annihilation of electrons in the gaseous state,
(vii) elastic collisions of electrons with ions, dust particles, and neutral atoms,
(viii) elastic collisions of ions with neutral atoms (elastic collisions with dust particles can be ignored on account of the slow speed of ions),
(ix) accretion of electrons and ions on the surface of dust particles,
(x) energy exchange between plasma constituents in the above processes,
(xi) number and energy balance of all the constituents,
(xii) charge balance on the dust particles,
(xiii) electrical neutrality of the complex plasma [it follows from (xi) and (xii) above], and (xiv) electromagnetic wave equation.

9.5.2 Net Flux of Electrons/Ions

Consider a complex plasma consisting of electrons, singly charged ions, neutral atoms, and uniform size spherical dust grains with radius a and charge Ze. The net flux of constituent species is given by

$$\boldsymbol{\Gamma}_e = -D_e \nabla n_e - n_e \mu_e \boldsymbol{E} - D_e' n_e (\nabla T_e / T_e), \tag{9.8a}$$

$$\Gamma_i = -D_i \nabla n_i + n_i \mu_i E - D_i' n_i (\nabla T_i / T_i), \tag{9.8b}$$

and

$$\boldsymbol{\Gamma}_d = -D_d \nabla n_d - n_d \mu_d \boldsymbol{E}, \tag{9.8c}$$

where $\boldsymbol{\Gamma}_h$, n_h, D_h, $(n_h D_h')$, μ_h, and $\boldsymbol{E}$ are the flux, density, diffusion coefficient, thermal diffusion coefficient, mobility, and ambipolar space charge field; the subscript h refers to electrons (e), ions (i), and charged dust particles (d), respectively.

In steady state, the net flux of positive and negative charges across a unit area should be zero, thus one should have

$$\boldsymbol{\Gamma}_i = \boldsymbol{\Gamma}_d + \boldsymbol{\Gamma}_e \tag{9.9a}$$

As a good approximation, the dust particles are assumed to be stationary on account of their large mass and hence D_d and μ_d may be put as zero. Thus (9.8c) gives $\Gamma_d = 0$ and (9.9a) reduces to

$$\boldsymbol{\Gamma}_e = \boldsymbol{\Gamma}_i = \boldsymbol{\Gamma} \tag{9.9b}$$

Using (9.8a), (9.8b), and (9.9b) the net flux of electrons (or ions) is given by

$$\begin{aligned} \Gamma &= -D_e \nabla n_e - D'_e n_e (\nabla T_e / T_e) - n_e \mu_e \left(\frac{D_i \nabla n_i - D_e \nabla n_e + D'_i n_i (\nabla T_i / T_i) - D'_e n_e (\nabla T_e / T_e)}{(n_e \mu_e + n_i \mu_i)} \right) \\ &= -\left(\frac{D_i [c_2 n_e \nabla n_i + c_1 n_i \nabla n_e] + n_e n_i D'_i [c_2 (\nabla T_i / T_i) + c_3 (\nabla T_e / T_e)]}{(n_e c_2 + n_i)} \right) \end{aligned}, \tag{9.10}$$

where $D'_h = (5kT_h/2m_h\nu_h)$, $\mu_h = (eD_h/kT_h)$, $D_h = (kT_h/m_h\nu_h)$, $c_1 = (D_e/D_i)$, $c_2 = (\mu_e/\mu_i)$, and $c_3 = \left(D'_e/D'\right)$; T_h, m_h, ν_h, and Q_{hn} correspond to the temperature, mass, effective collision frequency, and collision cross-section with neutrals of the charged species $'h'$ (e is the numerical value of electronic charge and k is Boltzmann's constant). In what follows D'_e, D'_i, D_e, D_i, μ_e and μ_i correspond to the parameters on the axis of the electromagnetic beam; thus the radial dependence of these parameters has not been taken into account.

9.5.3 Complex Plasma Kinetics

The kinetics of complex plasma can be described by the following set of equations.

9.5.3.1 Charging of Dust Grains

$$(dZ/dt) = n_{\text{ee}}(Z, T) + n_{\text{ic}}(Z, T_i) - n_{\text{ec}}(Z, T_e), \tag{9.11}$$

where $n_{\text{ee}}(Z, T)$ is the rate of electron emission from the surface of a particle, $n_{\text{ec}}(Z, n_e, T_e)$ and $n_{\text{ic}}(Z, n_i, T_i)$ are the accretion rates of electrons and ions on the surface of the particle, and T is the temperature of the dust particles and neutral atomic species (assumed to be constant on account of their large heat capacities).

9.5.3.2 Electron and Ion kinetics

$$(\mathrm{d}n_e/\mathrm{d}t) = \beta_i n_0 - \alpha_r n_e n_i - n_d(n_{\text{ec}} - n_{\text{ee}}) - \nabla \boldsymbol{\Gamma} \tag{9.12}$$

and

$$(\mathrm{d}n_i/\mathrm{d}t) = \beta_i n_0 - \alpha_r n_e n_i - n_d n_{\mathrm{ic}} - \nabla \boldsymbol{\Gamma}, \tag{9.13}$$

where

β_i	is the coefficient of ionization, due to processes, responsible for ionization in the absence (or presence) of dust particles
$\alpha_r(T_e) = \alpha_{r0} T_e^{-\kappa}\,\mathrm{cm}^3/\mathrm{s}$	is the coefficient of recombination of electrons and ions and
α_{r0} and κ	are constants.

The first two terms on the right-hand side of (9.5a), (9.5b), (9.5c), and (9.6a), (9.6b) correspond to the net gain in electron and ion density due to ionization of neutral species and recombination of electrons and ions in the plasma. The next term in both the equations represents the net electron and ion currents, respectively, accreting on the surface of the dust particles. The last term in both the equations refer to loss in electron and ion density on account of diffusion of electrons and ions, respectively.

9.5.3.3 Neutral Kinetics

On account of the low level of ionization, the density of neutral atoms n_0 may be considered to be constant, i.e.,

$$n_0 = n_{00}, \tag{9.14}$$

where n_{00} refers to the neutral atom density in the absence of beam and dust particles.

9.5.3.4 Charge Neutrality

The equation for charge neutrality may be expressed as

$$Zn_d + n_i - n_e = 0.$$

Since (9.8a), (9.8b,) and (9.8c) can be derived from the set of (9.4a), (9.4b) and (9.4c), (9.5a), (9.5b), (9.5c), and (9.6a), (9.6b), the equation of charge neutrality (9.8a), (9.8b) and (9.8c) has been omitted in further analysis.

9.5.3.5 Electron and Ion Energy Balance Equations

$$\begin{aligned}\frac{\mathrm{d}}{\mathrm{d}t}\left(\frac{3}{2}kn_eT_e\right) &= \beta_i n_0\varepsilon_e - \alpha_r n_e n_i(3kT_e/2) - n_d(n_{\mathrm{ec}}\varepsilon_{\mathrm{ec}} - n_{\mathrm{ee}}\varepsilon_{\mathrm{ee}}) \\ &\quad + \frac{1}{r}\frac{\partial}{\partial r}\left(\chi_e r\frac{\partial T_e}{\partial r}\right) + \left(\frac{e^2 n_e \nu_e}{4m_e(\nu_e^2+\omega^2)}\right)EE^*\end{aligned} \tag{9.15}$$

and

$$\frac{\mathrm{d}}{\mathrm{d}t}\left(\frac{3}{2}kn_iT_i\right) = \beta_i n_0\varepsilon_i - \alpha_r n_e n_i(3kT_i/2) - n_d n_{\mathrm{ic}}\varepsilon_{\mathrm{ic}} + \frac{1}{r}\frac{\partial}{\partial r}\left(\chi_i r\frac{\partial T_i}{\partial r}\right), \tag{9.16}$$

where small energy exchange in elastic collisions is neglected,

$\chi_h = \left(5k^2 n_h T_h / m_h \nu_h\right)$, corresponds to the coefficients of thermal conductivities of electrons and ions,

$\nu_{\mathrm{ei}} = \nu_{\mathrm{ei0}}(n_i/n_{e0})(T_e/T_{e0})^{-3/2}$, is the electron collision frequency corresponding to elastic collisions with ionic species,

$\nu_{\mathrm{ei0}} = \left(5.5n_{e0}/T_{\mathrm{eo}}^{3/2}\right)\ln\left[220T_{\mathrm{eo}}/n_{e0}^{1/3}\right]$

$\nu_{\mathrm{en}} = \nu_{\mathrm{en0}}(n_0/n_{00})(T_e/T_{e0})^{1/2}$ is the electron collision frequency corresponding to elastic collisions with neutral species,

$\nu_{\mathrm{en0}} = (8.3\times10^5)Q_0 n_{00}T_{e0}^{1/2}$,

$\nu_{\mathrm{in}} = \nu_{\mathrm{ino}}n_0(T_i+T)/n_{00}(T_{i0}+T)$,

$\nu_{\mathrm{in0}} = \frac{16\sqrt{(2kT/\pi)}}{3}Q_{\mathrm{in}}n_{00}(T_{i0}+T)^{1/2}$,

$\nu_e = (\nu_{\mathrm{ed}} + \nu_{\mathrm{ec}} + \nu_{\mathrm{ei}} + \nu_{\mathrm{en}})$, is the effective collision frequency of electrons,

$\nu_i = (\nu_{\mathrm{ic}} + \nu_{\mathrm{in}})$, is the effective collision frequency of ions,

$\nu_{\mathrm{ec}} = n_d(n_{\mathrm{ec}}/n_e)$, is the frequency of accretion of an electron on dust particles,

$\nu_{\mathrm{ic}} = n_d(n_{\mathrm{ic}}/n_i)$, is the frequency of accretion of an ion on dust particles,

$\nu_{\mathrm{ed}} = \nu_{\mathrm{ed0}}(T_e/T_{e0})^{-3/2}Z^2$ is the collision frequency corresponding to elastic collisions with dust particles[47],

$\nu_{\mathrm{ed0}} = (2.9\times10^{-6})n_d T_{e0}^{-3/2}\ln\Lambda$ (with $\ln\Lambda \approx 10$),

- Q_0 is the collision cross-section of the electron-neutral elastic collision
- Q_{in} is the collision cross-section of the ion-neutral elastic collision
- m_i is the mass of an ionic species
- m_0 is the mass of a neutral atom and
- T_{j0} is the temperature constituent species in the absence of the dust grains and field-free space, here j stands for electron (e) and ionic (i) species, respectively.

The first two terms in (9.9a), (9.9b), and (9.10) refer to the net gain in the mean energy of electron and ionic species due to ionization of neutrals and recombination of electrons and ions in the plasma. The next term in both the equations correspond to the net loss in energy of electrons and ions, due to their accretion on and electron emission from the surface of dust particles. The next term describes the power loss per unit volume of electrons and ionic species on account of the binary elastic collisions between electrons, ions, and neutral atomic species. The next term correspond to net power gain per unit volume of electrons and ions due to thermal conduction. The last term in (9.9a), (9.9b) refers to the power gained by electrons per unit volume due to *Ohmic heating* of electrons from the electric field of the beam; *Ohmic heating* of ions (which are much heavier) can be neglected. The radial dependence of χ_e and χ_i is not taken into account, for the sake of convenience; the values correspond to the parameters on the axis of the beam.

Due to the large thermal capacity of neutral atoms and dust particles and efficient energy exchange between them, it is a good approximation to assume their temperature to be the same viz. T (no energy exchange). The value of the ionization coefficient (β_i) may be determined by applying the electron kinetics in dust-free plasmas; thus

$$\beta_i n_{00} = \alpha_r n_{e0} n_{i0} = \alpha_r n_{e0}^2, \tag{9.17}$$

where the suffix zero refers to the absence of dust.

Further the mean energy of electrons and of ions generated due to ionization may be obtained by imposing the initial conditions for dust-free plasma in the energy balance equations; thus

$$\varepsilon_e = 3kT_{e0}/2 \tag{9.18a}$$

and

$$\varepsilon_i = 3kT_{i0}/2 \tag{9.18b}$$

9.5.3.6 Paraxial Approximation

The initial $(z = 0)$ irradiance distribution of a *Gaussian EM* beam can be expressed as

$$(E_0^2)_{z=0} = E_{00}^2 \exp[-(r^2/r_0^2)] \tag{9.19a}$$

where E_{00} is the amplitude of the electric vector and r_0 is the beam width.

In this approximation, the irradiance distribution in the complex plasma is given by (Sodha et al. [8])

$$A_0^2 = (E_0^2/f^2)\exp[-(r^2/r_0^2 f^2)], \tag{9.19b}$$

where the beam width parameter f is a function of the distance of propagation z.

As the expression suggests, most of the power of the beam is concentrated in a region around central axis $(r = 0)$; paraxial ray approximation [i.e., $(r/r_0 f) \ll 1$]

has been adopted to study the propagation of Gaussian *EM*-beams. In this context, all the relevant parameters can be expanded around central maximum ($r = 0$) as

$$X(r,z) = X_a(z) - (r^2/r_0^2)X_r(z), \tag{9.20}$$

where $X_a(z)$ and $X_r(z)$ are the axial and radial parts of the parameter $X(r,z)$.

Substituting equations like (9.20) in the system of (9.10) to (9.13) $n_a(z)$ and $n_r(z)$ and hence $\varepsilon_a(z)$ and $\varepsilon_r(z)$ can be evaluated.

9.5.3.7 Effective Dielectric Function of the Complex Plasma

The nonlinear complex dielectric function of the plasma is given by

$$\varepsilon(r,z) = \varepsilon_a(z) - (r^2/r_0^2)\varepsilon_r(z), \tag{9.21a}$$

where the subscripts a and r refer to axial and radial parts of the dielectric function.

$$\varepsilon_a = 1 - \left(\frac{\omega_p^2}{\omega^2 + v_{\text{ea}}^2}\right)(n_{\text{ea}}/n_{e0}) \tag{9.21b}$$

$$\varepsilon_r = \left(\frac{\omega_p^2}{\omega^2 + v_{\text{ea}}^2}\right)\left(\frac{(n_{\text{ea}}/n_{e0})v_{\text{ea}}v_{\text{er}}}{\omega^2 + v_{\text{ea}}^2} - (n_{\text{er}}/n_{e0})\right), \tag{9.21c}$$

where $\omega_p = \left(4\pi n_{e0}e^2/m_e\right)^{1/2}$ is the plasma frequency in the absence of dust.

9.5.4 *Propagation of Gaussian Electromagnetic Beam*

Consider the propagation of a linearly polarized *Gaussian EM-beam* along the *z-axis* in a complex plasma with its electric vector polarized along the *y-axis*; the electric field vector $\boldsymbol{E}$ for such a beam may be expressed in the cylindrical coordinate system with azimuthal symmetry as

$$\boldsymbol{E}(z) = \boldsymbol{j}E_0(r,z)\exp(i\omega t) \tag{9.22a}$$

where $(E_0)_{z=0} = E_{00}\exp[-(r^2/2r_0^2)]$, refers to the complex amplitude of the Gaussian beam of initial beam width r_0, E_{00} is a real constant characterizing the amplitude of the beam, $\boldsymbol{j}$ is the unit vector along the *y-axis*.

Starting from the wave equation and (9.21a), it can be shown (Sodha et al. [8]) that

$$E_0^2(r,z) = E_{00}^2[\varepsilon_a(0)/\varepsilon_a(z)]^{1/2}.\frac{1}{f^2}\exp\left(-r^2/r_0^2f^2(z)\right), \tag{9.22b}$$

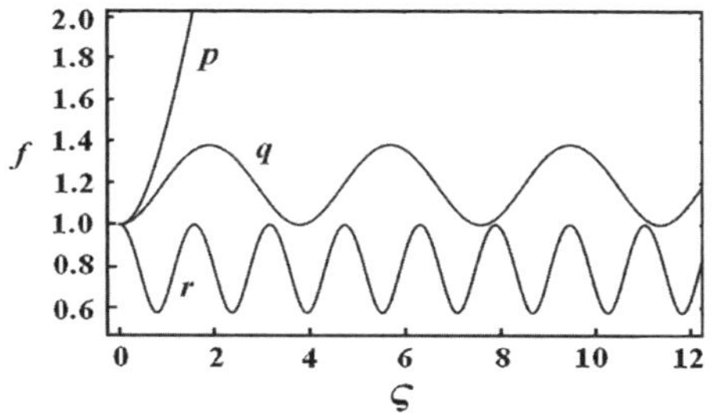

Fig. 9.3 Three regimes of beam propagation: Dependence of the dimensionless beam width parameter f on the dimensionless distance of propagation ξ for the standard set of parameters as stated in the text. The labels $p, q,$ and r refer to $\beta E_{00}^2 = 2{,}2500$ and 10^4, respectively ($\beta = e^2 m_0/6m_e\omega^2 k_B T$) (after, Sodha et al. [9], curtsey authors and publishers AIP)

where the beam width parameter f is given by

$$\varepsilon_a f \frac{\mathrm{d}^2 f}{\mathrm{d}\xi^2} = \left(\frac{1}{f^3} - \rho_0^2 f \varepsilon_r\right) - \frac{1}{2}\left(\frac{\mathrm{d}f}{\mathrm{d}\xi}\right)\left(\frac{\mathrm{d}\varepsilon_a}{\mathrm{d}\xi}\right), \tag{9.22c}$$

$\xi = \left(c/r_0^2\omega\right)z$ is the dimensionless distance of propagation; $\rho_0 = (r_0\omega/c)$ is the dimensionless initial beam width.

The critical curves are obtained by putting $\left(\mathrm{d}^2 f/\mathrm{d}\varsigma^2\right)$ and $(\mathrm{d}f/\mathrm{d}\varsigma)$ equal to zero and f equal to unity in (9.22c). Thus

$$\rho_{0c}^2 \varepsilon_r \left(E_{00}^2\right) = 1. \tag{9.22d}$$

Thus the critical curve expresses the relationship between the beam width (ρ_0) and initial axial irradiance (βE_{00}^2) for critical focusing.

The dependence of the beam width parameter f on the dimensionless distance of propagation ς can be obtained by the numerical integration of (25b) after putting suitable expressions for ε_a and ε_r, and using the initial boundary conditions $f = 1$, $(\mathrm{d}f/\mathrm{d}\xi) = 0$ at $\xi = 0$.

9.5.5 Numerical Results and Discussion

Standard set of parameters for which computations have been made are as follows:

$\left(\omega_p^2/\omega^2\right) = 0.8, \omega = 10^8\,\mathrm{s}^{-1}, n_{e0} = n_{i0} = 2.52 \times 10^6\,\mathrm{cm}^{-3}, n_{00} = 10^{12}\,\mathrm{cm}^{-3},$

$n_d = 100\,\mathrm{cm}^{-3}, a = 10^2\,\mathrm{nm}, T_{e0} = 1000\,\mathrm{K}, T_{i0} = 400\,\mathrm{K}, T = 300\,\mathrm{K}, m_0 = 30\,\mathrm{amu},$

$m_i = 30\,\mathrm{amu}, \kappa = 1.0, \alpha_{r0} = 5 \times 10^{-7}\,\mathrm{cm}^3/\mathrm{s}, Q_0 \approx Q_{in} = 1.5 \times 10^{-16}\,\mathrm{cm}^2,$

and $v_{e0} = 6.24 \times 10^3\,\mathrm{s}^{-1}$, and the suffix 0 refers to the absence of dust and beams.

Figure 9.3 illustrates the dependence of beam width parameter f on dimensionless distance of propagation ξ in a complex plasma; the figure indicates the propagation of Gaussian *EM*-beam in three modes viz. steady divergence self-focusing and oscillatory divergence and as the power of the beam increases.

References

1. P.B. Chilson, E. Belova, M.T. Rietvald, S. Kirkwood, U.P. Hoppe, Geophys. Res. Lett. **27**, 3801 (2000)
2. V.L. Ginzburg, A.V. Gurevich, Sov. Phys. Uspekhi **3**, 115 (1960)
3. A.V. Gurevich, *Nonlinear Phenomena in the Ionosphere* (Springer, Berlin, 1978)
4. G.H. Milikh, M.J. Freeman, L.M. Duncan, Radio Sci. **29**, 1355 (1994)
5. S.K. Mishra, S. Misra, M.S. Sodha, Phys. Plasmas **18**, 043702 (2011)
6. M. Rapp, J. Lubken, Geophys. Res. Lett. **27**, 3285 (2000)
7. M.S. Sodha, S.K. Mishra, S.K. Agarwal, IEEE Trans. Plasma Sci. **37**, 375 (2009)
8. M.S. Sodha, A.K. Ghatak, V.K. Tripathi, *Self Focusing of Laser Beams in Plasmas and Semiconductors in, Progress in Optics,* vol. 13, ed. by E. Wolf, (North Holland & Co, Amsterdam, 1976) p. 169
9. M.S. Sodha, S.K. Mishra, S. Misra, Phys. Plasmas **18**, 023701 (2011)
10. M.S. Sodha, S. Misra, S.K. Mishra, Phys. Plasmas **18**, 083708 (2011)

Chapter 10
Fluctuation of Charge on Dust Particles in a Complex Plasma

10.1 Introduction

The charge on a dust particle in a plasma can fluctuate for two reasons. The spatial, temporal, and turbulent changes in the properties of the plasma may induce [2, 3, 8] fluctuations in the charge on the dust particles. The more important reason is the discrete nature of charge on electrons, ions, and dust particles. Electrons and ions are incident on the charged particles at random times, causing fluctuation of charge Ze on the dust particles (where $-e$ is the electronic charge) even in a steady state uniform plasma. Morfill et al. [7] anticipated that corresponding to normal distribution of charge ζe on particles the root-mean square fluctuation of the charge is $\Delta\zeta = \sigma = (|Z|)^{1/2}$; however, using numerical simulation, Cui and Goree [1] concluded that $\Delta\zeta = 0.5(|Z|)^{1/2}$, where Z is the mean charge on the particles. Matsoukas and Russel [5] and Matsoukas et al. [6] considered the charging of a dust particle in a plasma as a one step *Markov process* [11] with the probability density governed by a master difference equation.

$$\begin{aligned}\frac{\mathrm{d}n_\zeta}{\mathrm{d}t} &= n_{\zeta+1}\, n_{\mathrm{ec}}(\zeta+1) + n_{\zeta-1}\left[n_{\mathrm{ee}}(\zeta-1) + n_{\mathrm{ic}}(\zeta-1)\right] \\ &\quad - n_\zeta\left[n_{\mathrm{ic}}(\zeta) + n_{\mathrm{ec}}(\zeta) + n_{\mathrm{ee}}(\zeta)\right],\end{aligned} \tag{10.1}$$

where n_ζ is the probability of a particle having a charge ζe (which is the same as the fraction of particles having a charge ζe), $n_{\mathrm{ee}}(\zeta)$ is the rate of electron emission by a particle having a charge ζe, and $n_{\mathrm{ec}}(n_{\mathrm{ic}})$ is the rate of electron/ion accretion on the particles.

The analytical solution of the master equation was obtained by the same workers as

$$n_\zeta = \left(1/\sigma\sqrt{2\pi}\right)\exp\left[-(\zeta - Z)^2\big/2\sigma^2\right] \quad \text{for } Z > 30. \tag{10.2}$$

For a plasma having *Maxwellian distribution* of speed of electrons and ions, the standard deviation σ is given by Matsoukas and Russel [5] and Matsoukas et al. [6]

M. S. Sodha, *Kinetics of Complex Plasmas*, Springer Series on Atomic, Optical, and Plasma Physics 81, DOI: 10.1007/978-81-322-1820-3_10,

$$\sigma^2 = \frac{1}{2}\left(\frac{n_{\text{ic}} + n_{\text{ec}}}{n'_{\text{ic}} - n'_{\text{ec}}}\right)_{\zeta=Z},$$

where n_{ec} and n_{ic} are the number of electrons and ions incident per unit time on a particle of charge Ze and the primes denote the differential coefficient with respect to ζ. In case the particles emit n_{ee} electrons per unit time (10.2) gets modified to [9]

$$\sigma^2 = \frac{1}{2}\left(\frac{n_{\text{ic}} + n_{\text{ec}} + n_{\text{ee}}}{n'_{\text{ic}} - n'_{\text{ec}} + n'_{\text{ee}}}\right)_{\zeta=Z}. \tag{10.3}$$

Khrapak et al. [4] have given an alternate derivation of (10.2), which they put in a different form. Some comments on *Khrapaks,* work are given later in the chapter.

Most of the work on the fluctuation of charge on the particles is based on the assumption that the dust does not significantly affect the electron/ion density and temperature; expressions for n_{ec} and n_{ic} are based on the magnitudes of parameters in the absence of dust. Recently, as may be seen from the previous chapters, considerable work on the kinetics of complex plasma has been conducted considering (i) ionization and electron–ion recombination in the gaseous phase, (ii) suitable expressions for the parameters, occurring in the kinetics equations, (iii) number cum energy balance of the constituents of the complex plasma, and (iv) charge balance on the particles. Thus, the effect of dust on the electron/ion number density and temperature gets accounted for and realistic values of n_{ee}, n_{ec}, n_{ic} and hence σ^2 can be obtained by using these values of n_e, n_i, T_e, T_i and Z thus obtained. Such an approach for the investigation of fluctuation of charge on particles has been adopted by Sodha et al. [9].

Another approach for determining the charge distribution on the particles in a dust cloud (in the absence of neutral atoms and ions), is based on the use of steady state (analogous to equilibrium) constants and takes account of the emission and accretion processes, applicable to the particles. Two cases of electron emission viz. thermionic emission in a dust cloud in thermal equilibrium and photoelectric emission in an irradiated dust cloud have been analyzed [10].

In the following sections, we consider the cases of (i) complex plasma with dust particles of uniform size and (ii) complex plasmas with size distribution of dust.

10.2 Fluctuation of Charge on Uniform Size Dust Particles in a Complex Plasmas

10.2.1 Methodology

The starting point is the determination of the mean charge Z on the particles and the electron/ion density and temperature in the complex plasma, following the logic and methodology in the earlier chapters.

Knowing n_e, n_i, T_e, T_i, and Z the parameters n_{ec}, n_{ic}, n_{ee} and n'_{ec}, n'_{ic} and n'_{ee} can be determined for $\zeta = Z$. Hence σ^2 can be evaluated by using (10.3) in the case when $Z > 30$. It is common to use a parameter $\alpha = \sigma/Z^{1/2}$ in discussions on the subject.

It should be remembered that (10.3) is valid only for $Z > 30$. Hence if Z comes out to be less than 30, another approach is followed viz solution of a system of equations analogous to (10.1) for $Z_1 < \zeta < Z_2$, along with the number and energy balance of electrons/ions where $Z_{1,2} = Z \mp 2Z^{1/2}$.

10.2.2 Numerical Results and Discussion

Sodha et al. [9, 10] have considered three cases; corresponding to $Z > 30$.

Case 1: *Absence of electron emission from the particles*
If there is no emission process then accretion of electron and ions are the only processes, which lead to the charging of the dust grains. Thus, one may obtain the mean charge and various relevant parameters by solving the equations of kinetics, as explained earlier in the book.

Case 2: *Thermionic emission*
In this case, all the plasma constituent species have the same temperature; hence, the energy balance equations are not relevant. The mean charge on dust grain and other parameters can be obtained by solution of the equations of number/energy balance, charging of particles and equation for electrical neutrality.

Case 3: *Photoelectric emission*
In case when photoelectric emission is the dominant electron emission process the mean charge on the dust particle and other parameters may be estimated by simultaneous solution of the relevant kinetics equations.

[On account of efficient energy exchange between abundant (high thermal capacity) neutral atoms and dust particles, it is a good approximation to assume their temperatures to be the same viz. $T_n = T_d = T$. The standard sets of parameters for which computations have been made are as follows:

Case 1: *Absence of electron emission from the particles*

$$n_{e0} = n_i \approx 10^{10}/\text{cm}^3,\ n_{00} = 10^3 n_{e0},\ T_{e0} = 1000\,\text{K},\ T_{i0} = 400\,\text{K},\ T = 300\,\text{K},$$
$$a = 1.0\,\mu,\ m_0 \approx m_i = 30\,\text{amu},\ \mu = 1.2,\ \text{and}\ \alpha_r = 5 \times 10^{-7}\,\text{cm}^3/\text{s}.$$

Case 2: *Thermionic emission*

$$n_{e0} = n_i \approx 10^{10}/\text{cm}^3,\ n_{00} = 10^3 n_{e0},\ T_{e0} = T_{i0} = T = 2000\,\text{K},\ n_d = 10^5\,\text{cm}^{-3},$$
$$a = 1.0\,\mu,\ m_0 \approx m_i = 30\,\text{amu},\ \mu = 1.2\ \text{and}\ \alpha_r = 5 \times 10^{-7}\,\text{cm}^3/\text{s}$$

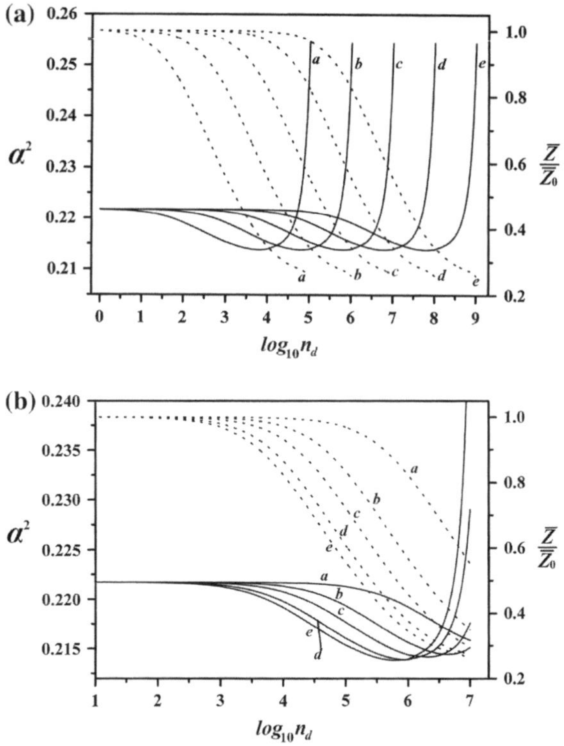

Fig. 10.1 **a** The dependence of the parameter $\alpha^2(=\sigma^2/Z)$ and mean charge (Z/Z_0) on the number density of the dust particle n_d, (in the absence of electron emission) for the standard set of parameters as stated in Case 1 in the text. The letters on the *curve a*, *b*, *c*, *d* and *e* refer to the initial electron density $n_{e0} = 10^8$, 10^9, 10^{10}, 10^{11} and 10^{12} cm^{-3}, respectively. Further the *solid* and *broken curves* correspond to *left hand side* scale (α^2) and *right hand side* scale, respectively (Z/Z_0) (after Sodha et al. [9], curtsey authors and publishers AIP). **b** The dependence of the parameter $\alpha^2(=\sigma^2/Z)$ and mean charge (Z/Z_0) on n_d, (in the absence of electron emission) for the standard set of parameters as stated in Case 1 in the text. The letters on the *curve a*, *b*, *c*, *d* and *e* refer to the radius of the dust particle $a = 0.1, 0.3, 0.5, 0.8$, and $1.0\ \mu$, respectively. Further the *solid* and *broken curves* correspond to *left hand side* scale (α^2) and *right hand side* scale, respectively (Z/Z_0) (after Sodha et al. [9], curtsey authors and publishers AIP)

Case 3: *Photoelectric emission*

$$n_{e0} = n_i \approx 10^8/\text{cm}^3,\ n_{00} = 10^3 n_{e0},\ T_{e0} = 1000\,\text{K},\ T_{i0} = 400\,\text{K},\ T = 300\,\text{K},$$
$$\xi = (h\nu - \emptyset)/k_B T = 20,\ a = 1.0\,\mu,\ n_d = 10^4\,\text{cm}^{-3},\ m_0 \approx m_i = 30\,\text{amu},\ \mu = 1.2 \text{ and}$$
$$\alpha_r = 5 \times 10^{-7}\,\text{cm}^3/\text{s}.]$$

The set of Fig. 10.1a, b illustrates the dependence of the parameter $\alpha^2(=\sigma^2/Z)$ (describing the charge fluctuation of a dust grain) and the mean charge on the particle (Z/Z_0) (Z_0 is the mean charge on the dust grain for $n_d = 1$) on the number density of the dust particles (n_d) in a complex plasma when the accretion of electrons and ions are the only processes responsible for the charge on the particles. Figure 10.1a indicates that the mean charge |Z| on the dust particles decreases with increasing n_d and increases with increasing initial electron density. Figure 10.1b shows the dependence of (Z/Z_0) and α on the size of the dust grains; the figure indicates smaller fluctuations for larger radii of the dust grains associated with larger values of |Z|. It is also significant to note that for the chosen parameters $0.45 < \alpha < 0.51$ this agrees with the results predicted by simulation techniques viz. $\alpha \approx 0.5$ [1].

The set of Fig. 10.2a–c, describes the dependence of the parameter $\alpha^2(=\sigma^2/Z)$ and the mean charge on the particle Ze on the work function of the material of the dust grains (φ) in a complex plasma when thermionic emission is the significant mechanism for electron generation; accretion of electron and ionic species on the particles has also been considered. Figure 10.2a indicates that the mean charge decreases with increasing φ and n_{e0}. Further α increases sharply with increasing n_{e0}, due to larger electron accretion on the surface of the dust grain. The figure excludes the region $|Z| \leq 10^{1.5}(\approx 30)$, ($n_{\mathrm{ec}} \approx n_{\mathrm{ic}}$) because in this region the derivation of (10.1) from the master difference equation is not valid. It can also be seen from the figure that for the case of negative charging of the dust grain (large φ), α is around 0.5. The dependence of Z and α^2 on size and dust temperature have been displayed in Fig. 10.2b, c, respectively.

The set of Fig. 10.3a, b illustrate the dependence of the parameter α^2 and the mean charge on the particles Ze, on the parameter n_p in a complex plasma irradiated by a monochromatic source of radiation when the photoemission is the significant charging process of the dust grains. The effect of size on the charging and hence on α^2 has been shown in Fig. 10.3a; the mean charge increases with increasing n_p and increasing radius of the dust grains on account of larger photoemission. Figure 10.3b indicates increasing charge of the dust grains with increasing ξ. Further α takes higher values in the case of positive charging with increasing ξ; this may also be understood in terms of charging characteristics of the dust grains.

It is of some interest to compare the results of the present analysis with those of Khrapak et al. [4]. For this purpose, it will be useful to take note of the difference in notation between the two analyses viz. $\Gamma^+ = n_{\mathrm{ic}} + n_{\mathrm{ee}}$, $\Gamma^- = n_{\mathrm{ec}}$, $\beta = (\mathrm{d}I/\mathrm{d}\zeta)_{\zeta=Z}$ and $\gamma = Ze^2/akT_{e0}$; in both cases $\alpha = \Delta\zeta/\bar{Z}^{1/2}$ has the same significance. Since Khrapak et al. [4] formulation does not take into account the dependence of plasma parameters on n_d, the plasma parameters corresponding to Khrapak et al. [4] have been taken to be the parameters in the absence of dust; this analysis ignores ions and thus conforms to a dust cloud (without gas or ions). In the absence of electron emission, a neutral gas number density $n_{00} = 10^{12}\ \mathrm{cm}^{-3}$ has been chosen for computational purpose. It is seen that the departure of our results with those of Khrapak et al. [4] decreases as $n_d \rightarrow 1$ or 0.

For the case of thermionic emission in dust cloud, the formulation is the same in both the analyses and hence the results are in agreement.

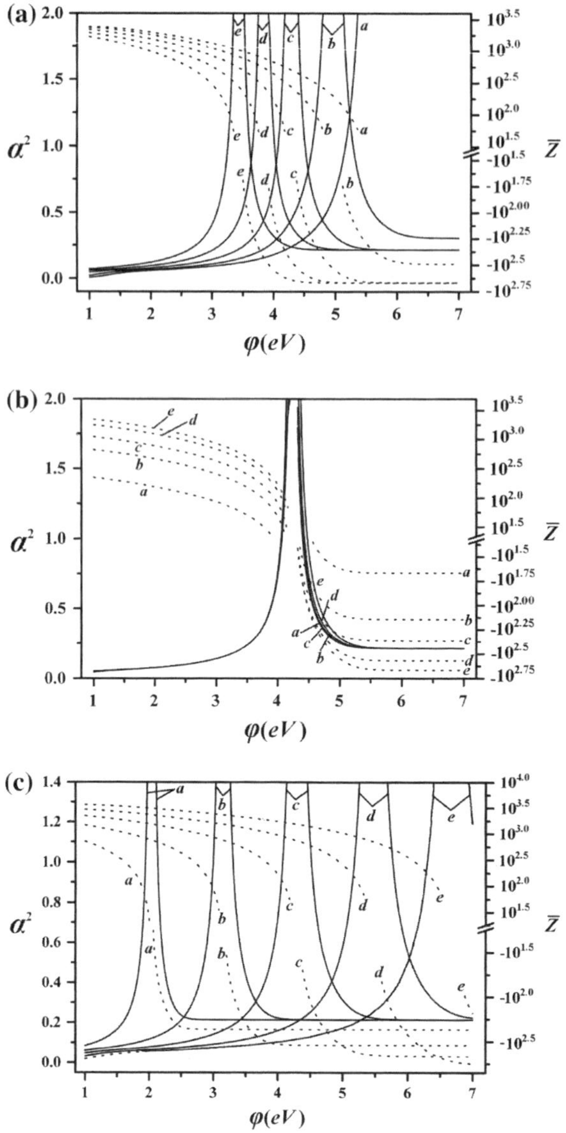

(a)
(b)
(c)
α²
φ(eV)
Z̄
a
b
c
d
e

Fig. 10.2 **a** The dependence of the parameter $\alpha^2(=\sigma^2/Z)$ and mean charge (*Z*) on the work function φ of the material, in the case of thermionic emission for the standard set of parameters as stated in Case 2 in the text. The letters on the *curve a, b, c, d* and *e* refer to the initial electron density $n_{e0} = 10^8$, 10^9, 10^{10}, 10^{11} and 10^{12} cm^{-3}, respectively. Further the *solid* and *broken curves* correspond to *left hand side* scale (α^2) and *right hand side* scale, respectively (*Z*) (after Sodha et al. [9]). **b** The dependence of the parameter α^2 and mean charge (*Z*) on the work function φ of the material, in the case of thermionic emission for the standard set of parameters as stated in Case 2 in the text. The letters on the *curve a, b, c, d* and *e* refer to the radius of the dust particle $a = 0.1, 0.3$, 0.5, 0.8, and 1.0 μ, respectively. Further the *solid* and *dotted curves* correspond to *left hand side* scale (α^2) and *right hand side* scale, respectively (*Z*) (after Sodha et al. [9]). **c** The dependence of the parameter $\alpha^2(=\sigma^2/\bar{Z})$ and mean charge (*Z*) on the work function φ of the material, in the case of thermionic emission for the standard set of parameters as stated in Case 2 in the text. The letters on the *curve a, b, c, d* and *e* refer to $T = 1000$, 1500, 2000, 2500, and 3000 K, respectively. Further, the *solid* and *broken curves* correspond to *left hand side* scale (α^2) and *right hand side* scale, respectively (*Z*) (after Sodha et al. [9], curtsey authors and publishers AIP)

10.3 Fluctuation of Charge on Dust Particles with a Size Distribution in a Complex Plasmas

Consider (Chap. 7) the simple case of complex plasma with dust particles, having a uniform surface potential V_s and size distribution given by the MRN power law viz.

$$n(a)\,\mathrm{d}a = Aa^{-s}\,\mathrm{d}a.$$

For a uniform potential V_s, $a = Ze/V_s$ and the, mean charge (*Z*) distribution can be put as

$$n(Z)\,\mathrm{dZ} = (e/V_s)^{1-s}Z^{-s}\,\mathrm{dZ}.$$

Hence, the fraction of particles with a mean charge between *Z* and *Z* + dZ is

$$\begin{aligned} f(Z)\,\mathrm{dZ} &= n(Z)\,\mathrm{dZ}\left[\int_{Z_1}^{Z_2} n(Z)\,\mathrm{dZ}\right]^{-1} = (1-S)(Z_2^{1-S} - Z_1^{1-S})^{-1}Z^{-S}\,\mathrm{dZ} \\ &= \Phi(s,\,Z_1,\,Z_2)Z^{-s}\,\mathrm{dZ} \end{aligned}$$

where $Z_2 = a_2V_s/e$ and $Z_1 = (a_1V_s/e)$.

Of the particles having a mean charge *Z* the fraction having a charge between ζ and $\zeta + \mathrm{d}\zeta$ is given by (for $Z > 30$)

$$\begin{aligned} F(\zeta, z)\,\mathrm{d}\zeta &= \left(1/\sigma\sqrt{2\pi}\right)\exp\left[-(\zeta - Z)^2/2\sigma^2\right]\,\mathrm{d}\zeta. \\ &= (2/\pi)^{1/2}(|Z|)^{-1/2}\exp\left[-2(\zeta - Z)^2/|Z|\right]\,\mathrm{d}\zeta \end{aligned} \tag{10.4}$$

and $\sigma = 0.5(|Z|)^{1/2}.$

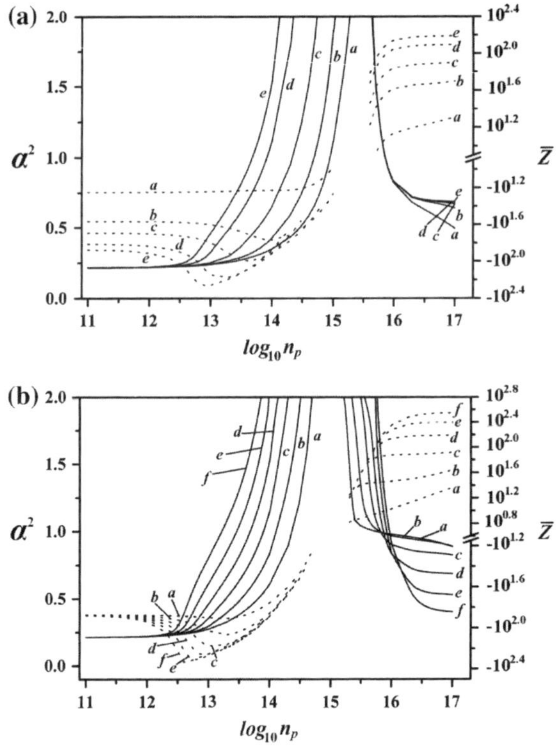

Fig. 10.3 **a** The dependence of the parameter $\alpha^2(=\sigma^2/Z)$ and mean charge (*Z*) on the parameter n_p, in the case of photoelectric emission for the standard set of parameters as stated in Case 3 in the text. The letters on the *curve a*, *b*, *c*, *d* and *e* refer to the radius $a = 0.1$, 0.3, 0.5, 0.8, and 1.0 μ, respectively. Further the *solid* and *broken curves* correspond to *left hand side* scale (α^2) and *right hand side* scale, respectively (*Z*) (after Sodha et al. [9]). **b** The dependence of the parameter $\alpha^2(=\sigma^2/Z)$ and mean charge (*Z*) on the parameter n_p, in the case of photoelectric emission for the standard set of parameters as stated in Case 3 in the text. The letters on the *curve a*, *b*, *c*, *d*, *e* and *f* refer to $\xi = 5$, 10, 15, 20, 25, and 30, respectively. Further the *solid* and *broken curves* correspond to *left hand side* scale (α^2) and *right hand side* scale, respectively (*Z*) (after Sodha et al. [9], curtsey authors and publishers AIP)

Hence, the fraction of particles having charge between ζ and $\zeta + \mathrm{d}\zeta$ is given by

$$
\begin{aligned}
\phi(\zeta)\mathrm{d}\zeta &= \int_{Z_1}^{Z_2} f(Z)F(\zeta, Z)\,\mathrm{d}Z.\mathrm{d}\zeta \\
&= \mathrm{d}\zeta.\Phi(S, Z_1, Z_2)\int_{Z_1}^{Z_2} Z^{-S}F(\zeta, Z)\,\mathrm{d}Z. \quad \text{for} \quad |Z_1| > 30
\end{aligned}
\tag{10.5}
$$

For given values of parameters s, Z_1 and Z_2 the distribution $\phi(\zeta)$ can be obtained by numerical integration of (10.5). For a typical set of parameters viz.

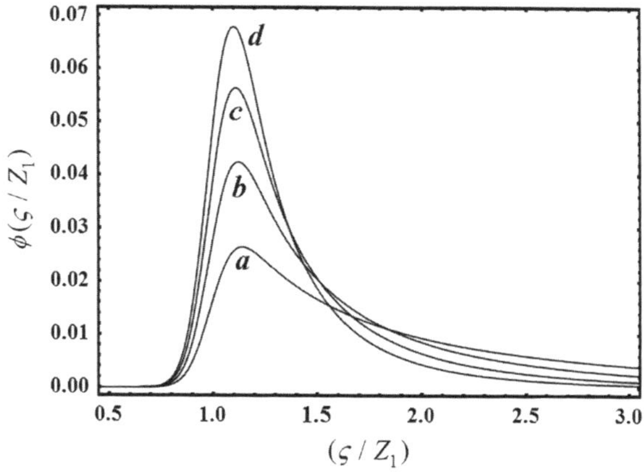

Fig. 10.4 Dependence of charge distribution on dust particles (ζ) as a function of charge ζe on the particles; letters a, b, c, d refer to $s = 2$, 3, 4, and 5, respectively $Z_1 = 30$ and $Z_2 = 200$ (curtsey Dr. S.K. Mishra, I.P.R., Ahmedabad, India)

$Z_1 = 30$ and $Z_2 = 200$, $\phi(\zeta)$ has been plotted in Fig. 10.4 as a function of ζ for various values of s; letters *a, b, c* and *d,* refer to $s = 2$, 3, 4 and 5 respectively. It may be pointed out that the (10.5) is meaningful only for $Z_1 > 30$ and $s > 1$. Further in obtaining $\phi(\zeta)$ only the magnitude of Z_1 and Z_2 is considered; hence, the distribution function is also valid for negatively charged particles.

References

1. C. Cui, J. Goree, IEEE Trans. Plasma Sci. **22**, 151 (1994)
2. C.K. Gorertz, Rev. Geophys. **27**, 271 (1989)
3. M.R. Jana, A. Sen, P.K. Kaw, Phys. Rev. E **48**, 3930 (1993)
4. S.A. Khrapak, A.E. Nefedov, O.F. Petrov, O.S. Vaulina, Phys. Rev. E **59**, 6017 (1999)
5. T. Matsoukas, M. Russel, J. Appl. Phys. **77**, 4292 (1995)
6. T. Matsoukas, M. Russel, M. Smith, J. Vac. Sci. Technol., A **14**, 624 (1996)
7. G.E. Morfill, E. Grun, T.V. Johnson, Planet. Space Sci. **28**, 1087 (1980)
8. T.G. Northrop, J.R. Hill, J. Geophys. Res. **88**, 01 (1983)
9. M.S. Sodha, S.K. Mishra, S. Misra, S. Srivastava, Phys. Plasmas **17**, 073705 (2010a)
10. M.S. Sodha, A. Dixit, S. Srivastava, S.K. Mishra, M.P. Verma, L. Bhasin, Plasma Sources Sci. Technol. **19**, 015006 (2010b)
11. N.G. Van Kampen, *Stochastic Processes in Physics and Chemistry* (North Holland, New York, 1990)

Part II
Applications

Chapter 11
Kinetics of Complex Plasmas in Space

11.1 Introduction

A variety of environments in space comprise of slightly to highly ionized plasma with a suspension of fine (nanometer to micrometer) dust. In this chapter such plasmas have been referred to as complex plasmas, regardless of the relative magnitudes of Debye radius and intergrain distance; the only criterion is that the dust carries a charge and significantly affects the properties of the plasma. Electron emission from the dust particles (usually photoelectric, caused by ultraviolet light) and electron/ion accretion lead to the charging of the dust particles. The physics and dynamics of the plasma are significantly affected by the presence of charged dust and the change in physical properties, caused by the dust. In this section, some typical environments, comprising of complex plasmas in space have been mentioned; in this context the author has benefitted much from a keynote talk by Mendis [32].

11.1.1 Planetary Magnetospheres

Dust–plasma interactions in the planetary magnetospheres have been studied for a long time. The period of vigorous research on the role of complex plasma in the magnetospheres started with the highly significant observations by spacecraft on interesting phenomena in the magnetospheres of giant planets in the early 1980s (*For a review see* Horanyi et al. [13]). It is instructive to take a look at some specific cases.

Saturn's Spokes

The approximately radial features (Spokes) across the dense B ring, observed intermittently by Voyager spacecraft 1 and 2 as they flew by Saturn have attracted a great amount of attention; several theories have been invoked to explain the observations.

Light scattering measurements indicated the presence of fine (micron and submicron) dust in the rings. This fact along with the observation that the grains

M. S. Sodha, *Kinetics of Complex Plasmas*, Springer Series on Atomic, Optical, and Plasma Physics 81, DOI: 10.1007/978-81-322-1820-3_11,

were levitated above the plane of the ring indicated the existence of a force normal to the plane. All theories have proposed a sporadic increase of the plasma density and subsequent charging of fine dust to a negative potential and consequent levitation. Thus the charging of the dust turns out to be an important part of the dynamics of the spokes.

High Speed Dust Streams from Jupiter and Saturn
The observation [9] of collimated quasi-periodic and high speed streams of fine dust particles, emanating from Jupiter by the Ulysses spacecraft highlighted the role of electrodynamic force in the dynamics of fine dust particles; the charge on the particles is an important parameter in the process. Subsequently such streams, emanating from Saturn were also discovered by the Cassini spacecraft [18].

Differential Collection of Charged Dust by Planetary Satellites
Mendis and Axford [33] have proposed that the two-tone appearance of the satellites could be explained in terms of the differential collection of charged dust by the leading and trailing faces of the satellite. Recent Cassini observations [60] support the model of Mendis and Axford [33]. As in other cases the charge on the dust is a critical parameter.

11.1.2 Cometary Magnetosphere

The solar radiation and solar wind cause a variety of phenomena, associated with dust–plasma interaction in the magnetosphere of a comet, approaching the sun. These include electrostatic levitation of dust from the bare cometary nucleus at large distance from the sun and electrostatic disruption and electrodynamic transport in the induced cometary magnetosphere as the distance from the sun gets decreased. For a review of these processes, the reader is referred to a review by Mendis [31]. The timely interest is on account of the incoming rendezvous of the Rosette spacecraft with Churiyumov—Gerasimenko comet in the spring of 2014. Needless to say the charging of dust particles is an important part of the process.

11.1.3 Interplanetary Dust

Particulates of size between a few centimeters to few nanometers in space with appreciable solar wind in the solar system are referred to as interplanetary dust. The main source of interplanetary dust in the inner region (within a distance of 5 A.U. from the sun) of the solar system are the asteroidal debris and ejecta from comets. In the outer region (at a distance of more than 5 A.U. from the sun), the interplanetary dust is contributed by the interstellar medium, volcanic satellites of giant planets, and the collisional debris of small satellites. The dominant forces which determine the dynamics of dust particles are gravitation for large particles, solar radiation (Poynting Robertson effect) and solar wind for intermediate size particles, and electrodynamic forces for small dust particles; the electrodynamic forces are of course dependent on the charge on the particles.

11.1.4 Interstellar Dust

The interstellar medium (ISM) is highly nonuniform and it contains a suspension of fine (mm to nm) dust in regions of very low (H I) and very high (H II) ionization. Dust plays an important role in the dynamics and thermodynamics of ISM and in the secondary star formation. The dust tends to be negatively charged by the impact of low energy cosmic rays while the ultraviolet radiation causes photoelectric emission from the dust particles, making these positively charged. Dust exists in the H I region and localized regions in H II.

11.1.5 Polar Mesospheric Clouds

Noctilucent clouds (NLC) and polar mesospheric summer echoes (PMSE, discussed in Chap. 9) are of considerable interest on account of their relevance to global warming. The low temperatures (110–130 K) at the low ionospheric altitudes in the polar region cause condensation of water vapor; the resulting system of suspended ice particles with sizes from 3 nm to 0.1 μm are known as polar mesospheric clouds (PMC); this term includes both the NLC (dust size from 20/30 nm to 0.1 μm) and PMSE regions (dust size from 3 to 20/30 nm). These dust particles get charged on account of accretion of ionospheric plasma electrons/ions on the surface of the particles and photoelectric emission from dirty ice particles, if present. PMCs are a well-known manifestation of dust–plasma interaction in the terrestrial atmosphere.

In the rest of the chapter, this kinetics of the complex plasma in the PMC, cometary environment, Saturn E ring, and interplanetary and interstellar media have been discussed as applications of the basics, presented in Part I of the book.

11.2 Kinetics of Polar Mesospheric Clouds (NLCs and PMSEs) (After Sodha et al. [55])

11.2.1 Basic Information

Based on earlier work [11] have listed typical values of plasma, neutral gas, and dust parameters in the mesopause region, associated with NLCs and PMSEs. Table 11.1 indicates the values of the parameters, and the range of values for

Table 11.1 Values of typical parameters for PMCs

Characterization	Measured and observed data	Present theory		Evaluated values
		Data used		
Height	82–90 km	≈85 km		
Temperature minimum				
Winter	200 K	130 K		–
Summer	130 K			
Neutral gas number density (30 a.m.u.)				
80 km	5×10^{14} cm^{-3}	3.5×10^{13} cm^{-3} (O_2)		–
90 km	6×10^{13} cm^{-3}	1.3×10^{14} cm^{-3} (N_2)		
Ionospheric electron density				
Day	10^3–10^4 cm^{-3}	10^3–10^4 cm^{-3}		–
Night	10–10^3 cm^{-3}	10–10^3 cm^{-3}		
Dust				
Radius (a)	50 nm	10–10^2 nm		–
Number density	10–1000 cm^{-3}	1–2500 cm^{-3}		–
PMC *electron density*				
Day time	Corresponds to Bite out	a	10–10^2 nm	20–100 % of ionospheric electron density
		n_d	1–2500 cm^{-3}	
		n_{e0}	4.5×10^3 cm^{-3}	
Dust charge (Pure ice)				
Frequent day time	−e, −2 e, −3 e	a	50 nm	−e, −2 e, −3 e
		n_d	100 cm^{-3}	
		n_{e0}	4.5×10^3 cm^{-3}	
Frequent nighttime	−e, 0, e	a	50 nm	−e, 0, e
		n_d	100 cm^{-3}	
		n_{e0}	10 cm^{-3}	
Dust charge (Dirty ice)				
Occasional day time	≤ 100 e	a	10–100 nm	Z_{max} = 100 e
		n_d	100 cm^{-3}	Z_{mean} = 85 e
		n_{e0}	4.5 x 10^3 cm^{-3}	

which computations have been made in this section; the agreement of the observed and computed charge distribution on pure ice particles and the condition for which the computed charge on dirty ice particles is around ≈80e are notable features of the present analysis.

Klumov et al. [21–23] have published three important papers on the charging of ice dust particles and formation of dust structures (PMCs) in the upper atmosphere. The models introduced the role of primary and proton hydrate (PHs) ions in the context of PMCs.

In contrast to the models, discussed in Part I of the book the attachment of electrons with O_2 molecules and the detachment (including photodetachment) of electrons from O_2^- ions have to be taken into account. The departure of Q_{abs}, the fraction of radiant energy absorbed on the surface of the particles from unity, as predicted by *Mie's* theory of scattering should also be accounted for. It is seen that for pure ice particles and a specific set of mesospheric parameters, the observed charge distribution (as given in Table 11.1) during day (−3e, −2e, −e) and night (−e, 0, e) gets explained; the corresponding decrease (*bite out*) in the electron density during the day is more than 30 %. Further with dirty ice particles of work function 2.3 eV (a value, which explains observations but has no other basis), large values of positive charge on the particles can also be explained; as an example for $a = 80$ nm and $n_d = 100\ \text{cm}^{-3}$, the mean charge on the particles has been computed as $\overline{Z} = 80$ which compares very well with the occasional value (<100 e), listed in Table 11.1. Corresponding values of electron density and temperature have also been computed.

11.2.2 Analytical Model

11.2.2.1 Master Equation

Consider the ionospheric region at mesospheric altitudes (≈82–90 km), illuminated by solar radiation; it consists [21, 22] of electrons, two different groups of singly charged ions (viz. *primary and cluster or* PHs *ions*), neutral species (mainly N_2 and O_2) and uniform size spherical pure (with work function $\varphi \approx 8.7$ eV) or dirty ice dust grains with an effective work function φ; the value $\varphi = 2.3$ eV explains the observations but has no other basis. The electron and ion accretion currents on and photoemission of electrons from the surface of dust grains are the basic processes determining the charging of dirty ice grains; hence the inherent charging processes may be expressed as

$$P_{Z-1} \rightleftarrows P_Z + e^-,$$

$$P_{Z+1} + e^- \leftrightarrows P_Z$$

$$P_{Z+1} + i^- \leftrightarrows P_Z$$

and

$$P_{Z-1} + i^+ \leftrightarrows P_Z \tag{11.1}$$

where P_Z refers to particles with charge *Ze* and e^- and $i^\pm$ represent electrons and ionic species respectively. It may be emphasized that this model is not limited by thermal equilibrium.

The population balance equation describing the generation and depletion of ice particles with charge *Ze* may be written as [28, 29]

$$(\mathrm{d}n_Z/\mathrm{d}t) = n_{Z+1}[n_{\mathrm{ec}}(Z+1) + n_{-\mathrm{ic}}(Z+1)] + n_{Z-1}[n_{\mathrm{ee}}(Z-1) + n^p_{+\mathrm{ic}}(Z-1) + n^h_{+\mathrm{ic}}(Z-1)] - n_Z[n_{\mathrm{ee}}(Z) + n^p_{+\mathrm{ic}}(Z) + n^h_{+\mathrm{ic}}(Z) + n_{-\mathrm{ic}}(Z) + n_{\mathrm{ec}}(Z)], \tag{11.2}$$

where $n_Z(=N_Z/n_d)$ is the fraction of particles carrying a charge Ze, n_{ec} and $n^y_{\pm\mathrm{ic}}$ refer to electron and ion accretion currents on the surface of the ice grains and n_{ee} refers to the photoemission rate of electrons from the surface of ice dust particles; here p, h stand for primary (p) and cluster or proton hydrate (h) positive ions.

The basic equations of the kinetics of a mesospheric layer with ice dust are as follows.

11.2.2.2 Approximate Constancy of Neutral Species

On account of the low level of ionization, the density of the neutral species can be considered to be a constant. Hence

$$n_{0,\mathrm{O}_2} = n_0 \text{ (constant)} \tag{11.3a}$$

$$n_{0,\mathrm{N}_2} = N_0 \text{ (constant)} \tag{11.3b}$$

At mesospheric altitudes, N_2 molecules do not take part in the formation of ions either by solar radiation or three body collision; here n refers to the O_2 molecules. The N_2 molecules contribute only to negligible energy exchange in elastic collisions with electrons and ions. There are no other natural gases, present in the mesosphere in significant amounts.

11.2.2.3 Charge Neutrality

$$n_d \sum_{Z_1}^{Z_2} Z n_Z = (n_e + n_{-i} - n^p_{+i} - n^h_{+i}), \tag{11.4}$$

where one restricts consideration to the values of Z lying between Z_1 and Z_2 ($Z_1 < Z < Z_2$) and n_d is the number density of ice dust particles and n_e is the electron density.

11.2.2.4 Electron Kinetics

$$\frac{\mathrm{d}n_e}{\mathrm{d}t} = \beta_i n_0 - n_e\left(\alpha^p_r n^p_{+i} + \alpha^h_r n^h_{+i}\right) + (\nu_d n_{-i} - \nu_a n_e) - n_d\left(\sum_{(Z_1+1)}^{Z_2} n_Z n_{\mathrm{ec}}(Z) - \sum_{Z_1}^{(Z_2-1)} n_Z n_{\mathrm{ee}}(Z)\right), \tag{11.5}$$

where

β_i is the coefficient of ionization due to the processes responsible for the ionization in the absence (also in presence) of dust particles,
$\alpha_r^y(T_e) = \alpha_{r0}^y T_e^{-\eta_y}$ cm^3 s^{-1} is the coefficient of recombination of electrons and ions $y = p, h$
$\nu_a(=k_a n_0^2)$ is the frequency of the electron attachment to the neutral species due to triple collisions,
$\nu_d = (\nu_{\rm ph} + \nu_{\rm ed})$ is the electron detachment frequency,
$\nu_{\rm ph}$ is the photodetachment frequency, under conditions of the day time ionosphere,
$\nu_{\rm ed}(=k_d n_0)$ is the collision detachment frequency,
$k_a = (1.4 \times 10^{-29}) \exp[-(600/T_e)]$ cm^6 s^{-1},

$$k_d = 10^{-20}\,\text{cm}^3\,\text{s}^{-1}$$

and T_e is the electron temperature and α_{r0}^y and η_y are constants.

The formulation and values of the parameters have been taken from the book of Gurevich [10].

11.2.2.5 Kinetics for Ionic Species (After Klumov et al. [21])

Primary Ion Kinetics

$$\frac{\mathrm{d}n_{+i}^p}{\mathrm{d}t} = \beta_i n_0 - \alpha_r^p n_e n_{+i}^p - \beta_c n_{+i}^p - n_d \sum_{Z_1}^{(Z_2-1)} n_Z n_{+\mathrm{ic}}^p(Z). \tag{11.6a}$$

Proton Hydrate Ion Kinetics

$$\frac{\mathrm{d}n_{+i}^h}{\mathrm{d}t} = \beta_c n_{+i}^p - \alpha_r^h n_e n_{+i}^h - n_d \sum_{Z_1}^{(Z_2-1)} n_Z n_{+\mathrm{ic}}^h(Z). \tag{11.6b}$$

Negative Ion Kinetics

$$\frac{\mathrm{d}n_{-i}}{\mathrm{d}t} = (\nu_a n_e - \nu_d n_{-i}) - n_d \sum_{(Z_1+1)}^{Z_2} n_Z n_{-\mathrm{ic}}(Z). \tag{11.6c}$$

It is seen that any one of (11.3a), (11.3b), (11.4), (11.5) and (11.6a), (11.6b), (11.6c) can be derived from the remaining three equations and hence one of the equations becomes redundant. As a convenience (11.4) has been omitted in further analysis.

11.2.2.6 Energy Balance for Electrons and Ions

$$\frac{\mathrm{d}}{\mathrm{d}t}\left(\frac{3}{2}kn_eT_e\right) = \beta_i n_0 \varepsilon_e - n_e[\alpha_r^p n_{+i}^p + \alpha_r^h n_{+i}^h](3kT_e/2) + [\nu_d n_{-i}\varepsilon_v - \nu_a n_e(3kT_e/2)]$$
$$- n_d\left[\sum_{(Z_1+1)}^{Z_2} n_Z n_{\mathrm{ec}}(Z)\varepsilon_{\mathrm{ec}}(Z) - \sum_{Z_1}^{(Z_2-1)} n_Z n_{\mathrm{ee}}(Z)\varepsilon_{\mathrm{ee}}(Z)\right] \quad (11.7)$$

and

$$\frac{\mathrm{d}}{\mathrm{d}t}\left(\frac{3}{2}k(n_{+i}^p + n_{+i}^h + n_{-i})T_i\right) = \beta_i n_0 \varepsilon_{+i}^p - n_e\left(\alpha_r^p n_{+i}^p + \alpha_r^h n_{+i}^h\right)(3kT_i/2)$$
$$+ (\nu_a n_e - \nu_d n_{-i})(3kT_i/2)$$
$$- n_d \sum_{Z_1}^{(Z_2-1)} \left[n_Z n_{+\mathrm{ic}}^p(Z)\varepsilon_{+\mathrm{ic}}^p(Z) + n_{+\mathrm{ic}}^h(Z)\varepsilon_{+\mathrm{ic}}^h(Z)\right.$$
$$\left. + n_{-\mathrm{ic}}(Z)\varepsilon_{-\mathrm{ic}}(Z)\right] \quad (11.8)$$

where

ε_e, ε_{+i}^p are the mean energies of electrons and primary ions produced by ionization of neutral atoms,
$\varepsilon_{\mathrm{ec}}$,$\varepsilon_{\pm\mathrm{ic}}^y$ are the mean energies (far away from the surface) of electrons and ions accreting on the surface of the ice grain,
ε_v is the mean energy of photodetached electrons from negative ions,
$\varepsilon_{\mathrm{ee}}$ is the mean energy of photoemitted electrons at a large distance from the surface of the dust grain,
T_i is the temperature of the ionic species,
and T_{e0}, T_{i0} and T represent the temperature of electrons, ions, and neutrals, respectively, in absence of dust; unless otherwise stated most of this data is from the book by Gurevich [10].

Energy exchange in elastic collisions has been neglected.

11.2.3 Model of the Mesosphere Without Dust

The mesosphere has been modeled by putting $n_d = 0$ in the kinetic equations, given as above. The value of β_i was adjusted to yield an electron density, equal to the observed value; this however gives an electron temperature, a little larger than the observed value in the day time environment (this is mainly because of inclusion of the process of photodetachment in the model).

Table 11.2 Observed and adjusted parameters in PMC

Time	Electron density (observed) n_{e0} (cm^{-3})	Ionization rate (adjusted) $\beta_i n_0$ (cm^{-3} s^{-1})	Electron temperature T_{e0} (K)		Ion temperature T_{i0} (K)	
			Observed	Computed	Observed	Computed
Day	4.5×10^3	141.5	130	145	130	130
Night	35	3.35	130	130	130	130

After Sodha et al. [55], courtesy authors and publishers AIP

The adjusted values of and the corresponding electron density and temperature are as in Table 11.2.

Further the mean energy of electrons and ions generated due to ionization may be obtained by imposing the initial conditions for dust-free plasma in the energy balance equations; thus

$$\varepsilon_e \approx (3kT_{e0}/2) - (\beta_i n_{00})^{-1} v_{\text{ph}}(\varepsilon_v - 3kT_{e0}/2) \tag{11.9}$$

and

$$\varepsilon_i^p = (3kT_{i0}/2), \tag{11.10}$$

where the suffix zero refers to the absence of dust. Equations (11.9) and (11.10) have been obtained by putting $n_d = 0$ in (11.7) and (11.8) in the steady state.

11.2.4 Computational Methodology

For an arbitrarily chosen range of Z values, one can easily obtain the steady-state charge distribution and other relevant parameters viz. n_e, n_i^p, n_i^h, $n_{i\text{-}}$, n_0, T_e, and T_i by simultaneous numerical integration of (11.2), (11.3a), (11.3b), (11.5), (11.6a), (11.6b), (11.6c), (11.7), and (11.8) for a given initial set (in the absence of dust) of parameters as $t \to \infty$; the steady state is indeed independent of the initial conditions. The arbitrary range of Z may be estimated around mean charge $\overline{Z}$ on the dust particles, obtained by average charge theory (Sodha et al. [52, 53]). For such a chosen range (11.2) yields a set of equations for n_Z.

11.2.5 Photoelectric Emission from Charged Dust Particles by Solar Radiation

Expressions for the rate of emission n_{ph}, and mean energy of photoelectrons, emitted by the incidence of white (sun) light on the dust particles have been given in Sect. 5.4.4 [(5.15) and (5.17)]. The effect of Mie scattering may be incorporated by having Q_{abs} in the integrand.

11.2.6 Other Parameters

The expressions for the relevant parameters are as per Chaps. 2 and 3.

The photodetachment frequency ν_{ph} and the mean energy of the electrons produced by photodetachment are given by

$$\nu_{\text{ph}} = \int_{E_d}^{E_{\nu m}} Q_d(E_\nu) dn_{\text{inc}}(E_\nu) \tag{11.11a}$$

and

$$\varepsilon_\nu = \nu_{\text{ph}}^{-1} \int_{E_d}^{E_{\nu m}} (E_\nu - E_d) Q_d(E_\nu) dn_{\text{inc}}(E_\nu), \tag{11.11b}$$

where $E_d = 0.5$ eV is the energy of dissociation of photodetachment.

Following McEwan and Philips [30], the photodetachment cross-section corresponding to O_2^- negative ions can be expressed as

$$Q_d = \left[\left(2.3 \times 10^{-18}\right)/2.5\right](E_\nu - 0.5)\,\text{cm}^2, \quad (\text{where } E_\nu \text{ is in } e\text{V})$$

or

$$Q_d = (2.3/4) \times 10^{-6}\left(E_\nu - 8 \times 10^{-13}\right)\text{cm}^2. \quad (\text{where } E_\nu \text{ is in } ergs) \tag{11.11c}$$

11.2.7 Numerical Results and Discussion

PMC *plasma with pure ice dust (Basic data)*

$n_{e0} = n_{i0} = 35\,\text{cm}^{-3}(\text{night time}), 4.5 \times 10^3\,\text{cm}^{-3}(\text{day time}), n_d = 100\,\text{cm}^{-3}, n_{O_2} = 3.5 \times 10^{13}\,\text{cm}^{-3}$
$n_{N_2} = 1.3 \times 10^{14}\,\text{cm}^{-3},\ a = 50\,\text{nm},\ T_{e0} = T_{i0} = T_n = T_d = 130\,\text{K}(\text{observed}) m_0 \approx m_i^p = 30\,\text{amu},$
$m_i^h = 5m_i^p, \eta_p = 0.7, \eta_h = 1.2, \alpha_{r0}^p = 10^{-7}\,\text{cm}^3/s, \alpha_{r0}^h = 10^{-5}\text{cm}^3/s, \beta_i n_{00}(\text{adjusted})$
$= 141.5\,s^{-1}(\text{day}),\ 3.35\,s^{-1}(\text{night})$ and Computed $T_{e0} = 130\,\text{K}(\text{night}),\ 145.45\,\text{K}(\text{day})$.

The set of Fig. 11.1 (corresponding to the above data) illustrates the charge distribution [i.e. ($n_z = N_Z/n_d$) *vs* Z] in day and night time mesospheric dust clouds (consisting of pure ice particles characterized by high work function $\varphi = 8.7$ eV) and corresponding density and temperature of electrons; since there are no photons available having energy more than the work function of pure ice there is no photoelectron emission from the pure ice dust. It may be appreciated that in the absence of electron emission from pure ice dust, a large change in the electron

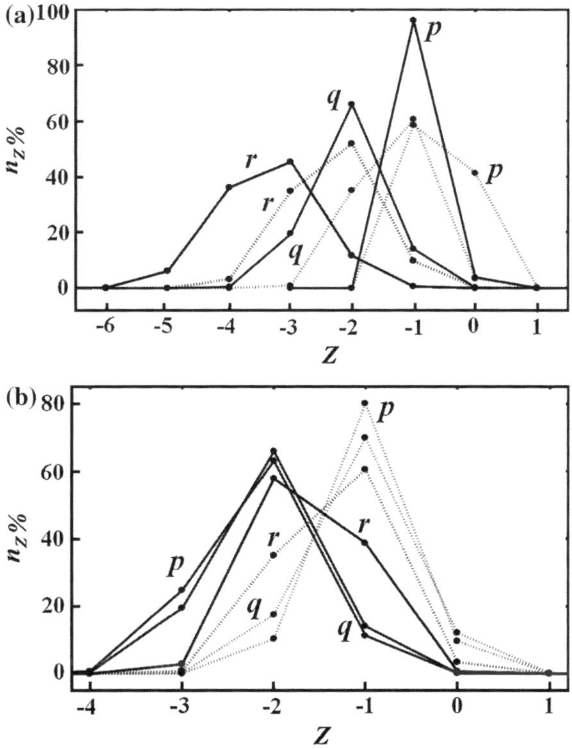

Fig. 11.1 **a** Dependence of the charge distribution [*i.e.* (N_Z/n_d) *vs* Z] (for pure ice dust and in absence of electron emission), on the grain size (a), for standard set of parameters as stated in the text. The letters p, q, and r on the *curves* refer to $a = 10, 50$, and 100 nm, respectively. *Solid* and *broken lines* correspond to day and night time conditions, respectively. Only points on the *curves* which correspond to integral values of Z are significant. **b** Dependence of the charge distribution on the number density of ice grains (n_d) for the standard set of parameters as stated in the text. The letters p, q, and r on the *curves* refer to $n_d = 10$, 100, and 1,000 cm^{-3}, respectively. *Solid* and *broken lines* correspond to day and night time conditions, respectively. Only points on the *curves* which correspond to integral values of Z are significant. **c** Effect of the dust-free mesospheric electron density (n_{e0}) on the charge distribution for the same standard set of parameters as stated in text. The letters p, q, and r on the *curves* correspond to $n_{e0} = 10$, 100, and 10^3 cm^{-3}. *Solid* and *broken lines* correspond to day and night time. Only points on the *curves* which correspond to integral values of Z are significant. **d** Dependence of electron density (n_e/n_{e0}) and electron temperature (T_e/T_{e0}) on the ice grain size (a) as a function of n_d, for standard set of parameters as stated in text (day time). The letters p, q, and r on the *curves* correspond to $a = 10, 50$, and 100 nm, respectively. The *solid* (n_e/n_{e0}) and *broken* (T_e/T_{e0}) *lines* correspond to *left-* and *right-hand side scales*, respectively (after Sodha [55], courtesy authors and publishers AIP)

density can be caused by the presence of very few ice particles on account of electron/ion accretion on dust surface. This reduction in electron density at mesospheric layers leads to *electron bite out* phenomenon. The analysis incorporates both day and nighttime mesospheric data and the results have been displayed separately in the figures.

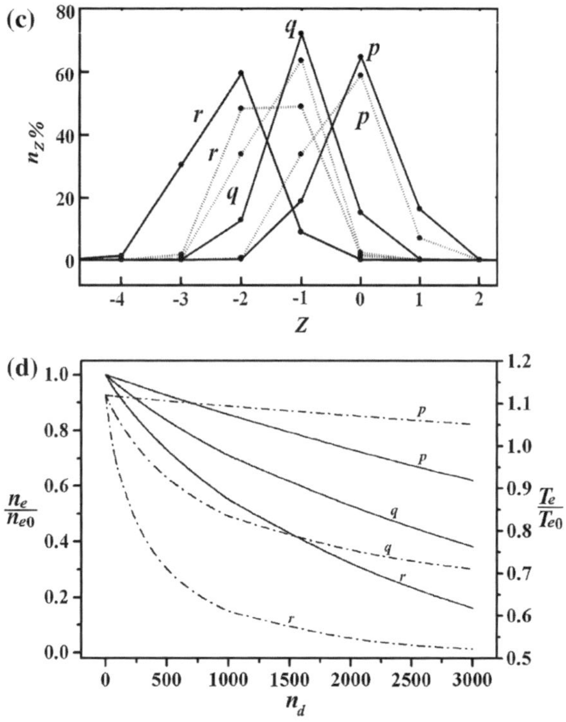

Fig. 11.1 (continued)

Figure 11.1a indicates that the charge distribution shifts toward larger values of negative charge with increasing radius of the particles. The broadening of the distribution may be qualitatively understood in terms of the earlier result [3] that $\Delta Z \approx 0.5(\bar{Z})^{1/2}$. Curves with solid lines correspond to day time mesospheric conditions while broken lines refer to nighttime situation; the difference between the two may be ascribed to difference in day and nighttime mesospheric electron densities. The dependence of the charge distribution (n_Z) on the number density of ice particles n_d and the initial electron density n_{e0} have been displayed in Fig. 11.1b and c, respectively. Figure 11.1b indicates that the charge distribution [viz. day (−3e, −2e, −1e) and night (−2e, −1e, −0e) time NLCs] remains almost unaffected by the number density of dust grains; however, the nature of the distribution varies significantly. The charge distribution dependence on n_{e0} may be understood in terms of smaller number of electrons/ions, available for accretion on the dust particles for small values of n_{e0}. Figure 11.1d illustrates the dependence of corresponding number density of electrons (n_e/n_{e0}) and temperature (T_e/T_{e0}) of electrons as a function of number density of pure ice particles (n_d).

Some interesting features extracted from Fig. 11.1.

(i) From Fig. 11.1a, it is seen that the number of possible values of charge on the ice grains for which there are significant number (probability more than 5 %) of ice grains, increases with increasing radii of the ice particles (from two to seven). However, for $a = 50$ nm this number (three) appears to be independent of n_d and n_{e0} (as illustrated in Fig. 11.1b and c).

(ii) Figure 11.1a and b reflect the fact that for $n_d = 100\ \text{cm}^{-3}$ and $a = 50$ nm, the computed charge is distributed over the range $(N_{-3} \approx 19\ \text{cm}^{-3}, N_{-2} \approx 66\ \text{cm}^{-3}, N_{-1} \approx 14\ \text{cm}^{-3})$ for the day time mesospheric parameters (viz. $n_{e0} = 4.5 \times 10^3\ \text{cm}^{-3}$), while the charge distribution shifts toward less negative charging over the range $(N_{-2} \approx 35\ \text{cm}^{-3}, N_{-1} \approx 60\ \text{cm}^{-3}, N_0 \approx 4\ \text{cm}^{-3})$ for $n_{e0} = 35\ \text{cm}^{-3}$, corresponding to night time mesospheric layer at NLC altitude. Figure 11.1c indicates that in night time NLCs [with $n_d = 100\ \text{cm}^{-3}$, $a = 50$ nm and $n_{e0} = 10\ \text{cm}^{-3}$] the computed charge is distributed over the range $(N_{-1} \approx 33\ \text{cm}^{-3}, N_0 \approx 59\ \text{cm}^{-3}, N_1 \approx 7\ \text{cm}^{-3})$. This provides an explanation for the observed/estimated data of Havnes et. al (Table 11.1).

(iii) The nature of dependence of the electron density on n_d (as shown in Fig. 11.1d) is a clear indication of the occurrence of *electron bite* out, the electron density gets reduced by at least ≈ 30 % (corresponding to $n_d = 1000$) from its initial concentration viz. $n_{e0} = 4.5 \times 10^3\ \text{cm}^{-3}$ and $a = 50$ nm during the day time mesospheric layers.

Thus it may be concluded that the main features of NLCs/PMSEs (like *electron bite out* and *charge distribution*) can be explained by considering the mesospheric ice dust plasma layer as an open system [52, 53] consisting of pure ice grains; these pure ice particles act as a sink of electrons and ionic species. It may be appreciated that the predicted numerical outcome of the present analytical approach is in excellent agreement with the observed data for charge distribution (Table 11.1) in PMCs.

Computations also show that high positive charge on dust can be explained on the basis of a work function of 2.3 eV, for dirty ice; however, there is no other evidence for such a work function.

11.3 Cometary Plasma (After Sodha et al. [54])

11.3.1 Basic Information

It is well known that the dust grains in cometary plasmas significantly affect the number density and temperature of the constituents on account of the accretion of electrons and ions and emission of electrons by the grains. The solar radiation causes ionization of the gaseous species. The photoelectric emission of electrons from the grains acts as an additional source of electrons.

The charging of the regolith of the comet on account of solar wind and electromagnetic radiation from the sun has been the main theme of most publications in the field of cometary plasma. Qualitative aspects of the role of dust in the plasma of cometary tails have been frequently mentioned. Possibly the most comprehensive model of the coma plasma, which enables an analysis of the impact of dust (comprising ice particles) on the plasma chemistry of an inner comet coma has been developed by Klumov et al. [21, 23, 25] who considered the following processes:

(i) Photoionization of water molecules by extreme UV solar radiation with wavelength less than 98.4 nm (the sun was assumed to be at a distance of 1 A.U.).
(ii) Ionization of atoms by solar radiation.
(iii) Photoelectric emission from ice particles (work function $\varphi = 8.7$ eV) by the incidence of solar radiation (Lyman-α radiation); dust of Fe, Na, K, Al, and Si may also be present (amount not specified).
(iv) Recombination of molecular ions with electrons.
(v) Charge exchange between atomic ions and neutral molecules, dominating over recombination of atomic ions with electrons, which has been neglected.
(vi) Accretion of electrons and ions on the ice particles on account of efficient charge exchange process; the recombination of atomic ions with electrons was ignored.

Sodha et al. [54] have improved the analysis by incorporating the following additional features in the analysis.

(i) The energy balance of the various constituents has been taken into account.
(ii) Appropriate expressions (Chap. 2) for the rate of photoelectric emission of electrons from the particles and the corresponding mean energy have been used.
(iii) The recombination of atomic ions with electrons has also been taken into account.

The effect of solar radiation on the gaseous species is to produce electrons, atomic ions, and molecular ions at the rate of q_e, q_i^a and q_i^m per unit volume, respectively, with energies $(3kT_{e0}/2)$ (electrons) and $(3kT_{i0}/2)$ (ions).

Further since the number density of the neutral species is very large, their temperature and number density can be justifiably assumed to remain unchanged; hence the number of electron/ion pairs created per unit time per unit volume by the ionization of molecules/atoms due to solar radiation is also constant.

11.3.2 Analytical Model for Electronic Processes in a Cometary Coma Plasma

Consider a cometary coma, consisting of water vapor plasma and spherical ice grains with radius a and charge Ze (all the grains have been assumed to have the same charge), located at a heliocentric distance of 1.0 A.U. Photoemission from ice (work function 8.7 eV) particles, caused by the extreme UV solar radiation (Lyman-α) is assumed to be the prime mechanism of electron/ion generation in addition to photoionization of water molecules by EUV radiation from the sun. It may be appreciated [1] that 80 % of the incident photons in EUV (122.5–4.0 nm) part of solar spectrum are accounted for by Lyman-α radiation; further the photoelectric efficiency drops rapidly with decreasing wavelength. Hence Lyman-α radiation alone has been considered as the source of photoelectron emission from the ice dust. The basic equations, governing the kinetics of the cometary plasma are as follows:

Charge Neutrality

$$Zn_d + n_i^a + n_i^m = n_e, \tag{11.12}$$

where Ze is the charge on an ice particle and n_d is the number density of dust particles.

Charging of the Dust Particles

$$(\mathrm{d}Z/\mathrm{d}t) = n_{\mathrm{ic}}^a + n_{\mathrm{ic}}^m + n_{\mathrm{ph}} - n_{\mathrm{ec}}, \tag{11.13}$$

where $n_{\mathrm{ic}}^j(j - a, m)$ is the atomic and molecular ion accretion current to a dust particle and n_{ph} is the rate of photoemission of electrons per unit time from the surface of the dust particle.

Electron Kinetics

$$(\mathrm{d}n_e/\mathrm{d}t) = q_e + n_{\mathrm{ph}}n_d - \alpha_r^a n_e n_i^a - \alpha_r^m n_e n_i^m - n_{\mathrm{ec}}n_d. \tag{11.14}$$

In general q_e is proportional to the number density of neutral species n_0; however, as pointed out earlier on account of the low level of ionization it has been treated as constant.

Ion Kinetics

Here two distinct groups of ions have been considered viz. atomic (O^+, H^+, etc.) and molecular (H_2O^+, OH^+, CO_2^+, etc.) ions; the kinetics is as follows:

For Atomic Ions

$$\left(\mathrm{d}n_i^a/\mathrm{d}t\right) = q_i^a - \beta_a n_0^m n_i^a - \alpha_r^a n_e n_i^a - n_{\mathrm{ic}}^a n_d, \tag{11.15}$$

For Molecular Ions

$$\left(\mathrm{d}n_i^m/\mathrm{d}t\right) = q_i^m + \beta_a n_0^m n_i^a - \alpha_r^m n_e n_i^m - n_{\mathrm{ic}}^m n_d \tag{11.16}$$

where β_a is a coefficient, characterizing the charge exchange between atomic ions and neutral molecules.

It is necessary to mention that any one of (11.2), (11.3a), (11.3b), (11.4), (11.5) and (11.6a), (11.6b), (11.6c) can be derived from the remaining four equations and hence one of the equations becomes redundant. Thus (11.2) has been omitted in further analysis.

Energy Balance of Electrons

$$\frac{\mathrm{d}}{\mathrm{d}t}\left(\frac{3}{2}kT_e n_e\right) = q_e\varepsilon_e + n_{\mathrm{ph}}n_d\varepsilon_{\mathrm{ph}} - (\alpha_r^a n_i^a + \alpha_r^m n_i^m)n_e(3kT_e/2) - n_{\mathrm{ec}}n_d\varepsilon_{\mathrm{rec,e}} \tag{11.17}$$

where

T_i^j is the temperature of the ionic species,
T_0 is the temperature of the neutral species,
ε_e is the mean energy of electrons produced by ionization of neutral species,
$\varepsilon_{\mathrm{ph}}$ is the mean energy of the photoemitted electrons at large distance from the surface of the dust particles,
$\varepsilon_{\mathrm{rec},e}$ is the mean energy of the electrons, (at large distance from the surface) collected by the dust particles,
and the small energy exchange in elastic collisions is neglected.

Energy Balance of Ionic Species
For Atomic Ions

$$\frac{\mathrm{d}}{\mathrm{d}t}\left(\frac{3}{2}kT_i^a n_i^a\right) = q_i^a\varepsilon_i^a - \alpha_r^a n_i^a n_e(3kT_i^a/2) - n_{\mathrm{ic}}^a n_d\varepsilon_{\mathrm{rec,i}}^a \tag{11.18}$$

For Molecular Ions

$$\frac{\mathrm{d}}{\mathrm{d}t}\left(\frac{3}{2}kT_i^m n_i^m\right) = q_i^m\varepsilon_i^m - \alpha_r^m n_i^m n_e(3kT_i^m/2) - n_{\mathrm{ic}}^m n_d\varepsilon_{\mathrm{rec,i}}^m \tag{11.19}$$

where ε_i^j is the mean energy of ions, produced by ionization due to the ionization agency and the energy exchange in the charge exchange process is neglected on account of same temperature of ions and neutral atoms.

The rate of ionization $\left(q_e, q_i^j\right)$ and mean energy $\left(\varepsilon_e, \varepsilon_i^j\right))$ of electrons and ionic species, generated due to ionization may be obtained by imposing the initial conditions for dust-free coma plasma in the number and energy balance equations, respectively.

Thus

$$\begin{aligned}
q_e &= \left(\alpha_r^a n_{i0}^a + \alpha_r^m n_{i0}^m\right)n_{e0},\\
q_i^m &= \left(\alpha_r^a n_{i0}^a n_{e0} + \beta_a n_0^m n_{i0}^a\right),\\
q_i^a &= \left(\alpha_r^a n_{i0}^m n_{e0} - \beta_a n_0^m n_{i0}^a\right),\\
q_e\varepsilon_e &= (3k/2)n_{e0}[(\alpha_r^a n_{i0}^a + \alpha_{r0}^m n_{i0}^m)T_{e0}\\
q_i^a\varepsilon_i^a &= (3k/2)n_{e0}\alpha_r^a n_{i0}^a T_{i0}^a
\end{aligned}$$

and

$$q_i^m \varepsilon_i^m = (3k/2) n_{e0} \alpha_i^m n_{i0}^m T_{i0}^m$$

The above equations are in conformance with the basic information. Expressions for $n_{\rm ph}$, $\varepsilon_{\rm ph}$, $n_{\rm ee}$, $\varepsilon_{\rm ec}$, $n_{\rm ic}$, and $\varepsilon_{\rm ic}$ may be seen in Chaps. 2 and 3.

11.3.3 Numerical Results and Discussion

The contribution of the EUV radiation (from 122.5–4.0 nm) to photoelectron emission by ice dust may be approximated [1] by 3.25×10^{11} photons/cm^{2}s corresponding to Lyman-α radiation with the wavelength 121.57 nm.

$$\begin{aligned}
&n_{e0} = n_{i0} \approx 10^4/\mathrm{cm}^3,\ n_{00} = 10^{13}/\mathrm{cm}^3, T_{e0} = 1000\,\mathrm{K},\ T_{i0}^m = T_{i0}^a = T_{00} = 400K \\
&T_d = 250\,\mathrm{K},\ \text{photoelectric efficiency}\ Y = 0.03,\ n_{i0}^a = 0.3 n_{e0},\ a = 1.0\,\mu\,\mathrm{m} \\
&m_i^m = m_0 \approx 30\,\mathrm{amu},\ m_i^a 0^a \approx 10\,\mathrm{amu},\ \alpha_r^a = 10^{-8}\mathrm{cm}^3/s, \alpha_r^m = 10^{-7}\mathrm{cm}^3/s \\
&\beta_a = 10^{-4}\,\mathrm{s}^{-1}\ \text{and}\ \Lambda_0 = 3.25 \times 10^{11}/\mathrm{cm}^2\mathrm{s}
\end{aligned}$$

The values of parameters n_{e0}, n_{i0}, n_{00}, T_{e0}, T_{i0}, T_{i0}^a, T_{i0}^m, α_r^m, β_a, a and Y have been chosen from the range of values, given by Klumov et al. [25]. The effect of change in one of these parameters can be studied by computation of the result of change of that parameter, keeping others the same, as in the standard set.

Simultaneous solution of (11.3a, 11.3b), (11.4), (11.5), (11.6a, 11.6b, 11.6c), (11.7), (11.8), and (11.9) has been obtained by using the MATHEMATICA software.

Figure 11.2a illustrates the dependence of the charge on the particle Ze and mean free path of electrons λ_c on the number density of the ice particles (n_d) in the coma environment. It is seen that the character of positive charging of the ice particles increases with increasing number density and the particles acquire a maximum positive charge at a certain value of n_d; the charge finally gets saturated to a small value for large n_d.

The dependence of electron density n_e and electron temperature T_e on n_d has been shown in Fig. 11.2b. The curves indicate that the electron density falls with increasing number density of ice grains and is higher for smaller radii; this is due to the fact that larger size and larger n_d, both lead to larger accretion of electrons and ionic species.

Atomic and molecular ion densities decrease with increasing a and n_d, on account of their larger accretion to the ice particles; this nature has been shown in Fig. 11.2c.

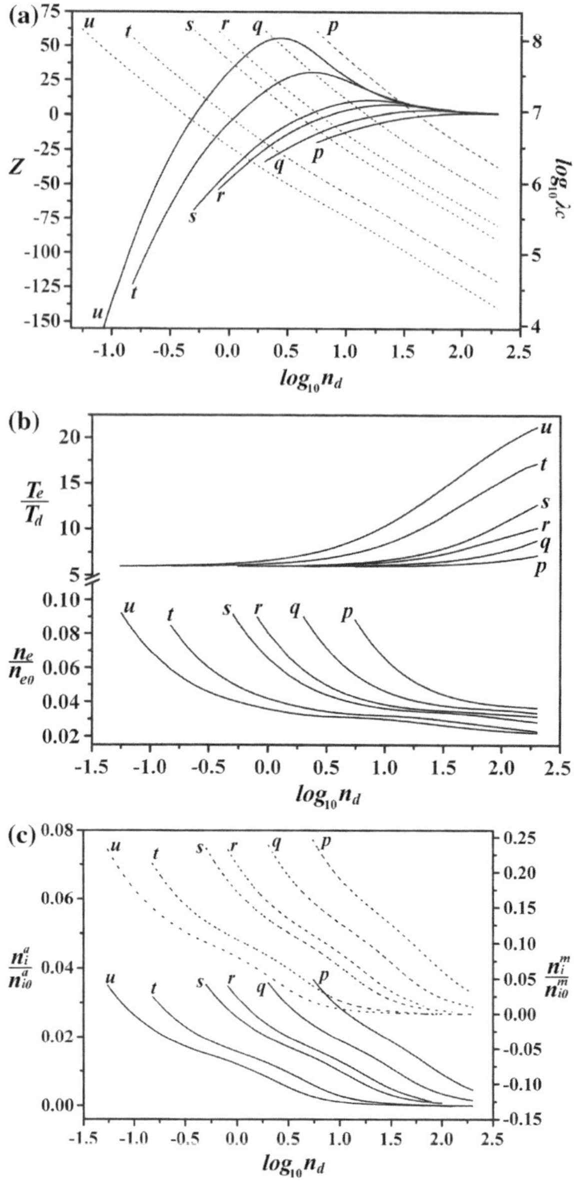
(a)
Z
$log_{10} \lambda c$
$log_{10} n_d$
u
t
s
r
q
p
(b)
$\frac{T_e}{T_d}$
$\frac{n_e}{n_{e0}}$
$log_{10} n_d$
(c)
$\frac{n_i^a}{n_{i0}^a}$
$\frac{n_i^m}{n_{i0}^m}$
$log_{10} n_d$

Fig. 11.2 **a** Dependence of charge of dust particles *Ze* and mean free path of the electrons λ_c (cm), on the number density of ice particles n_d (cm^{-3}) in a coma environment; the relevant parameters are $T_{e0} = 1000$ K, $T_{i0}^a = T_{i0}^m = 400$ K, $T_d = 250$ K, $n_{e0} \approx n_{i0} \approx 10^4$/cm^3, $n_{00} = 10^{13}$/cm^3, $n_{i0}^m = 0.3n_{e0}$, $Y = 0.03$, and $\Lambda_0 \approx 3.25 \times 10^{11}$/cm^{2}s for varying radii a of the ice particles; the letters on the *curves* *p*, *q*, *r*, *s*, *t*, and *u* refer to $a = 0.3$, 0.5, 0.8, 1.0, 2.0, and 3.0 μm, respectively. Further the *solid* and *dotted curves* correspond to *left-hand side scale* (*Ze*) and *right-hand side scale* ($log_{10}\lambda_c$), respectively. **b** Dependence of electron density (n_e/n_{e0}) and the electron temperature (T_e/T_d), on the number density of ice particles n_d(cm^{-3}) for different radii; the relevant parameters and the meaning of the letters is as in the caption of Fig. 11.2a. Further the scale of (n_e/n_{e0}) and (T_e/T_d) has been indicated in the figure. **c** Dependence of atomic ion density (n_i^a/n_{i0}^a) and molecular ion density (n_i^m/n_{i0}^m), on the number density of ice particlesn_d (cm^{-3}); the relevant parameters and the meaning of the letters is as in the caption of Fig. 11.2a. Further the *solid* and *dotted curves* correspond to *left-hand side scale* for (n_i^a/n_{i0}^a) and *right-hand side scale* for (n_i^m/n_{i0}^m), respectively (after Sodha et al. [54], courtesy authors, and publishers IOP)

11.4 Charging of Ice Grains in Saturn E Ring (After Misra et al. [37])

11.4.1 Model of Complex Plasma Environment

Saturn's tenuous outer E ring lies at a distance 1.8×10^5 to 6.4×10^5 km from the planet. The ring's brightness peaks near the orbit of Saturn's moon Enceladus, where the thickness of the ring is least. The most unusual characteristic of the ring is the prominent presence of micron-sized particles with a distinctly blue color.

Misra et al. [37], have on the basis of numerous studies, including the conclusions of the spacecraft missions arrived at the following model of the complex plasma environment in the *Saturn's E* ring ($\approx 3 - 8\ R_s$). The plasma is illuminated by solar radiation and consists of cold (C) ($\approx$5 eV) and hot (H) ($\geq$50 eV) electrons, two different groups of singly charged ions viz. protons (*p*) and water group heavy ions ($w \equiv O^+$, OH^+, H_2O^+, etc.) and neutral species (mainly OH and H) along with pure and dirty spherical ice dust particles; dirty ice particles develop on account of inherent impurities of sodium and silicates and are characterized by an effective work function lower than that of pure ice [i.e., φ (=8.7 eV)]. The ice grains acquire a charge on account of the accretion of the cold electrons and the ionic species on the dust and the photoelectric emission due to Lyman-α radiation (see Sect. 11.3.2) and secondary electron emission (due to hot electron impact) from the surface of the ice grains.

The data (based on model calculations) available on the size distribution characterizing the ice grains in *E ring plasma,* given by Juhasz and Horanyi [15] (also provided in the review by Grasp et al. [8]) is summarized in Table 11.3. It is seen that there are three regions (*j*) of radii with different characteristic distributions. The parameters $a^2_{\mathrm{rms},j}$ and $a_{m,j}$ for the whole range of the radii of the particle size distribution may thus be evaluated.

Table 11.3 Size distribution of dust in Saturn E ring plasma

	Range of a (μm)	Max. n_d (m^{-3})	a_{rms} (μm)			a_m (μm)		
			$s = 4$	$s = 5.0$	$s = 6.0$	$s = 4$	$s = 5.0$	$s = 6.0$
1.	0.1 to 0.5	7.0	0.156	0.139	0.129	0.145	0.133	0.125
2.	0.5 to 1.0	1.2	0.655	0.633	0.614	0.643	0.622	0.605
3.	1.0 to 3.0	0.3	1.441	1.342	1.270	1.385	1.300	1.241
4.	0.1 to 3.0	8.5	0.392	0.369	0.352	0.259	0.243	0.232

After Misra et al. [37], courtesy authors and publishers Oxford University Press

11.4.2 Mathematical Modeling of Kinetics

11.4.2.1 Charging Kinetics

It has been shown in Chap. 7 that the charging of dust grains (with a size distribution) of the *j*th variety can be represented by

$$\begin{aligned}(\mathrm{d}V_{s,j}/\mathrm{d}t) = e\pi a_{m,j}\big[f_{\mathrm{ph,j}}\left(V_{s,j},T_d\right) + n_{\mathrm{eH}}f_{\mathrm{se,j}}\left(V_{s,j},T_{\mathrm{eH}}\right) + n_{\mathrm{ip}}f_{\mathrm{icp,j}}\left(V_{s,j},T_{\mathrm{ip}}\right) \\ + n_{\mathrm{iw}}f_{\mathrm{icw,j}}\left(V_{s,j},T_{\mathrm{iw}}\right) - n_{\mathrm{ec}}f_{\mathrm{ecC,j}}\left(V_{s,j},T_{\mathrm{ec}}\right)\big],\end{aligned} \tag{11.20}$$

where the suffix “se” indicates secondary electron emission.

It should be appreciated that although $V_{s,j}$ is independent of a, only in the steady state, this property has been used in writing (11.20). Since the aim is to analyze the steady state ($t \to \infty$) only, this simplification does not lead to any error.

Further analysis follows the methodology in Part I of the book.

11.4.2.2 Constancy of Neutral Plus Ionic Species

$$n_p + n_{\mathrm{ip}} = n_{p0} + n_{\mathrm{ip}0} = N_{t_1}(\mathrm{Constant}) \tag{11.21a}$$

$$n_w + n_{\mathrm{iw}} = n_{w0} + n_{\mathrm{iw}0} = N_{t_2}(\mathrm{Constant}) \tag{11.21b}$$

where n_{iy} refers to the density of protons (p) and water group (w) ions while n_y corresponds to the neutral densities of hydrogen atom and water group molecules. The subscript ‘0’ refers to the corresponding number densities in the absence of ice grains.

11.4.2.3 Electron Number Balance

$$(\mathrm{d}n_{\mathrm{ec}}/dt) = \left[\left(\beta_{\mathrm{ip}}n_p + \beta_{\mathrm{iw}}n_w\right) - n_{\mathrm{ec}}\left(\alpha_{\mathrm{rp}}n_{\mathrm{ip}} + \alpha_{\mathrm{rw}}n_{\mathrm{iw}}\right)\right] + \sum_j \pi a^2_{\mathrm{rms},j}n_j\left(f_{\mathrm{ph,j}} + n_{\mathrm{eH}}f_{\mathrm{seC,j}} - n_{\mathrm{ec}}f_{\mathrm{ecC,j}}\right), \tag{11.22}$$

where β_{iy} is the mean value of the coefficient of ionization due to the processes, responsible for the ionization of neutral species (viz. photo and electron impact ionizations), α_{ry} is the mean value of the recombination coefficient of electrons and respective ions, n_j is the number density of ice dust grains corresponding to jth variety of ice dust, and $n_d = \sum_j n_j$. For simplification of the analysis, the temperature dependence of β_{iy} and α_{ry} has been ignored.

11.4.2.4 Number Balance for Ionic Species

Proton Kinetics

$$\left(\mathrm{d}n_{\mathrm{ip}}/\mathrm{d}t\right) = \left(\beta_{\mathrm{ip}}n_p - \alpha_{\mathrm{rp}}n_{\mathrm{ec}}n_{\mathrm{ip}}\right) - \beta_c n_{\mathrm{ip}} - \sum_j \pi a^2_{\mathrm{rms},j}n_j n_{\mathrm{icp}}f_{\mathrm{icp,j}} \tag{11.23a}$$

Water Group Ion Kinetics

$$(\mathrm{d}n_{\mathrm{iw}}/\mathrm{d}t) = (\beta_{\mathrm{iw}}n_w - \alpha_{\mathrm{rw}}n_{\mathrm{ec}}n_{\mathrm{iw}}) + \beta_c n_{\mathrm{ip}} - \sum_j \pi a^2_{\mathrm{rms},j}n_j n_{\mathrm{icw}}f_{\mathrm{icw,j}}. \tag{11.23b}$$

where β_c is the mean value of the conversion rate [46] of protons to water group of ions on account of the charge exchange process.

Since charge neutrality may be derived from (11.21a), (11.21b), (11.22), and (11.23a), (11.23b) the equation of charge neutrality has been excluded from further analysis.

11.4.2.5 Energy Balance for Electrons

$$(3k/2)[\mathrm{d}(n_{\mathrm{eC}}T_{\mathrm{eC}})/\mathrm{d}t] = \left(\beta_{\mathrm{ip}}n_p + \beta_{\mathrm{iw}}n_w\right)\varepsilon_e - n_{\mathrm{eC}}\left(\alpha_{\mathrm{rp}}n_{\mathrm{ip}} + \alpha_{\mathrm{rw}}n_{\mathrm{iw}}\right)(3kT_{\mathrm{eC}}/2) + \sum_j {}^2_{\mathrm{rms},j}n_j\left(f_{\mathrm{ph}}\varepsilon_{\mathrm{ph,j}} + n_{\mathrm{eH}}f_{\mathrm{se,j}}\varepsilon_{\mathrm{se,j}} - n_{\mathrm{eC}}f_{\mathrm{ecC,j}}\varepsilon_{\mathrm{ecC,j}}\right), \tag{11.24}$$

where ε_e is the mean energy of electrons produced by ionization of neutral species, $\varepsilon_{\mathrm{ecC},\,j}$ is the mean energy (far away from the grain surface) of cold electrons, accreting on the surface of the ice grains, $\varepsilon_{\mathrm{ph},j}$ and $\varepsilon_{\mathrm{sc},j}$($\approx$2–3 eV) are the mean energies of the photo and secondary emitted electrons (far away from the surface of the ice grain), and T_{ec} is the temperature of the cold electrons.

11.4.2.6 Energy Balance for Ions

Proton Energy Balance

$$\begin{aligned}(3k_B/2)\left[\mathrm{d}\left(n_{\mathrm{ip}}T_{\mathrm{ip}}\right)/\mathrm{d}t\right] =& \beta_i n_p \varepsilon_{\mathrm{ip}} - \alpha_{\mathrm{rw}} n_{\mathrm{ec}} n_{\mathrm{ip}}\left(3k_B T_{\mathrm{ip}}/2\right)\\ &- \beta_c n_{\mathrm{ip}}\left(3k_B T_{\mathrm{ip}}/2\right) - \sum_j \pi a^2_{\mathrm{rms},j} n_j n_{\mathrm{icp}} f_{\mathrm{icp,j}} \varepsilon_{\mathrm{icp,j}}\end{aligned} \tag{11.25a}$$

Energy Balance Water Group Ions

$$\begin{aligned}(3k_B/2)[\mathrm{d}(n_{\mathrm{iw}}T_{\mathrm{iw}})/\mathrm{d}t] =& \beta_{\mathrm{iw}} n_w \varepsilon_{\mathrm{iw}} - \alpha_{\mathrm{rw}} n_{\mathrm{ec}} n_{\mathrm{iw}}(3k_B T_{\mathrm{iw}}/2)\\ &+ \beta_c n_{\mathrm{ip}}(3k_B/2)\left(T_{\mathrm{ip}} - T_w - T_p\right)\\ &- \sum_j \pi a^2_{\mathrm{rms},j} n_j n_{\mathrm{icw}} f_{\mathrm{icw,j}} \varepsilon_{\mathrm{icw,j}}\end{aligned} \tag{11.25b}$$

where $\varepsilon_{\mathrm{ip}}$ and $\varepsilon_{\mathrm{iw}}$ are the mean energies of protons and water group of ions, respectively, produced by ionization of corresponding neutral species, $\varepsilon_{\mathrm{icp},j}$ and $\varepsilon_{\mathrm{icw},j}$ are the mean energies (far away from the surface of the grain) of protons and water group ions accreting on the surface of the ice grain, T_{iy} and T_y are the temperatures of corresponding ions and neutral species, respectively.

11.4.2.7 Computational Highlights

As a simplification it is assumed that in the steady state the temperature of the neutral species remains unaffected. Further the temperature of the dust particles may be estimated by equating the solar radiation absorbed per unit time by the particle to the power lost by thermal radiation; other electronic processes do not play a significant role. For computational purposes a typical value of $T_d = 130$ K (*for ice grains*), corresponding to $(\alpha_a/\varepsilon) \approx 1.17$ has been assumed (α_a is the absorptivity and ε is emissivity of the dust); the results depend only slightly on T_d.

Further the mean energy of electrons and ions generated due to ionization of neutral species may be obtained by imposing the initial conditions for dust-free plasma in the energy balance equations; thus one has

$$\varepsilon_e = \left[n_{\mathrm{eC0}}\left[\alpha_{\mathrm{rp}} n_{\mathrm{ip0}} + \alpha_{\mathrm{rw}} n_{\mathrm{iw0}}\right](3k_B T_{\mathrm{eC0}}/2)\right]/\left(\beta_{\mathrm{ip}} n_{p0} + \beta_{\mathrm{iw}} n_{w0}\right) \tag{11.26}$$

$$\varepsilon_{\mathrm{ip}} = \left[\alpha_{\mathrm{rw}} n_{\mathrm{eC0}} n_{\mathrm{ip0}}\left(3k_B T_{\mathrm{ip0}}/2\right) + \beta_c n_{\mathrm{ip0}}\left(3k_B T_{\mathrm{ip0}}/2\right)\right]/\left(\beta_{\mathrm{ip}} n_{p0}\right) \tag{11.27}$$

$$\varepsilon_{\mathrm{iw}} = \left[\alpha_{\mathrm{rw}} n_{\mathrm{eC0}} n_{\mathrm{iw0}}(3k_B T_{\mathrm{iw0}}/2) - \beta_c n_{\mathrm{ip0}}(3k_B/2)\left(T_{\mathrm{ip0}} + T_w - T_p\right)\right]/(\beta_{\mathrm{iw}} n_{w0}) \tag{11.28}$$

where the suffix zero refers to plasma parameters in the absence of ice dust particles.

For a chosen set of plasma parameters, one can obtain the steady-state potential on ice grains ($V_{s,j}$) and other parameters viz. n_{ec}, n_{iy}, n_y, T_{ec}, and T_{iy} by simultaneous numerical integration of (11.3a), (11.3b), (11.5), (11.6a), (11.6b), (11.6c), (11.7), (11.9), and (11.10) along with (11.11a), (11.11b), (11.11c), and appropriate expressions for other relevant parameters as $t \to \infty$; the steady state is indeed independent of the initial conditions.

Expressions for the parameters, occurring in these equations can be seen in Part I of the book.

Further for $V_{\mathrm{sj}} < 0$

$$f_{\mathrm{se,j}}\left(V_{s,j}\right) = 3.7\delta_M J_{0\mathrm{eH}} \exp\left(V_{5,j}\alpha_{\mathrm{eH}}\right) F_5(E_M/4kT_{\mathrm{eH}}) \tag{11.29a}$$

and

$$f_{\mathrm{ph,j}}\left(V_{s,j}\right) = Q_{\mathrm{abs}} Y(v) \Lambda_s(v) \tag{11.30a}$$

where

$$F_5(\eta) = \eta^2 \int_0^\infty u^5 \exp[-(\eta u^2 + u)]\mathrm{d}u.$$

and for $V_{s,\ j} \geq 0$:

$$f_{\mathrm{se,j}}\left(V_{s,j}\right) = 3.7\delta_M J_{0\mathrm{eH}}\left(1 + V_{s,j}\alpha_{\mathrm{ec}}\right) \exp\left[V_{s,j}(\alpha_{\mathrm{eH}} - \alpha_{\mathrm{ec}})\right] f_{5,B}(E_M/4kT_{\mathrm{eH}}) \tag{11.29b}$$

$$f_{\mathrm{ph,j}}\left(V_{s,j}\right) = Q_{\mathrm{abs}} Y(v) \Lambda_s(v) \left[\Psi\left(\xi, V_{s,j}\alpha_d\right)/\Phi\left(\xi_j\right)\right] \tag{11.30b}$$

where

$$F_{5,B}(\eta) = \eta^2 \int_B^\infty u^5 \exp\left[-(\eta u^2 + u)\right] \mathrm{d}u, B_j = -\left(4kT_{\mathrm{eH}} V_{s,j}\alpha_{\mathrm{eH}}/E_M\right),$$

for $V_{s,j} < 0$:

Equation (11.29a) and (11.29b) have been given by Mayer-Vernet [34].

The flux $\Lambda_s(v)$ of the Lyman-α photons on the *Saturn* atmosphere is given by

$$\Lambda_s(v) = (D_e/D_s)^2 \Lambda_e(v)$$

where D_e and D_s are the distances of *Earth* and *Saturn* from *Sun* and, $\Lambda_e(v)$ and $\Lambda_s(v)$ denotes the flux on the *Earth and Saturn* atmosphere, respectively.

11.4.3 Numerical Results and Discussion

The present study considers the plasma regions in the *E belt* near the orbits of icy satellites (*Mimas* (3.08 R_s), *Enceladus* (3.95 R_s), *Tethys* (4.89 R_s), *Diane* (6.26 R_s) *and Rhea* (8.75 R_s)), which consist of cold and hot electrons, hydrogen and water group neutral/ionic species, and ice grains characterized by the MRN power law of the size distribution (see Table 11.3). In view of the recent understanding of ice grain composition in the *E ring plasma* environment viz. 10 % *Sodium rich* impure ice grains along with remaining 90 % pure ice particles [17, 18, 39, 40], this analysis considers both pure and dirty ice grains.

For computations the relevant plasma parameters (listed in Table 11.4, have been picked from available literature and data based on *Cassini/Voyager/Pioneer* space missions and HST observations [16, 43, 44, 46–48, 50]. The number densities of hydrogen and water group neutrals and corresponding ionic species have been taken from the *E ring* plasma environment model based on HST observation, proposed by Richardson et al. [47]. The number density of the cold and hot electrons has been picked from an elegant paper by Richardson [44]. The coefficient of ionization and recombination has been extracted from one of the basic papers by Richardson et al. [46].

The size distribution of ice grains following MRN power law and corresponding number densities of ice grains (as listed in Table 11.3) have been used for computational purpose. The incident EUV radiation, causing photoelectric emission from ice can be approximated (See Sect. 11.3.2) as photon flux $\Lambda_e \approx 3 \times 10^{11}/\text{cm}^2$, corresponding to Lyman-α radiation, just outside the earth's atmosphere (see Sect. 11.3.2). The photoelectric yield is taken to be 0.1 [8]. As discussed earlier the *Mie coefficient* (Q_{abs}) is almost unity for the whole range of size of the dust particles. The data corresponding to secondary electron emission viz yield ($\delta_m = 1.5$), optimum impact energy of electrons ($E_m = 400$ eV), and characteristic mean energy of the secondary emitted electrons ($E_{\text{se}} \approx 2.5$ eV) has been collected from a recent review by Graps et al. [8]. The dataset, used for computational purpose to evaluate the potential of ice grains corresponding to the orbits of *Saturnian moons* has been tabulated in Table 11.4.

Figure 11.3a illustrates the dependence of the surface potential (V_s) of pure and dirty ice grains on their number density ($\gamma = n_d/n_{d,max}$) in the plasma environment of the orbits of *E ring Saturnian moons*. Increase in the photoelectric yield significantly leads to more positive values of dust potential as displayed in Fig. 11.3b. Due to strong photoemission current from positively charged grains [$\xi = (h\nu - \varphi)/k_BT_d$], the dust potential is significantly enhanced with the lowering in the work function of dirty ice, as displayed in case of *Rhea environment* in Fig. 11.3c.

Table 11.4 Characteristic plasma parameters and evaluated potential over ice grains in Saturn E ring plasma environment

Orbit of satellite	Mimas (3.08 R_s)	Enceladus (3.95 R_s)	Tethys (4.89 R_s)	Dione (6.26 R_s)	Rhea (8.75 R_s)
Plasma parameters					
n_{ec0} (cm^{-3})	111.80	99.80	44.03	22.67	3.893
n_{eh0} (cm^{-3})	0.20	0.20	0.40	0.40	0.30
n_{ip0} (cm^{-3})	8.0	20	3.59	1.53	1.41
n_{iw0} (cm^{-3})	104	80	40.84	21.54	2.78
n_{p0} (cm^{-3})	8.23	45.82	69.44	24.53	24.59
n_{w0} (cm^{-3})	358.23	620.22	540.23	105.48	42.52
T_{ec0} (eV)	1.0	3.0	4.0	6.0	14
T_{eh0} (eV)	80	100	120	155	230
$T_{ip0} \approx T_{p0}$ (eV)	8	11	14	16.75	21.5
$T_{iw0} \approx T_{w0}$ (eV)	13	40	80	109	210
a (μm)	0.1–3	0.1–3	0.1–3	0.1–3	0.1–3
Observed Potential (Volts) [8]	−2 to −1	−2 to −1	−2 to −1	−7 to +5	≈+3.0
Overall E ring observed Potential (Volts)	Lies between −7 to +5 (in a good agreement with estimated values, as indicated in figures)				
Estimated potential, based on data of Grasp et al. [8] (Volts) for $s = 4$, $Y = 0.3$, $\varphi_2 = 7$ eV and $\gamma = 0.1$–1.0	$V_{s1} \approx V_{s2} \approx -1.5$ to -0.9	$V_{s1} \approx V_{s2} \approx -4.5$ to -2.2	$V_{s1} \approx V_{s2} \approx -3.4$ to -1.4	$V_{s1} \approx -1.2$ to 0.2 $V_{s2} \approx -1.2$ to 0.3	$V_{s1} \approx 0.5$ to 1.2 $V_{s2} \approx 0.7$ to 2.1

After Misra et al. [37], courtesy authors and publishers Oxford University Press

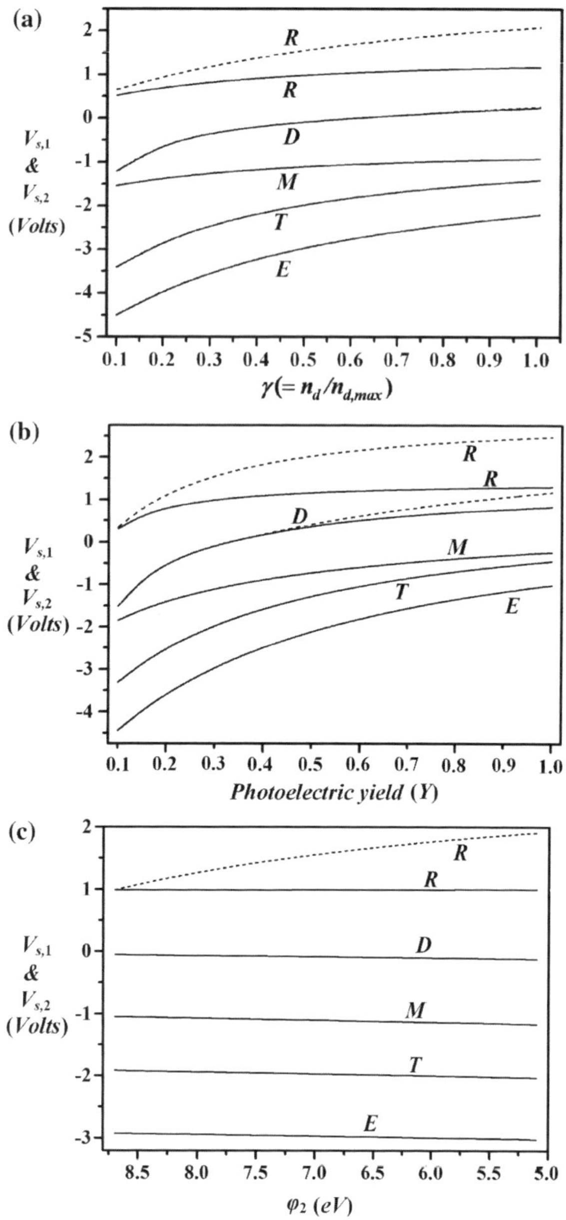

Fig. 11.3 **a** Dependence of ice dust surface potential (V_s) on number density (γ) for the parameters $s = 4$, $Y = 0.3$, $\varphi_2 = 7$ eV, and tabulated plasma parameters in Table 11.4, in the orbit of Saturnian satellite in E ring environment; the labels *M*, *E*, *T*, *D*, and *R* on the curve correspond Mimas, Enceladus, Tethys, Dione, and Rhea orbits, respectively. The *solid* and *broken curves* correspond to pure (V_{s1}) and dirty (V_{s2}) ice, respectively. **b** Dependence of ice dust surface potential (V_s) on parameter n_p; the parameters and labels on the curve are the same as in Fig. 11.3a. **c** Dependence of ice dust surface potential (V_s) on dirty ice work function φ_2; the parameters and labels on the curve are the same as in Fig. 11.3a (after Misra et al. [37], courtesy authors and publishers Oxford University Press)

11.5 Kinetics of Interplanetary Medium (After Misra and Mishra [36])

11.5.1 The Interplanetary Medium

The interplanetary medium comprises of highly ionized, high temperature solar wind [usually described as a continuous flow of electrons, ions (90 % protons and 10 % α particles) and neutral atoms] from the sun and interplanetary dust particles [IDP], irradiated by the sun. The speed of the solar wind is 300–800 km/s and the average plasma density is 5×10^6 m^{-3} at a distance of 1 A.U. from the sun. The solar wind carries a magnetic field of 3 nT, which is too weak to affect the kinetics. The medium also has a suspension of fine (mm to nm) dust particles of silicates and graphite; in literature, the particles are commonly referred to as interplanetary dust particles (IDP).

A number of investigations on the charging of IDP, based on the charge balance on the particles have been published. Recently, Misra and Mishra [36] have made a careful analysis of the kinetics of the interplanetary medium taking into account:

(i) Emission of photoelectrons due to solar irradiation and secondary electron emission (caused by protons in the solar wind) from the surface of the IDP.
(ii) Electron generation and annihilation in the gaseous phase.
(iii) Electron/ion accretion on the surface of IDPs.
(iv) Energy exchange in the three processes, enumerated above
(v) Fluctuation of charge on the IDPs.
(vi) Size distribution of IDPs
(vii) Uniform potential theory (Chap. 8) and
(viii) A mean value of Q_{abs} for the specific size distribution.

The investigation proceeds on the basis of the energy balance and number balance of the constituents of the medium and the charge balance on IDPs.

The size distribution may be represented by [12, 26]

$$f(a)da = Aa^{-s}da, \tag{11.31}$$

where $f(a)da$ is the number of particles per unit volume having radii between a and $a + da$, $s = 3$–5 [12, 26] and A is normalizing constant, which may be obtained by integrating $f(a)$, within the limits of a (*say* a_1 and a_2).

11.5.2 Analysis

Uniform Electric Potential

As per Chap. 7, the uniform electric potential V_s on the surface of the dust particle is given by

$$\begin{aligned}(\mathrm{d}V_{sj}/\mathrm{d}t) = {} & e\pi a_{mj}\left[f_{\mathrm{phj}}\left(V_{sj}, T_e\right) + (1-\eta_e)n_e f_{\mathrm{se}}(\upsilon_s, T_e) + n_e f_{th}\left(V_{sj}, T_d\right)\right. \\ & \left. + \eta_{\mathrm{ij}} n_{\mathrm{ij}} f_{\mathrm{ic}}\left(V_{sj}, T_i\right) - \eta_{ej} n_e f_{\mathrm{ec}}\left(V_{sj}, T_e\right)\right]\end{aligned} \tag{11.32}$$

where η_e and η_i denote the sticking coefficient of electrons and ions on the surface of dust; the suffix se refers to secondary emission and other symbols conform to Chap. 7.

11.5.3 Constancy of Neutral Plus Ionic Species

Since an atom gives rise to an ion on ionization and an ion converts to an atom after recombination with an electron the sum of number densities of ions and neutral atoms is constant; hence

$$n_0 + n_i = n_{00} + n_{i0} = N_t \text{ (Constant)}, \tag{11.33}$$

where the symbols are as in Chap. 7.

11.5.3.1 Electron and Ion Number Balance

The number balance of electrons and ions can be expressed as

$$(\mathrm{d}n_e/\mathrm{d}t) = (\beta_i n_0 - \alpha_r n_e n_i) + \pi a_{\mathrm{rms}}^2 n_d\left[f_{\mathrm{ph}} + (1-\eta_e)n_e f_{\mathrm{se}} - \eta_e n_e f_{\mathrm{ec}}\right] \tag{11.34}$$

and

$$(\mathrm{d}n_i/\mathrm{d}t) = (\beta_i n_0 - \alpha_r n_e n_i) - \pi a_{\mathrm{rms}}^2 n_d n_i f_{\mathrm{ic}}, \tag{11.35}$$

where β_i is the mean value of the coefficient of ionization due to the processes, responsible for the ionization of neutral species (viz. photo and electron impact ionizations), $\alpha_r(T_e) = \alpha_{r0} T_e^{-\theta}\,\mathrm{cm}^3/\mathrm{s}$ is the recombination coefficient of electrons and ions [10], and α_{r0} and θ are constants.

The value of the ionization coefficient may be determined from (11.34) by putting $n_d = 0$ and $\mathrm{d}/\mathrm{d}t = 0$, corresponding to the steady-state plasma in the absence of dust. Thus

$$\beta_i n_{00} = \alpha_r(T_{e0}) n_{e0} n_{i0} = \alpha_r(T_{e0}) n_{e0}^2. \tag{11.36}$$

It may be noticed that the electrical neutrality of the complex plasma is inherent in the number balance equations and can be easily derived from (11.32), (11.33), (11.34), and (11.35).

11.5.4 Energy Balance

The energy balance of solar wind plasma, with suspended dust particles may be expressed as

$$\begin{aligned}
&\frac{\mathrm{d}}{\mathrm{d}t}\left(\frac{3}{2}k_B(n_e+n_i+n_0)T\right) \\
&\quad=\left[\left[\beta_i n_0(\varepsilon_i+\varepsilon_e)+\alpha_r n_e n_i I_p-(3k_BT/2)\beta_i n_0\right]\right. \\
&\qquad+\pi a_{\mathrm{rms}}^2 n_d\left[f_{\mathrm{ph}}\varepsilon_{\mathrm{ph}}+(1-\eta_e)n_e f_{\mathrm{se}}\varepsilon_{\mathrm{se}}-\eta_e n_e f_{\mathrm{ec}}\varepsilon_{\mathrm{ec}}-\eta_i n_i f_{\mathrm{ic}}\varepsilon_{\mathrm{ic}}\right] \\
&\qquad\left.+(3k_BT_d/2)\left(\pi a_{\mathrm{rms}}^2\right)n_d\eta_i n_i f_{\mathrm{ic}}\right],
\end{aligned} \tag{11.37}$$

where ε_e (ε_i) is the mean energy of electrons (ions) produced by ionization of neutral species, I_p is the ionization potential of the neutral atoms, $\varepsilon_{\mathrm{ec}}(\varepsilon_{\mathrm{ic}})$ is the mean energy (far away from the grain surface) of the electrons (ions), accreting on the surface of the grains, $\varepsilon_{\mathrm{ph}}$ and $\varepsilon_{\mathrm{se}}$ correspond to the mean energy of the photoelectrons and secondary emitted electrons (far away from the surface of the grain), and T is the mean temperature of the plasma species. Further, the mean energy of electrons and ions produced due to ionization may be obtained by imposing the initial conditions for dust-free plasma $n_d = 0$ in the energy balance equations. The radiative and convective energy exchange between the dust and the gaseous component has been neglected.

11.5.5 Dust Temperature

Due to nonavailability of reliable values of absorptivity α_0 and emissions ε of the surface of the particles, a typical value of the dust particle temperature i.e., $T_d \approx 250$ K may be chosen.

11.5.6 Methodology

For a chosen set of initial ($t = 0$) values of parameters corresponding to dust-free plasma, one can obtain the steady-state potential on grains (V_s) and other parameters viz. n_e, n_i, n_0, and T by simultaneous numerical integration of (11.32–11.37) along with appropriate expressions for the electron/ion accretion and electron emission; the steady state ($t \to \infty$) is indeed independent of the initial conditions, since as $t \to \infty$, ($\mathrm{d}/\mathrm{d}t = 0$) the set of differential equations reduces to a set of algebraic equations, independent of the values at $t = 0$.

11.5.7 Photoelectric Efficiency

The photoelectric efficiency can be represented [38] as $\chi(\varepsilon_v) = y_o[1 - (b_o/12.5)]$, $[10^{0.6(\varepsilon_v - 12.5)}]$ which is applicable to the spectral regime $\varepsilon_v \leq 12.5$ eV; here (y_o, b_o) are constants and valued as (0.5, 8 eV) and (0.05, 7 eV) for *silicate* and *graphite* particles respectively. The *Mie scattering coefficient* (Q_a) for the *graphite and silicate* particles may be taken from the paper of Draine and Lee [6] in which the spectral dependence of Q_a for different size grains is illustrated graphically. This figure can be enlarged and the values of $Q_a(a, \lambda)$ read at intervals of 10 nm in wavelength for particles of radii 3, 10, 30, 100, 300, and 1000 nm. The mean spectral absorption efficiency $\overline{Q_a(a)}$ was evaluated by the relation

$$\overline{Q_a(a)} = \left(\sum_{\lambda_2}^{\lambda_1} Q_a(a, \lambda) F(E_v) \middle/ \sum_{\lambda_2}^{\lambda_1} F(E_v) \right), \tag{11.38a}$$

where $E_v = 300\ (ch/e\lambda)$ is the energy of a photon of wavelength λ in eV.

The mean absorption efficiency Q_a^m corresponding to a size distribution of particles and spectral irradiance distribution of incident radiation is given by

$$Q_a^m = \left[\sum \overline{Q_a(a)} f(a)\right] \Big/ \left[\sum f(a)\right]. \tag{11.38b}$$

Using the curves for $Q_a(a,\lambda)$ [6] and expressions for $F(E_v)$ and $f(a)$, the mean absorption efficiency Q_a^m can be evaluated. The mean values $Q_a^m = 0.198$ and 0.351 corresponding to *silicate* and *graphite* particles, has been used for the computations;

11.5.8 Secondary Electron Emission

The rate of secondary electron emission [34] is given by (11.29a) and (11.29b). Thus

For $\upsilon_s < 0$

$$f_{\text{se}}(\upsilon_s) = \left(n_{\text{se}}/\pi a^2\right) = 3.7\delta_M J_{0e,i}(1 - p_s\upsilon_s)\exp[-(p_e - p_s)\upsilon_s]F_{5,B}(E_M/4k_BT_e), \tag{11.39a}$$

where

$$J_{0e,i} = \left(8k_BT_{e,i}/m_{e,i}\pi\right)^{1/2},$$

$F_{5,B}(\mu) = \mu^2 \int_B^\infty u^5 \exp[-(\mu u^2 + u)]\mathrm{d}u$, $B = (4k_BT_d v_s/E_M)$, $p_s = (T_d/T_{\text{se}})$, $p_e = (T_d/T_e)$, and T_{se} corresponds to the temperature (mean energy) of the secondary electrons, emitted from the dust particles.

For $\upsilon_s \geq 0$

$$f_{se}(\upsilon_s) = \left((1-\eta)n_{se}/\pi a^2\right) = (1-\eta)3.7\delta_M J_{0e,i}\exp(-p_e\upsilon_s)F_5(E_M/4k_BT_e), \tag{11.39b}$$

where

$$F_5(\mu) = \mu^2 \int_0^\infty u^5 \exp\left[-\left(\mu u^2 + u\right)\right] du.$$

The data corresponding to secondary electron emission viz. (i) yield $\delta_m = 2.4$ (*Silicate*) and 1.43 (*Graphite*), (ii) optimum impact energy ($E_m = 400$ eV) of electrons [7, 20], and (iii) characteristic mean energy of the secondary emitted electrons ($\varepsilon_{se} \approx 2.5$ eV, [58]) has been chosen for the computations.

11.5.8.1 Accretion of Electrons/Ions

Well-known expressions for the electron/ion accretion current over the dust particles in a complex plasma based on the OML approach have been used for computations. Hence
For $\upsilon_s < 0$

$$\begin{aligned}
f_{ec}(\upsilon_s) &= \left(n_{ec}/\pi a^2\right) = J_{0e}(1 - p_e\upsilon_s), \\
p_e &= T_d/T_e, \\
f_{ic}(\upsilon_s) &= \left(n_{ic}/\pi a^2\right) = J_{0i}\exp(p_i\upsilon_s), \\
\varepsilon_{ec}(\upsilon_s) &= \left(\frac{2 - p_e\upsilon_s}{1 - p_e\upsilon_s}\right)kT_e \text{ and } \varepsilon_{ic}(\upsilon_s) = k_BT_i(2 - p_i\upsilon_s) \text{ with} \\
p_i &= (T_d/T_i).
\end{aligned}$$

For $\upsilon_s > 0$

$$\begin{aligned}
f_{ec}(\upsilon_s) &= \left(n_{ec}/\pi a^2\right) = J_{0e}\exp(-p_e\upsilon_s), \\
f_{ic}(\upsilon_s) &= \left(n_{ic}/\pi a^2\right) = J_{0i}(1 + p_i\upsilon_s), \\
\varepsilon_{ec}(\upsilon_s) &= k_BT_e(2 + p_e\upsilon_s)
\end{aligned}$$

and

$$\varepsilon_{ic}(\upsilon_s) = \left(\frac{2 + p_i\upsilon_s}{1 + p_i\upsilon_s}\right)kT_i.$$

11.5.9 Numerical Results and Discussion

For a numerical appreciation of the kinetics of the charging of IDPs in the interplanetary space plasma, the following standard set of parameters have been used for computations; the effect of various parameters on the dust potential has been studied by varying one parameter and keeping others the same.

$n_{e0} = n_{i0} = n_{00} = 5\ \text{cm}^{-3}$, $T_s \approx 5800$ K, $T_0 = 2 \times 10^5$ K, $T_d = 250$ K, $s = 3$, $a_1 = 10$ nm, $a_2 = 1000$ nm, $\eta_e = 0.5$, $\eta_i = 1$, $E_M = 400$ eV, $\kappa = 1.2$, $m_i \approx m_p = 1.67 \times 10^{-24}$ g, $\alpha_{r0} = 10^{-7}\ \text{cm}^3$/s, and total dust mass density $M_d = 10^{-23}$ g/cm^3 corresponding to $n_d = 1 \times 10^{-7}\ \text{cm}^{-3}$and $a_{\text{rms}} \approx 30$ nm, respectively; $r_s(\approx 6.96 \times 10^{10}$ cm) and $r_d(\approx 1.45 \times 10^{13}$ cm). *For Silicates:* $\varphi = 3.8$ eV, $\delta_m = 2.4$, $(y_o, b_o) = (0.5, 8$ eV), and $Q_a^m = 0.198$. *For Graphite:* $\varphi = 4.8$ eV, $\delta_m = 1.43$, $(y_o, b_o) = (0.05, 7$ eV), and $Q_a^m = 0.35$.

Using the initial set of parameters, the steady-state dust surface potential (V_s) can be obtained by simulations solution of the set of differential (11.32–11.37) as $t \to \infty$.

The set of Fig. 11.4a–c illustrates the dependence of the steady-state surface potential (V_s) of the (i) silicate and (ii) graphite particles on the number density (n_d) in the interplanetary space plasma. It is seen that V_s increases with increasing n_d; this can be understood on the basis of increasing photoemission of electrons from the dust surface with increasing n_d. Figure 11.4a reflects that the dust potential increases with decreasing work function (φ) of the dust material. The dependence of the surface potential of *silicate* and *graphite* particles on the size distribution parameter s has been illustrated in Fig. 11.4b. It is seen that V_s decreases with increasing s; this is because of decrease in a_{rms} with increasing s which leads to a smaller cross section available for the photoemission. Figure 11.4c illustrates the effect of sticking coefficient (η_e) on the surface potential of IDPs of *silicate* particles. The figure indicates that the surface potential increases with decrease in η_e; this can be explained by the fact that for a low value of η_e the negative charge transfer to the dust due to electron accretion will be small.

11.6 Temperature of Interstellar Warm Ionized Medium (After Misra et al. [37])

11.6.1 Interstellar Warm Ionized Medium

Warm ionized medium (WIM) regarded as diffused ionized gas region extends up to 1 kpc from the mid galactic plane [42]; this region usually consists of neutral and ionized H (about 99 %) gas along with small proportions of a few heavy gaseous elements (like C, N, O, etc., about 1 %). High energy radiation of the order of 10 nm wavelength from O-stars causes photo ionization of *H atoms* (ionization potential 13.6 eV) and subsequent heating of the gas through the

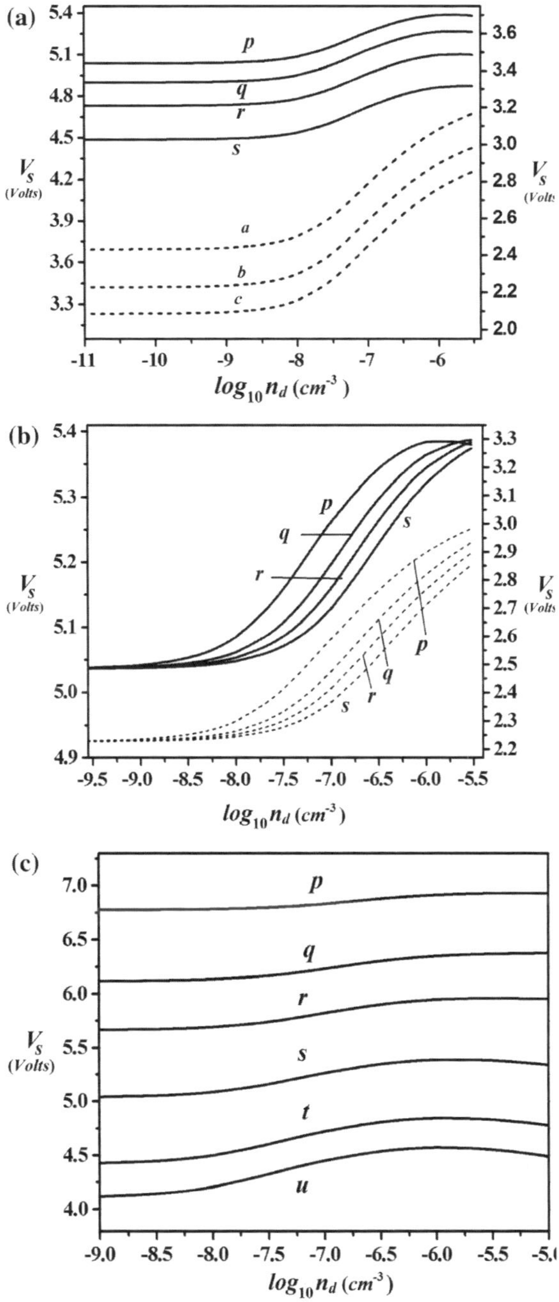

Fig. 11.4 **a** Surface potential (V_s, *solid lines* (silicates, L.H.S scale) and *broken lines* (graphite R.H.S scale)) of the dust particles as a function of the number density of dust particles n_d for different values of φ for the standard set of parameters, as stated in the text. The labels p, q, r, and s refer to $\varphi = 3.8$, 4.0, 4.2, and 4.5 eV while a, b, and c refer to $\varphi =$ 4.5, 4.8, and 5.0 eV. **b** Surface potential (V_s, *solid lines* (silicates, *L.H.S scale*) and *broken lines* (graphite *R.H.S scale*)) of the dust particles as a function of the number density of dust particles n_d for different values of s and the standard set of parameters, as stated in the text. The labels p, q, r, and s refer to $s = 3$, 3.5, 4.0, and 5.0. **c** Surface potential (V_s) of the silicate dust particles as a function of the number density of dust particles n_d for different values of η_e and the standard set of parameters, as stated in the text. The labels p, q, r, s, t, and u refer to $\eta_e = 0.1$, 0.2, 0.3, 0.5, 0.8, and 1.0 (after [36], courtesy authors and publishers Oxford University Press)

energy gained by the free electrons; the electron–ion plasma recombination, accounts for the maintenance of WIM region plasma [59]. On account of weak abundance of heavy elements (i.e., C, N, O, etc.) their contribution to WIM plasma heating is not significant, in comparison to that due to *H* atoms and may be ignored. Photoemission of electrons from the dust particles (specifically *graphite*) is reckoned as a major source of WIM environment heating on account of a large density of high energy photoelectrons. Cooling in WIM region occurs on account of electron-neutral collisional excitation of forbidden/semi-forbidden transitions, followed by a radiative decay; high energy neutral species also transfer their kinetic energy to excite the heavy element atoms (specifically C *and* O, at ground state) to radiate in the far IR region. In the context of heating, other plasma energy transfer mechanisms are insignificant in the energy balance of the WIM plasma environment.

Mishra et al. [35] investigated the WIM kinetics; the main features of the analysis are.

(i) Number and energy balance of ions/electrons and neutral species.
(ii) Use of appropriate expressions for the rate of emission and mean energy of emitted photoelectrons from a positively charged dust particle, considering the parametric dependence of the absorption efficiency, continuous spectral irradiance distribution of radiation and photoelectric efficiency.
(iii) Uniform potential approach in the context of the size distribution of dust.
(iv) The (heating/cooling) kinetic processes, enumerated before.

The main aim of the analysis was to explain the observed/measured/estimated consensus values of WIM temperature [5] $T = 8000$ K, the electric potential on the surface of graphite dust particles V_s(=2–4 V), and the electron density $n_{e0} = 0.01\ \mathrm{cm}^{-3}$ and neutral *H atom* density $n_{n0} = 0.1\ \mathrm{cm}^{-3}$on the basis of plausible values of the initial parameters.

11.6.2 Analysis

Charging Kinetics

From Chap. 7 one has, in the steady state

$$(\mathrm{d}V_s/\mathrm{d}t) = e\pi a_m\left[4f_{\mathrm{ph}}(V_s, T_d) + n_i f_{\mathrm{ic}}(V_s, T_i) - n_e f_{\mathrm{ec}}(V_s, T_e)\right], \tag{11.40}$$

where the symbols have usual meanings.

This equation assumes Q_{abs} to be independent of *a,* which is true of large particles $(2\pi a/\lambda) > 10$ or $a/\lambda > 1.6$ and when a mean value can be used.

Conservation of Neutral Plus Ionic Species

The conservation of atoms/ionic species implies

$$n_n + n_i = n_{n0} + n_{i0} = n_t(\mathrm{Constant}) \tag{11.41}$$

where n_n/n_i corresponds to the neutral/ion density and the subscript '0' refers to the corresponding number densities in the absence of the grains.

Electron Number Balance

The number balance of electrons may be expressed as

$$(\mathrm{d}n_e/\mathrm{d}t) = (\beta_i n_n - \alpha_r n_e n_i) + \pi a_{\mathrm{rms}}^2 n_d \left(4f_{\mathrm{ph}} - n_e f_{\mathrm{ec}}\right), \tag{11.42a}$$

where β_i is the mean value of the coefficient of ionization due to the processes, responsible for the ionization of neutral gaseous species (*specifically* photoionizations), $\alpha_r[\propto T^{-0.7}]$ is the recombination coefficient [57] of electrons and ions and T is the mean temperature of the species (electrons/ions/neutral atoms). For simplification of the analysis, the temperature dependence of β_i and α_r has been ignored.

Number Balance for Ionic Species

The number balance of ions may be represented by

$$(\mathrm{d}n_i/\mathrm{d}t) = (\beta_i n_n - \alpha_r n_e n_i) - \pi a_{\mathrm{rms}}^2 n_d n_i f_{\mathrm{ic}} \tag{11.42b}$$

Energy Balance

The important source for the energy gain in *warm ionized medium* plasma is the ionization of neutral atoms and photoemission from dust particles; all gaseous species may, following earlier investigations be assumed to have the same temperature T. Thus the energy balance in WIM may be expressed as

$$\begin{aligned}
&\frac{\mathrm{d}}{\mathrm{d}t}\left(\frac{3}{2}k_B(n_e + n_i + n_n)T\right) \\
&= \left[\beta_i n_n(\varepsilon_i + \varepsilon_e) + \alpha_r n_e n_i I_p - \beta n_n\left(\frac{3}{2}k_B T\right)\right] \\
&\quad - \alpha_{\mathrm{ex}} n_e n_n\left(f_{\mathrm{ex}}\frac{3}{2}k_B T\right) - f_{\mathrm{de}}\alpha_{\mathrm{de}} n_n^2 \exp[-92/T] \\
&\quad + \left[{}^{2}_{\mathrm{rms}} n_d[4f_{\mathrm{ph}}\varepsilon_{\mathrm{ph}} - f_{\mathrm{ec}} n_e \varepsilon_{\mathrm{ec}} - f_{\mathrm{ic}} n_i \varepsilon_{\mathrm{ic}}] + \left(\frac{3}{2}k_B T_d\right) n_d n_i f_{\mathrm{ic}}\right],
\end{aligned} \tag{11.43}$$

where $\varepsilon_e/\varepsilon_i$ is the mean energy of electrons/ions produced by ionization of neutral species, I_p is the ionization potential of neutral atom, f_{ex} corresponds to the fractional energy transfer to the neutrals by electrons which depends on energy difference between transition levels, f_{de} is the depletion factor of C with respect to neutrals, $\varepsilon_{\mathrm{ec}}/\varepsilon_{\mathrm{ic}}$ are the mean energy (far away from the grain surface) of the electrons/ions, accreting on the surface of the grains, $\varepsilon_{\mathrm{ph}}$ corresponds to the mean energy of the photoemitted electrons (far away from the surface of the grain), and T is the mean temperature of the plasma species.

The first term on the right-hand side in (11.43) represents the net power gained per unit volume by the plasma species on account of ionization and recombination processes. The second term in the right-hand side represents the energy loss per unit volume in the process of neutral atomic excitation, specifically by electron impact (due to higher mobility than ions). The next term is the power loss due to collisions of C atoms with other neutral atoms resulting in loss of power by radiation from C atoms [2]. The last term represents the net gain in the mean energy of plasma species on account of emission and accretion of electron/ionic species. The power loss due to elastic collisions between gaseous plasma constituents (electron/ion/neutral) have been ignored because this does not lead to any net gain/loss of energy by the system. The term $[(3/2)k_BT_d]n_dn_if_{ic}$ corresponds to the power carried by neutral atoms, produced from accretion of ions on the dust particles.

Temperature of Dust Particles
On account of the nonavailability of reliable values of *the parameters* the temperature of dust $T_d = 20$ K, corresponding to an idealized environment [14] has been used.

Computation Strategy
For a chosen set of initial parameters, one can obtain the steady state potential on grains (V_s) and other parameters viz. n_e, n_i, n_n, and T by simultaneous numerical integration of (11.40), (11.42a), (11.42b) and (11.43) along with appropriate expressions for other relevant parameters as $t \to \infty$; the steady state is indeed independent of the initial conditions, since as $t \to \infty$, ($\mathrm{d}/\mathrm{d}t = 0$) the set of differential equations reduces to a set of algebraic equations, independent of the values at $t = 0$. However the differential form of equations is retained, because these are amenable to numerical solution.

Photoemission from Dust Particles
Stellar Radiation
Based on available data from measurement/observations, Draine [5] has formulated an empirical fit for $F(E_v)dE_v$, the number of photons with energy between E_v *and* $E_v + dE_v$, incident on a unit area per unit time per steradian, which is in a good agreement with the data over the range 5.0 eV $\le E_v \le$ 13.6 eV. The empirical fit is

$$F(E_v) = (1.658 \times 10^6 E_v - 2.152 \times 10^5 E_v^2 + 6.919 \times 10^3 E_v^3)\, dE_v, \tag{11.44}$$

where E_v is expressed in eV.

Hence the number of photons with energy between E_v *and* $E_v + \mathrm{d}E_v$, incident on the surface $4\pi a^2$of a particle of radius a, per unit time is

$$\mathrm{d}n_{\mathrm{inc}}(E_v) = 4\pi a^2 \times 2\pi F(E_v)\mathrm{d}E_v = 8\pi^2 a^2 F(E_v)\mathrm{d}E_v, \tag{11.45}$$

where $F(E_v)$ is given by (11.44); it may be remembered that interstellar radiation is isotropic.

Mean Absorption Efficiency $\langle Q_a \rangle$
Please see Sect. 11.5.7.

Photoemission

After the photoemission of an electron from a dust particle of charge $(Z - 1)\,e$, the charge on the particle becomes Ze; the number of the emitted photoelectrons per unit time from a dust particle (of radius a) on account of photons with energy lying between $E_v(=hv)$ to $(E_v + \mathrm{d}E_v)$, is thus given by Sodha et al. [52, 53]

$$\mathrm{d}n_{\mathrm{ph}}(\upsilon_s) = \langle Q_a\rangle 8\pi^2 a^2 \chi(E_v)[\Psi(\xi,\upsilon_s)/\Phi(\xi)]F(E_v)\mathrm{d}E_v, \quad \text{for } Z \geq 0 \qquad (11.46a)$$

and

$$= \langle Q_a\rangle 8\pi^2 a^2 \chi(E_v)F(E_v)\mathrm{d}E_v, \quad \text{for } Z < 0, \qquad (11.46b)$$

where $\Psi(\xi, \upsilon_s) = \Phi(\xi - \upsilon_s) + 2\upsilon_s ln[1 + \exp(\xi - \upsilon_s)]$, $\chi(E_v)$ is the photoelectric efficiency of dust material, $\upsilon_s = Z\alpha = \mathrm{e}V_S/k_BT_d$ is the magnitude of the dimensionless potential energy (negative) of an electron at the surface with an electric potential (positive) V_s, T_d is the temperature of the dust particle, (−e is the electronic charge. $\pi \mathrm{a}^2$ is the cross-sectional area of the particle, $\beta(E_v)$ is the probability of absorption of a photon by an electron hitting the surface of the dust particle from inside per unit time, $\langle Q_a\rangle$ is the *mean absorption efficiency,* over the range of incident radiation of interest ($\varphi < E_v < 13.6$ eV), v is the frequency of the incident radiation, $\xi = (E_v - \varphi)/k_BT_d$, $\alpha = (e^2/ak_BT_d)$, and $\Phi(\kappa) = \int_0^{\exp\kappa} \ln(1+\Omega)/\Omega\, \mathrm{d}\Omega$.

The rate of photoemission due to radiation of frequency v from a positively charged particle is

$$(n_{\mathrm{ph}}/n_{\mathrm{ph0}}) = [\Psi(\xi,\upsilon_s)/\Phi(\xi)], \qquad (11.47)$$

where the symbols have usual meaning.

From (11.46a) and (11.46b) the number of photoelectrons, n_{ph} emitted from a dust particle per unit time due to the continuous spectrum of incident radiation, may be expressed as

$$n_{\mathrm{ph}}(V_s) = \int dn_{\mathrm{ph}}(V_s) = \langle Q_a\rangle 8\pi^2 a^2 \int_{E_{v0}}^{E_{vm}} \chi(E_v)[\Psi(\xi, ev_s/k_BT_d)/\Phi(\xi)]f(E_v)dE_v$$

$$\text{for } V_s \geq 0 \qquad (11.48a)$$

$$n_{\mathrm{ph}}(V_s) = \langle Q_a\rangle 8\pi^2 a^2 \int_{E_{v00}}^{E_{vm}} \chi(E_v)f(E_v)dE_v \quad \text{for } V_s < 0, \qquad (11.48b)$$

where $\chi(E_v)$ the photoelectric efficiency of the dust material may be expressed as either $\chi(E_v) = (729\chi_m/16)(E_{v00}/E_v)^4[1 - (E_{v00}/E_v)]^2$ [57] or $\chi(E_v) = \chi_m[1 - (E_{v00}/E_v)]$ [5], $E_{v_{00}} = hv_{00} = \varphi$ is the threshold energy of incident radiation,

$E_{v0} = h\nu_0 = (\varphi + eV_s)$, E_{vm} is the upper limit of the spectrum (13.6 eV), and χ_m is the maximum value of χ. As discussed before a mean value $\langle Q_a \rangle = 0.33$ has been adopted.

Similarly the mean energy [52, 53] of the photoemitted electrons (at a large distance from the dust surface) may be expressed as

for $V_s \geq 0$

$$\left(\frac{\varepsilon_{\text{ph}}(V_s)}{k_B T_d}\right) = \left[\left(\frac{\langle Q_a \rangle 8\pi^2}{n_{\text{ph}}}\right) \int_{E_{v0}}^{E_{vm}} \left(\frac{\chi(E_v)}{\Phi(\xi)} \int_{eV_s/k_B T_d}^{\infty} y^2[1+\exp(y-\xi)]^{-1} dy\right) f(E_v) dE_v\right] - (eV_s/k_B T_d) \tag{11.49a}$$

and for $V_s < 0$

$$\left(\frac{\varepsilon_{\text{ph}}(V_s)}{k_B T_d}\right) = \left[\left(\frac{\langle Q_a \rangle 8\pi^2}{n_{\text{ph}}}\right) \int_{E_{v00}}^{E_{vm}} \left(\frac{\chi(E_v)}{\Phi(\xi)} \int_{0}^{\infty} 2\eta ln(1+\exp(\xi-\eta) d\eta\right) f(E_v) \mathrm{d}\mathrm{E}_v\right] - (\mathrm{e}V_s/k_B T_d). \tag{11.49b}$$

Other Parameters

The expressions for other relevant parameters are as follows (Chap. 3):

$$\begin{aligned}
\text{for } V_s \geq 0: \; f_{\text{ic}} &= (8k_B T/m_i\pi)^{1/2} \exp(-eV_s/k_B T), \\
f_{\text{ec}} &= (8k_B T/m_e\pi)^{1/2}[1+(eV_s/k_B T)] \\
\varepsilon_{\text{ic}} &= [(2+eV_s/k_B T)/(1+eV_s/k_B T)]k_B T \text{ and } \varepsilon_{\text{ic}}(V_s) = 2k_B T + eV_s.
\end{aligned}$$

$$\begin{aligned}
\text{for } V_s < 0: \; f_{\text{ec}} &= (8k_B T/m_e\pi)^{1/2} \exp[eV_s/k_B T], \\
f_{\text{ic}} &= (8k_B T/m_i\pi)^{1/2}(1-eV_s/k_B T) \\
\varepsilon_{\text{ec}} &= (2-eV_s/k_B T)k_B T \quad \text{and} \\
\varepsilon_{\text{ic}} &= [(2-eV_s/k_B T)/(1-eV_s/k_B T)]k_B T.
\end{aligned}$$

11.6.3 Numerical Results and Discussion

For computations, the authors have chosen a standard set of parameters and studied the effect of change of one parameter, keeping others the same. The standard set of parameter for WIM (including spherical graphite particles) is as follows:

$F(E_\nu) = (1.658 \times 10^6 E_\nu - 2.152 \times 10^5 E_\nu^2 + 6.919 \times 10^3 E_\nu^3)\, dE_\nu$ [5], s = 3.5, $a_{rms} = 20$ nm corresponding to $a_{min} = 10$ nm and $a_{max} = 1{,}000$ nm [6], average absorption efficiency (*Mie coefficient of scattering*) computed before in this paper $\langle Q_a \rangle = 0.33$, $\beta_i = 10^{-16}/s$ [5], $\alpha_r = 1.75 \times 10^{-10} \times T^{-0.70}$ [4], $n_d = 10^{-7}$ cm^{-3} [49], $T_d = 20$ K [14], $\chi_m = 0.1$, $T_0 = 8000$ K, $I_p = 13.6$ eV (*hydrogen*), work function of graphite $\varphi = 5$ eV [27] and $n_t = 0.11$ cm^{-3} [5] and $\alpha_{de} \approx 8 \times 10^{-39}$/cm^{3}s, $f_{ex}\alpha_{ex} \approx 10^{-9}$/cm^{3}s, $f_{de} = 0.1$ [2].

In the steady state as $t \to \infty$, d/d$t = 0$, the set of differential Eqns. (11.40), (11.41), (11.42a), (11.42b), and (11.43) reduces to a set of algebraic equations, which give steady-state values of n_0, n_e, n_i, V_s, and T; these values are independent of the values at $t = 0$. Since the simultaneous numerical solution of the set of algebraic equations is formidable, a numerical solution of the set of differential equations has been obtained with plausible initial values of the parameters n_0, n_e, n_i, V_s and T at $t = 0$ viz. $n_{n0} = 0.10/cc$, $n_{e0} = n_{i0} = 0.01/cc$, $V_s = 0$, $T = 8000$ K. Since the values of n_0, n_e, n_i, V_s and T, so obtained are independent of the initial conditions, the accuracy of the assumed initial conditions $t = 0$ is irrelevant. The normalized values of n_e, n_i, n_0 and T viz. (n_e/n_{e0}), (n_i/n_{i0}), (n_n/n_{n0}), (T/T_0) and the surface potential V_s have been discussed later.

Figure 11.5a displays the surface potential (V_s) dependence of the photoemission current and corresponding mean energy of the photoelectrons (far away from the surface), from spherical graphite particles ($\varphi = 5.0$ eV) of radius a at a temperature of $T_d = 20$ K, irradiated by photons with energy distribution, given by (11.9). On account of the fact that with increasing (V_s) electrons in the particle need more energy to be emitted n_{ph} and E_{ph} get consequently decreased.

Figure 11.5b shows the dependence of the dust surface potential V_s and mean temperature of the plasma species T on the number density of dust particles for different values of root mean square radius a_{rms}. It is seen that as a_{rms} increases, both V_s and T/T_0 increase; this may be ascribed to larger number of emitted photoelectrons because of increased effective surface area of the particles and smaller number of electrons available per particle for accretion on the dust.

Figure 11.5c shows the corresponding dependence of the electron/ion density on dust particle density n_d; it is seen that the electron density decreases with increasing number density of dust particles (n_d). This is due to enhancement in the rate of accretion of electrons on dust particles due to increased temperature of the plasma and the surface potential of dust particles with increasing dust particle density n_d; the rate of photoelectric emission also decreases with increasing dust particle density n_d on account of simultaneous increase in V_s.

Figure 11.5d and e display the effect of increasing χ_m on the surface potential V_s, mean temperature T and electron/ion density as a function of dust particle density n_d.

Figure 11.5f displays the dependence of the temperature of the WIM plasma and the surface potential on the graphite particles on the number density of dust particles for different values of $f_{ex}\alpha_{ex}$. It is seen that both V_s and (T/T_0) increase with increase in dust particle density n_d, as discussed earlier. However with

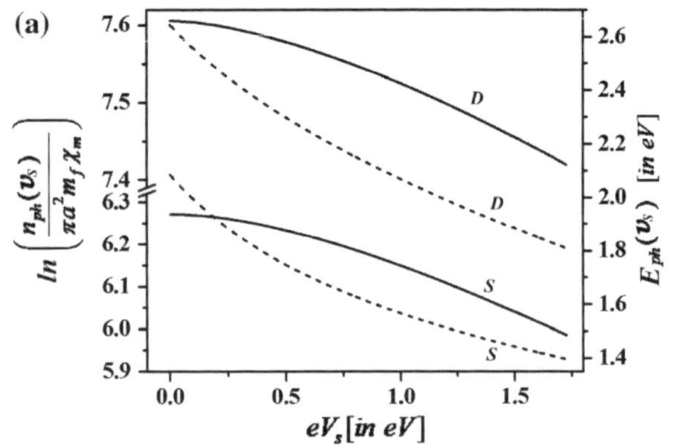

Fig. 11.5 **a** Photoemission current (*solid*) and corresponding mean energy far from the surface (*broken lines*) as a function of surface potential V_s. The labels D and S on the curves correspond to different dependences of χ on $E_{\upsilon,}$ as given by Draine [5] and Spitzer 57]. **b** Surface potential (V_s, *solid lines*) of the dust particles and the mean temperature (*T*, *broken lines*) of the plasma species as a function of the number density of dust particles n_d for different values of $a_{\rm rms}$ and the standard set of parameters, as stated in the text. The labels *a*, *b*, *c*, *d*, *e*, *f*, *g*, and *h* refer to $a_{\rm rms}$ = 1, 5, 10, 20, 30, 50, 80, and 100 nm, respectively. **c** Corresponding electron (n_e, *solid*) and ion (n_i, *broken*) densities as function of n_d for different values of $a_{\rm rms}$. The standard set of the parameters and labels are the same as in Fig. 11.2a. **d** Surface potential (V_s, *solid lines*) of the dust particles and the mean temperature (*T*, *broken lines*) of the plasma species as function of n_d for different values of χ_m and the standard set of parameters, as stated in the text. The labels *p*, *q*, *r*, *s*, and t refer to χ_m = 0.05, 0.08, 1.0, 0.3, and 0.5. **e** Corresponding electron (n_e, solid) and ion (n_i, broken) densities as function of n_d for different values of χ_m. The standard set of the parameters and labels are the same as in Fig. 11.3a. **f** Surface potential (V_s, *solid lines*) of the dust particles and the mean temperature (*T*, *broken lines*) of the plasma species as function of n_d for different values of parameter $f_{\rm ex}\alpha_{\rm ex}$ and the standard set of parameters, as stated in the text. The labels *a*, *b*, *c*, *d*, *e*, *f*, and *g* refer to $f_{\rm ex}\alpha_{\rm ex}$ = (0.01, 0.05, 0.1, 0.5, 1.0, 5, 10) × 10^{-9}/cm^3 *s*, respectively. **g** Corresponding electron (n_e, *solid*) and ion (n_i, *broken*) densities as function of n_d for different values of $f_{\rm ex}\alpha_{\rm ex}$. The standard set of the parameters and labels are the same as in Fig. 11.4a. **h** Surface potential (V_s, *solid lines*) of the dust particles and the mean temperature (*T*, *broken lines*) of the plasma species as function of n_d for different values of n_t and the standard set of parameters as stated in the text. The labels *p*, *q*, *r*, *s*, *t*, and *u* refer to n_t = (0.11, 0.22, 0.33, 0.44, 0.55, and 1.1/cm^3, respectively. **i** Corresponding electron (n_e, *solid*) and ion (n_i, *broken*) densities as function of n_d for different values of n_t. The standard set of the parameters and labels are the same as in Fig. 11.5a. **j** Values of n_d and $f_{\rm ex}\alpha_{\rm ex}$, which lead to consensus values of WIM parameters viz. $T = 8000\ K$, $V_s = 2.3\ V$, $n_e = 0.01$ cm^{-3}, $n_i = 0.01$ cm^{-3}, $n_n = 0.1$ cm^{-3} for the standard set of the parameters in the text (after Misra et al. [35], courtesy authors and publishers Springer)

increasing value of $f_{\rm ex}\alpha_{\rm ex}$ and hence associated cooling the temperature (T/T_0) and the surface potential decrease. The corresponding dependence of the electron/ion density on dust particle density n_d is displayed in Fig. 11.5g.

Figure 11.5h and i display the effect of change of n_t on the temperature T/T_0, surface potential V_s, electron density n_e/n_{e0}, ion density n_i/n_{i0}.

From Fig. 11.5f and g it is seen that the consensus measured/observed/estimated values of the temperature of the WIM plasma 8000 K, the surface potential

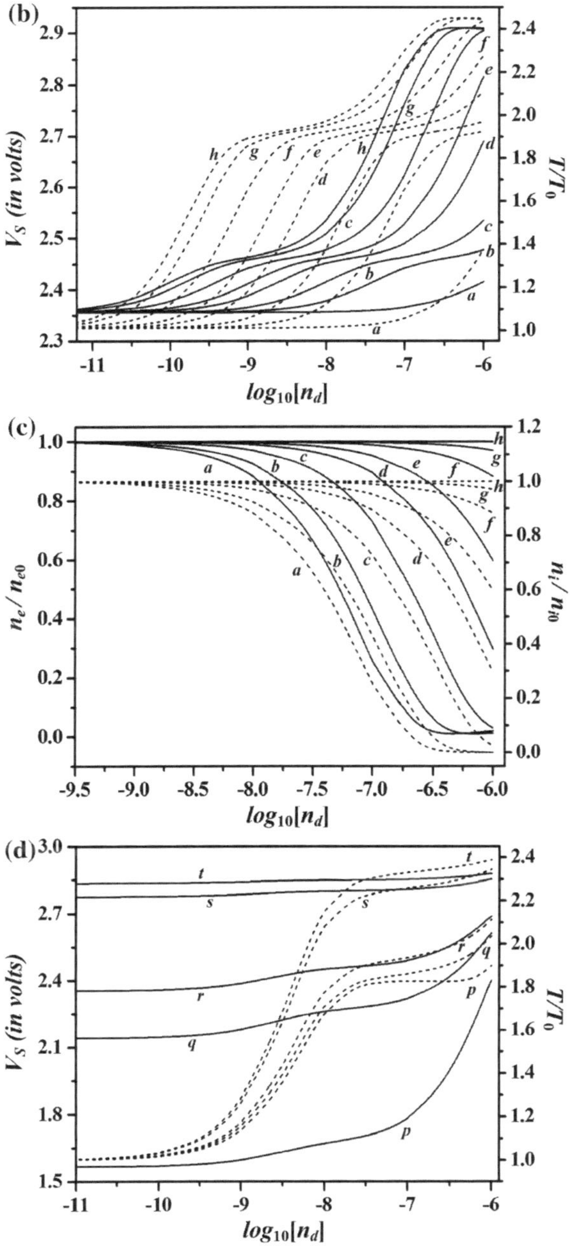

Fig. 11.5 (continued)

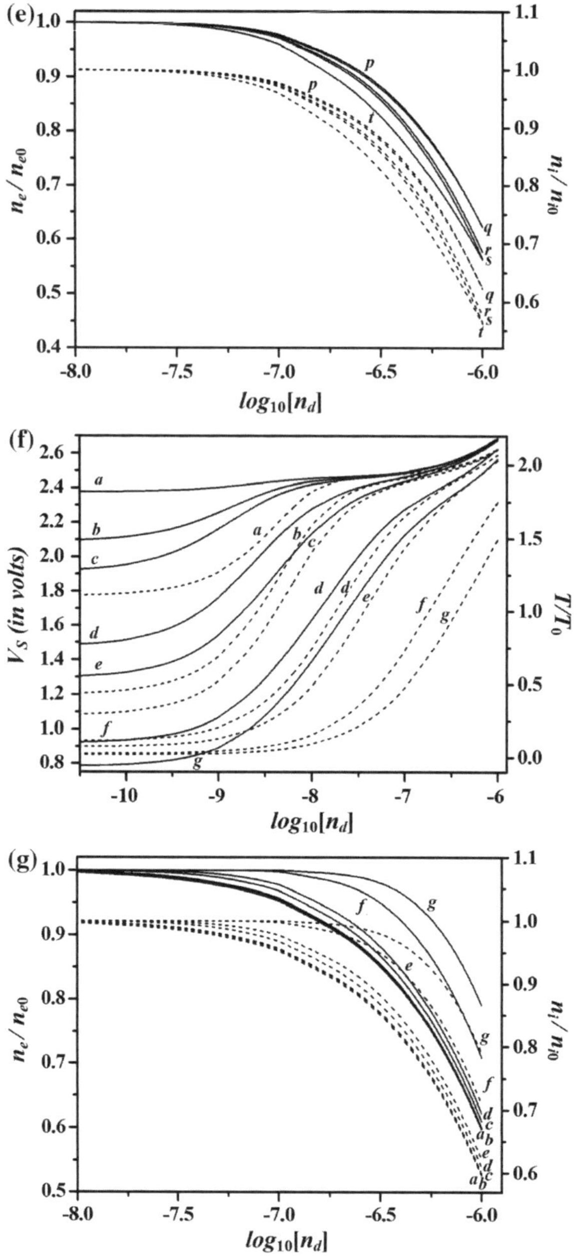

Fig. 11.5 (continued)

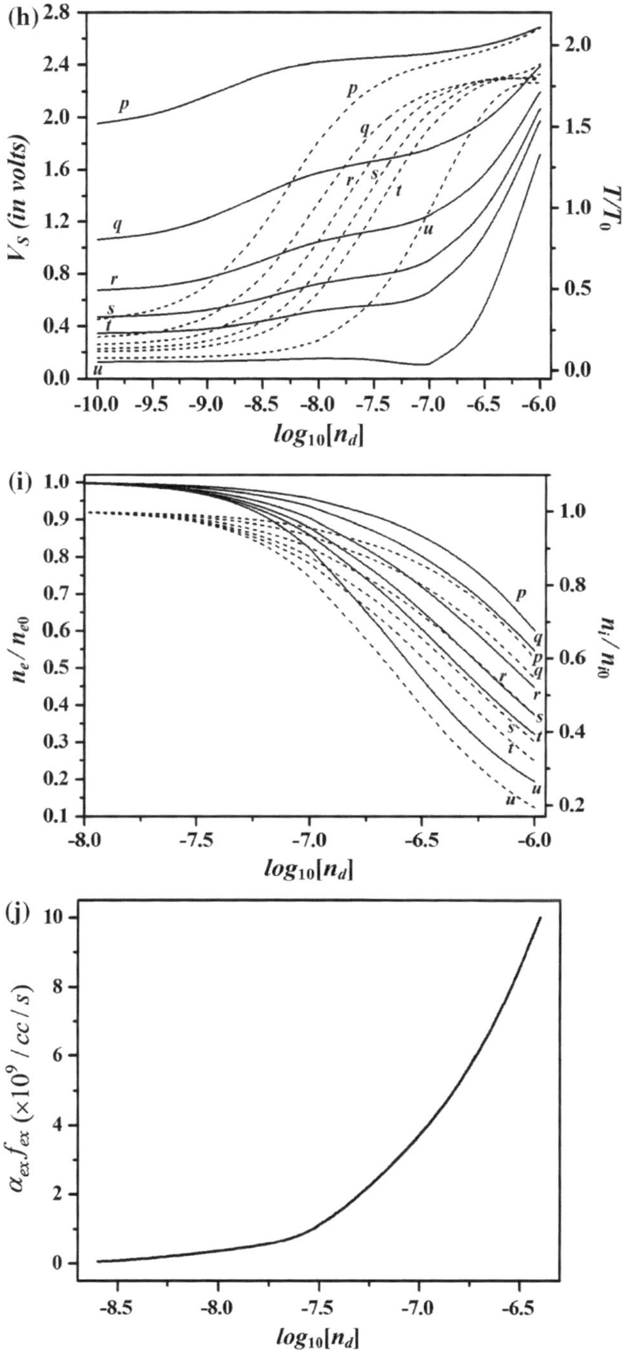

Fig. 11.5 (continued)

(2.3 V) on graphite particles [59]; ($2 < V_s < 4$ V) the electron/ion densities ($n_e = n_i = 0.01$ cm^{-3}) and density of neutral H atoms ($n_n = 0.1$ cm^{-3}) [5] can be explained on the basis of plausible combination of values of the number density of particles and the parameter $f_{ex}\alpha_{ex}$ as illustrated by Fig. 11.5j.

A precise determination of f_{ex} and α_{ex} in the laboratory will thus lead to a definite value of the density of graphite particles. The other parameters are as per Sect. 11.6.3.

Since all the particles are at the same electric potential $V_s = Ze/a$, the charge distribution may be obtained by putting $a = (Ze/V_s)$ in (11.1) and may be expressed as

$$F(Z)\mathrm{d}Z = (e/V)F(Ze/V_s)\mathrm{d}Z. \tag{11.50}$$

11.7 Conclusions

1. A parametric analysis of the kinetics and temperature of the WIM complex plasma, including number and energy balance of the constituents and size distribution of dust particles has been presented. The results have been graphically illustrated.
2. It is seen that the consensus values of the WIM plasma temperature, surface potential on the graphite dust particles, and electron/ion/neutral atom densities can be explained on the basis of plausible combinations of dust particle density n_d and the parameter $f_{ex}\alpha_{ex}$.

References

1. S.J. Bauer, *Physics of planetary ionospheres* (Springer, New York, 1973)
2. R.L. Bowers, T. Deeming, *Astrophysics* (Jones & Bartlett, Boston, 1984)
3. C. Cui, J. Goree, IEEE Trans. Plasma Sci. **22**, 151 (1994)
4. T. deJong, Astron. Astrophys. **55,** 137 (1977)
5. B.T. Draine, Astrophys. J. Suppl. Ser. **36**, 595–619 (1978)
6. B.T. Draine, H.M. Lee, Astrophys. J. **285**, 89–108 (1984)
7. B.T. Draine, E.E. Salpeter, Astrophys. J. **231**, 77 (1979)
8. A.L. Graps, G.H. Jones, A. Juhasz, M. Horanyi, O. Havnes, Space Sci. Rev. **137**, 435 (2008)
9. E. GrÜn et al., Nature **362**, 428 (1992)
10. A.V. Gurevich, *Nonlinear Phenomena in the Lonosphere* (Springer, New York, 1978)
11. O. Havnes, T. Aslaksen, A. Brattli, Phys. Scr. **T89**, 133 (2001)
12. H. Hirashita, T. Nozawa, Earth, Planets and Space **00**, 1–10 (2012)
13. M Horanyi et al, Rev. Geophys. **42,** RG4002 (2004)
14. T. Huang, B.T. Draine, A. Lazarian, Astrophys. J. **715**, 1462–1485 (2010)
15. A. Juhasz, M. Horanyi, J. Geophys. Res. **107**, A6 (2002)
16. S. Jurac, Johnson, J.D. Richardson, Icarus **149**, 384 (2001)
17. S. Kempf et al., Nature **433**, 289 (2005)

18. S. Kempf, R. Srama, M. Horanyi, M.E. Burton, S. Helfert, G. Moragas-Klostermeyer, M. Roy, E. Grun, Nature **433**, 289 (2005)
19. S. Kempf, R. Srama, F. Postberg, M. Burton, S.F. Green, S. Helfert, J.K. Hillier, N. Mcbride, M.J. McDonnell, G. Moragas-Klostermeyer, M. Roy and E. Grun, Science **307**, 1274 (2005b)
20. H. Kimura, I. Mann, Astron. J. **499**, 454 (1998)
21. B.A. Klumov, S.V. Vladimirov, G.E. Morfill, JETP Lett. **82**, 632–637 (2005)
22. B.A. Klumov, S.I. Popel, R. Bingham, JETP Lett. **72**, 364 (2000)
23. B.A. Klumov, G.E. Morfill, S.I. Popel, J. Expr, Theor. Phys. **100**, 152 (2005)
24. B.A. Klumov, S.V. Vladimirov, G.E. Morfill, JETP Lett. **82**, 632 (2005)
25. B.A. Klumov, S.V. Vladimirov, G.E. Morfill, JETP Lett. **85**, 478 (2007)
26. D. Le Sargeant, L.B. Hendecourt, P.H.L. Lamy, ICARUS **43**, 350 (1980)
27. D.R. Lide, *Handbook of Chemistry and Physics*. (CRC press, London, 2008)
28. T. Matsoukas, M. Russel, J. Appl. Phys. **77**, 4285 (1995)
29. T. Matsoukas, M. Russel, M. Smith, J. Vac. Sci. Tecnol. **A14**, 624 (1996)
30. M. McEwan, L.F. Phillips, *Chemistry of the Atmosphere*. (Edward Arnold, London, 1975)
31. D.A. Mendis, Ann. Rev. Astron. Astrophys. **26**, 11 (1988)
32. D.A Mendis, Keynote talk in ICPDN 6, (May 2011)
33. D.A. Mendis, W.I. Axford, Rev. Earth Planetar. Sci. **2**, 419 (1974)
34. N. Meyer-Vernet, Astron. Astrophys. **105**, 98 (1982)
35. S.K. Mishra, M.S. Sodha, S. Srivastava, Astrophys. Space Sci. **344**, 193 (2013)
36. S. Misra, S.K. Mishra, Mon. Not. Roy. Astron. Soc. **432**, 2985 (2013)
37. S. Misra, S.K. Mishra, M.S. Sodha, Mon. Not. Roy. Astron. Soc. **423**, 176 (2012)
38. T. Mukai, Astron. Astrophys **99,** 1 (1981)
39. F. Postberg, S. Kemf, J. Schmidt, N. Brilliantov, A. Beinsen, B. Abel, U. Burk, R. Srama, Nature **459**, 2009 (1098)
40. F. Postberg, J. Schmidt, J. Hillier, S. Kempf, R. Srama, Nature **474**, 620 (2011)
41. G.C. Reid, Adv. Space Res. **20**, 1285 (1997)
42. R.J. Reynolds, L.M. Heffner, S.L. Tufte, Evidence for an additional heat source in the warm ionized medium of galaxies. Astrophys. J. **525**, L21–L24 (1999)
43. J.D. Richardson, J. Geophys. Res. **97**, 13705 (1992)
44. J.D. Richardson, Geophys. Res. Lett. **22**, 1177 (1995)
45. J.D. Richardson, A. Eviatar, J. Geophys. Res. **93**, 7297 (1988)
46. J.D. Richardson, A. Eviatar, G.L. Siscoe, J. Geophys. Res. **91**, 8749 (1986)
47. J.D. Richardson, A. Eviatar, M.A. McGrath, V.M. Vasyliunas, J. Geophys. Res. Planets **103**, 245 (1998)
48. J.D. Richardson, E.C. Sittler Jr, J. Geophys. Res. **95**, 12019 (1990)
49. P.K. Shukla, A.A. Mamun, *Introduction to dusty plasma physics* (Institute of Physics, Bristol, 2002)
50. E.C. Sittler Jr., M. Thomsen, D. Chornay, M.D. Shappirio, D. Simpson, R.E. Jhonson, H.T. Smith, A.J. Coates, A.M. Rymer, F. Crary, D.J. Mc.Comas, D.T. Young, D. Reisenfeld, M.Dougherty, N. Andre, Geophys. Res. Lett. **32,** L14S07 (2005)
51. M.S. Sodha, S. Guha, Physics of colloidal plasma. Adv. Plasma Phys, eds. by A. Simon, W.B. Thomas, (Interscience) **4,** 219 (1971)
52. M.S. Sodha, S.K. Mishra, S. Misra, Phys. Plasma **16**, 123701 (2009)
53. M.S. Sodha, S. Misra, S.K. Mishra, Charging Phys. Plasmas **16**, 123705 (2009)
54. M.S. Sodha, S. Misra, S.K. Mishra, Plasma Sources Sci. Technol. **19**, 045022 (2010)
55. M.S. Sodha, S. Misra, S.K. Mishra, A Dixit, Phys. Plasmas. **18**, 083708 (2011)
56. L. Spitzer Jr., M.G. Tomasko, Astrophys. J. **152**, 971–986 (1968)
57. L. Spitzer, Astrophys. J. **107**, 6–33 (1948)
58. E.J. Sternglass, *The Theory of Secondary Electron Emission, Sci.* ***Paper*** **1772** (Westinghouse Res. Lab, Pittsburgh, 1954)
59. J.C. Weingartner, B.T. Draine, Astrophys. J. Suppl. Ser. **134**, 263–281 (2001)
60. M. Wilson, Phys. Today, **p.15** (Feb. 2010)

Chapter 12
Complex Plasma as Working Fluid in MHD Power Generators

12.1 Introduction

The efficiency of mechanical (and hence electrical) power generation in a steam turbine system is limited by the temperature of the superheated steam and not by the high temperature of the gases obtained from combustion or flow of a carrier gas through a nuclear reactor; some loss of efficiency is also suffered in the conversion of mechanical into electrical power. Hence, it is desirable to have a system that can utilize the high temperature gases, corresponding to combustion or flow through nuclear reactor, for direct conversion of thermal into electrical power. The magnetohydrodynamic (MHD) power generation system is a powerful means to this end. Figure 12.1 illustrates the basic configuration of an MHD generator. It consists of a stagnation ($v = 0$) chamber, containing hot gases (obtained by combustion or flow through nuclear reactor) and the MHD duct, where the electrical power gets generated; the gas exiting from the duct is fed into a conventional steam power generation unit to produce further electrical power. The electrodes are placed in the opposite walls of the duct as shown in Fig. 12.1 (segmented electrode configuration); in such a configuration, no current can flow along the direction (x axis) of flow of the gas (since it is an open circuit in the x direction) and along the magnetic field (z axis) because there is no force on the charge carriers along the z direction.

The parameters in the stagnation chamber and at the entrance to the duct are characterized by the suffixes 0 and 1.

12.2 Basic Equations

Following earlier works (e.g. [13]), the basic (viz., momentum, energy, continuity, state, and modified Ohm's Law) equations corresponding to the duct may be expressed as

M. S. Sodha, *Kinetics of Complex Plasmas*, Springer Series on Atomic, Optical, and Plasma Physics 81, DOI: 10.1007/978-81-322-1820-3_12,

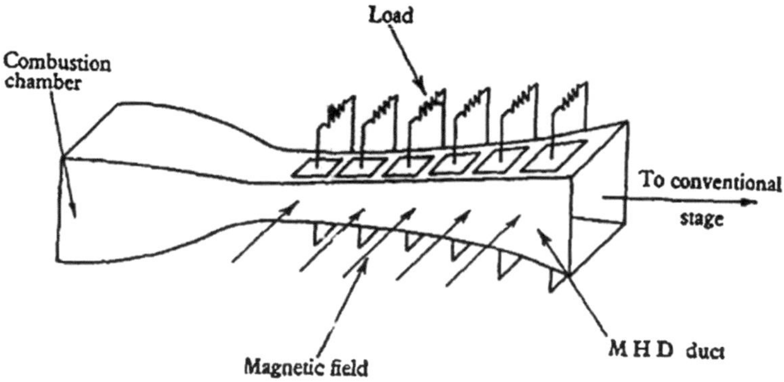

Fig. 12.1 MHD duct configuration (after Swifthook and Wright [13])

$$\text{Momentum } \rho\upsilon(\mathrm{d}\upsilon/\mathrm{d}x) + (\mathrm{d}p/\mathrm{d}x) + JB = 0, \tag{12.1}$$

$$\text{Energy: } \left[\rho\upsilon\,\mathrm{d}\left(C_pT + \upsilon^2/2\right)/\mathrm{d}x\right] + JE = 0 \tag{12.2}$$

and

$$\text{Continuity: } \rho\upsilon A = \text{constant} = \rho_1\upsilon_1A_1, \tag{12.3}$$

where $\mathrm{d}/\mathrm{d}t = \upsilon\mathrm{d}/\mathrm{d}x$,

ρ is the density of the gas
T is the temperature of the gas
υ is the flow velocity of the gas (along x direction)
p is the pressure of the gas
c_p is the specific heat at constant pressure of the gas
$\boldsymbol{J}$ is the current density
$\boldsymbol{B}$ is the magnetic field (along z direction)
$\boldsymbol{E}$ is the electric field in the gas
$\boldsymbol{J} \times \boldsymbol{B} =$ iJB is the force per unit volume exerted by the magnetic field on the flowing gas
JE is the Ohmic loss per unit volume, and
$A(x)$ is the area of cross section of the duct.

Recalling that

$$c_p - c_\upsilon = R/\mu$$

and

$$c_p/c_\upsilon = \gamma,$$

the equation of state for unit mass of the gas is

$$(p/\rho) = (R/\mu)T$$

can be expressed as

$$p = \rho c_p T(\gamma - 1)/\gamma, \tag{12.4}$$

where c_υ is the specific heat of gas at constant volume and μ is the molecular weight of the gas.

The modified Ohm's law, which takes into account the induced $(\boldsymbol{\upsilon} \times \boldsymbol{B})\,emf$ per unit volume may be expressed as

$$J = \sigma(\upsilon B - E), \tag{12.5}$$

where σ is the electrical conductivity of the gas and E is electric field along the y direction (perpendicular to $\boldsymbol{\upsilon}$ and $\boldsymbol{B}$). The systems of (12.1)–(12.4) have been solved by various workers (Coe and Eisen [1], Huth [5], Neuringer [8], Sutton [12], Way [16]), corresponding to one of the parameters $(A, p, T, \boldsymbol{\upsilon})$ being kept constant. Swifthook and Wright [13] solved this system of equations keeping the Mach number $M = \upsilon/c$ constant, where c is the speed of sound in the gas. This analysis is convenient for analyzing the performance of a combined cycle of MHD generation and generation by a steam turbine system; since the conductivity of the gas falls to low values after traversing some distance in the duct, the remaining energy of the gas can be used to extract further electrical power by the use of exit gas from the MHD duct as an input in the conventional steam turbine cycle.

From (12.4), the Mach number M of the flow is given as

$$M^2 = \upsilon^2/c^2 = \upsilon^2/(\gamma p/\rho) = \upsilon^2/(\gamma - 1)c_p T. \tag{12.6}$$

12.3 Analysis for Constant Mach Number (After Swifthook and Wright [13])

12.3.1 Efficiency

Using (12.2) the electrical power, extracted per unit length is given as

$$\mathrm{JEA} = -\rho\upsilon A\,\mathrm{d}\big(c_p T + \upsilon^2/2\big)/\mathrm{d}x.$$

Substituting for υ from (12.6) and $\rho\upsilon A$ from (12.3) into the above equation one obtains

$$\begin{aligned}\frac{\mathrm{d}Q}{\mathrm{d}x} &= \mathrm{JEA} = -\rho_1\,\upsilon_1\,A_1\,\mathrm{d}\big[c_p T\{1 + (\gamma - 1)M^2/2\}\big]/\mathrm{d}x \\ &= -\,\rho_1\,\upsilon_1\,A_1\,c_p \mathrm{d}T_0/\mathrm{d}x,\end{aligned}$$

where the stagnation temperature T_0 is defined as

$$c_p T_0 = c_p T + \upsilon^2/2 == c_p T\left[1 + (\gamma - 1)M^2/2\right].$$

Thus $\frac{\mathrm{d}Q}{\mathrm{d}x} = \mathrm{JEA} = \rho_1\, \upsilon_1 A_1\, \mathrm{d}(T/T_1)\,\mathrm{d}x = Q_1 \mathrm{d}(T_0/T_{01})/\mathrm{d}x$, where Q_1 is the enthalpy of the gas entering the duct ($x = 0$) per unit time.

Integrating the above equation

$$Q/Q_1 = 1 - (T_0/T_{01}) = 1 - (T/T_1) = \eta \tag{12.7}$$

where η is the efficiency of electrical power extractions.

It is interesting to note that the above expression corresponds to the efficiency of a reversible engine.

12.3.2 Other Parameters

From (12.1)–(12.4) and (12.6), it can be shown [13] that

$$(p/p_1) = (T/T_1)^{\beta}, \tag{12.8}$$

$$(\rho/\rho_1) = (T/T_1)^{\beta-1}, \tag{12.9}$$

$$(\upsilon/\upsilon_1) = (T/T_1)^{1/2} \tag{12.10}$$

and

$$(A/A_1) = (T/T_1)^{-(\beta-1/2)}, \tag{12.11}$$

where

$$\beta = \left\{\left[1 + (1-K)(\gamma-1)\,M^2/2\gamma\right]/(\gamma-1)\,K\right\} \tag{12.12a}$$

and

$$K = E_y/\upsilon B \text{ is the load factor.} \tag{12.12b}$$

The duct length x corresponding to a temperature T can be shown to be given by

$$(x/x_0) = -[K(1-K)]^{-1}\int_1^{T/T_1} (\sigma_1/\sigma)(T/T_1)^{\beta-3/2}\mathrm{d}(T/T_1) \tag{12.13a}$$

where

$$x_0 = (Q_1/A_1)/\sigma_1\, \upsilon_1^2 B^2. \tag{12.13b}$$

An analytical expression for x can be obtained when the electrical conductivity of the gas can be expressed as

$$\sigma = ST^y p^{-z}. \tag{12.14a}$$

In this case, it can be shown that

$$(x/x_2) = 1 - (T/T_1)^{\omega}, \tag{12.14b}$$

where

$$\omega = (\beta - 1/2) - (y - \beta z) \tag{12.14c}$$

and

$$x_2 = x_0/K(1 - K)\omega. \tag{12.14d}$$

12.4 Complex Plasmas as Working Fluid in MHD Generators (After Sodha and Bendor [10, 11])

12.4.1 The Need

The large-scale use of MHD power generation (particularly in closed cycles with nuclear reactor) is mainly constrained by the low electrical conductivity of the gas (even after seeding with cesium) at temperatures feasible with technology (particularly in the 1960s, when MHD power generation was pursued as a serious option). Hence, many interesting concepts were put forward to enhance the conductivity of the gas at temperatures compatible with the then technology. One of the approaches proposed at that time was to have a suspension of low work function dust in a gas and use this complex plasma as the working fluid. Honna and Fushima [2], Honna et al. [3], Hooper et al. [4], Mori et al. [7], Sodha and Bendor [10, 11], Waldie and Fells (1967) have investigated the use of this concept for sytems compatible with gas cooled reactor. Experimental [17], values of the electrical conductivity of Argon–BaO complex plasmas were found to be in good agreement with theory [14, 15].

12.4.2 Equivalent Parameters

The parameters of an equivalent gas (for study of gas dynamics) having a suspension of dust with a fraction λ by mass are given as

$$\rho = (1 + \lambda)\rho_g \tag{12.15a}$$

$$R = R_g/(1 + \lambda) \tag{12.15b}$$

$$c_p = \left(c_{\mathrm{pg}} + \lambda c_{\mathrm{pp}}\right) / (1 + \lambda) \tag{12.15c}$$

$$\gamma = \left(c_{\mathrm{pg}} + \lambda c_{\mathrm{pp}}\right) / \left(\lambda c_{\mathrm{pp}} + c_{\mathrm{pg}} / \gamma_g\right) \tag{12.15d}$$

where c_{pg} and c_{pp} are the specific heats per unit mass at constant pressure of the gas and the dust, R_g is the gas constant of the pure gas and $\gamma_g = c_{\mathrm{pg}} / c_g$.

12.4.3 Steam Turbine-Magnetohydrodynamic Topping Closed Cycle

The cycle is illustrated in Fig. 12.2. The suffixes 1, 2, 3, and 4 refer to inlet/outlet from various components of the cycle; the suffix 0 indicates the stagnation values. For a mass flow rate m of the complex plasma, the power input/output corresponding to different components is given by

$$W_M = \eta_M \, m c_p (T_{01} - T_{02}), \tag{12.16a}$$

$$W_T = \eta_T \, m c_p (T_{02} - T_{03}), \tag{12.17a}$$

$$W_c = m c_p (T_{04} - T_{03}) / \eta_c \tag{12.18a}$$

and

$$H = m c_p (T_{01} - T_{04}), \tag{12.19a}$$

where η_M is the efficiency of conversion of D.C. power (produced by the MHD generator) into A.C. power, η_T is the efficiency of conversion of heat absorbed from the complex plasma to electrical power by the steam turbine cum associated equipment, and η_c is the mechanical efficiency of the compressor.

We have neglected the energy needed to maintain the magnetic field, including refrigeration for superconductors, in case of superconducting magnets.

For a constant Mach number duct, from (12.8) one has

$$(p_2/p_1) = (T_{02}/T_{01})^{\beta}, \tag{12.16b}$$

further,

$$p_3 = K_T p_2 \tag{12.17b}$$

and

$$p_4 = K_H^{-1} p_1, \tag{12.18b}$$

where K_T and K_H indicate the pressure drop in the heat exchanger and the reactor.

The temperature ratio across the compressor is

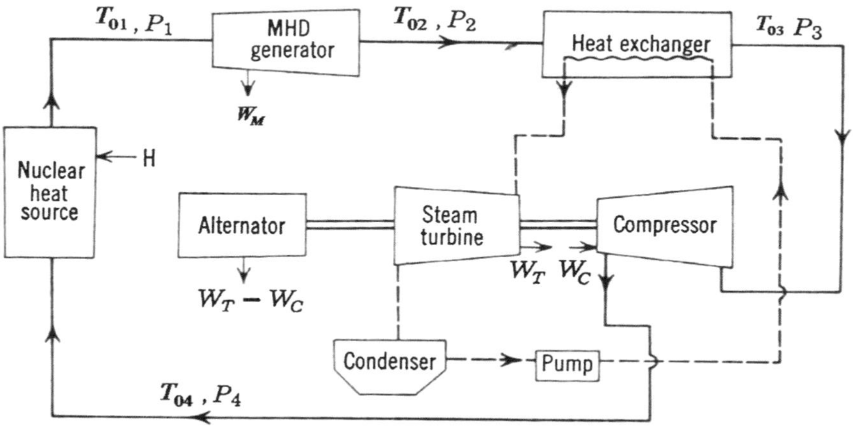

Fig. 12.2 Magnetohydrodynamic generator steam turbine topping cycle—steam water—gas particle suspended (after Sodha and Bendor [11], curtsey authors and publishers IOP). The complex plasma cycle is represented by *continuous line*, while the water stream cycle conforms to *dashed lines*

$$\frac{T_{04}}{T_{03}} = 1 + \left\{ (p_4/p_3)^{(\gamma-1)/\gamma} \right\} \eta_p^{-1}, \tag{12.18c}$$

where η_p is the polytrophic efficiency.

In all components except the MHD duct, the flow velocity is not large enough to make T_0 substantially different from T.

Thus, W_M/m, W_T/m, W_c/m and H/m may be evaluated by a choice of T_{02}/T_{01}, T_{03}/T_{02} and p_4/p_3; the cycle efficiency is given by

$$\eta = (W_m + W_T - W_c)/H.$$

For a numerical appreciation of the dependence of the overall efficiency η on T_2/T_1 and T_3/T_1, the following typical parameters have been assumed:

$$c_{pg} = 522\,\mathrm{J\,kg^{-1}K^{-1}},\ c_{pp} = 334\,\mathrm{J\,kg^{-1}K^{-1}},\ \gamma_g = 1.667,\ K = 0.8,\ \eta_M = 0.95,\ \eta_c = 0.98,$$
$$\eta_p = 0.9,\ \eta_T = 0.45,\ K_T = 0.96,\ K_H = 0.95.$$

For a suspension of BaO dust in Argon and λ the ratio of mass of dust to gas equals to 0.1. The dependence of the efficiency η on T_2/T_1 and T_3/T_1 is illustrated in Fig. 12.3 for $M = 0.7$.

Sodha and Bendor [11] evaluated the electrical conductivity of a suspension of BaO dust in Argon as function of pressure, temperature, and λ. The following parameters were used:

Diameter of particles 0.05 μ; density of BaO = 5.72×10^3 kg/m^3; work function of BaO = 1.7 eV; ionization potential of Argon = 15.68 eV; electron cross collision cross-section of Ar atoms = 3×10^{-10} cm^2.

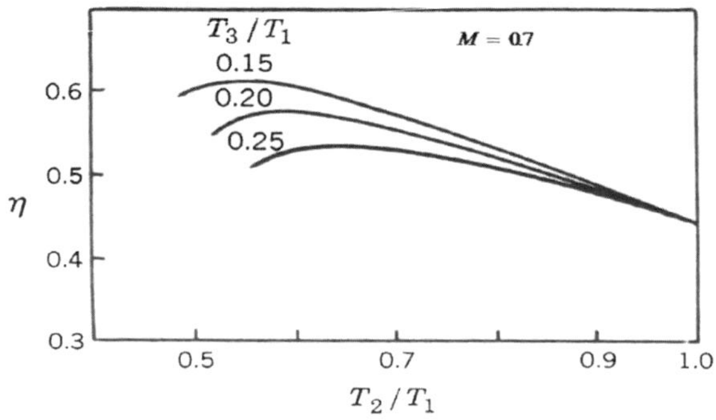

Fig. 12.3 Variation of efficiency of a steam turbine–MHD generator topping cycle with T_3/T_1 and T_2/T_1 (after Sodha and Bendor [11], curtsey authors and publishers IOP)

Table 12.1 Results for a selected cycle: $T_2/T_1 = 0.7, M = 1.0, \lambda = 0.3$ (after Sodha and Bendor [11], curtsey authors and publishers IOP)

T_{01} (K)	T_{02} (K)	T_{03} (K)	T_{04} (K)	P_1 (atm)	P_2 (atm)	P_3 (atm)	P_4 (atm)	σ_1 (mho m^{-1})	σ_2 mho m^{-1}	x_2 (m)
1,669	1,167	367	635	7.00	1.73	1.65	7.28	6.85	2.95	6.35

$\frac{W_M}{W}$	$\frac{W_T}{W}$	$\frac{W_c}{W}$	$\frac{H}{W}$	$\left(\frac{A_1}{W}\right) \times 10^{10}$ m^2/W	$\left(\frac{m}{W}\right) \times 10^7$ kg/J	$\left(\frac{\Omega}{W}\right) \times 10^8$ m^3/W	η
0.845	0.652	0.490	1.832	39.0	37.0	4.67	0.543

Ω Volume of the duct; W net out part

The evaluated conductivity (as per Part I of the book) can be approximately fitted to the relation $\sigma = 1.39 \times 10^{-8} \lambda^{0.66722}\ T^{2.92191}\ p^{-0.13130}$, where σ is expressed in mhos/m, T in K and p in atmospheres. Further, for duct B = 10 Wb/m^2 and $V = 11{,}505$ V.

Detailed results for a cycle, corresponding to $M = 1$ and $\lambda = 0.3$ are given in Table 12.1.

Corresponding dependence of the duct length on λ is illustrated in Fig. 12.4.

12.4.4 Feasibility

The interest in the commercial exploitation of the concept of MHD power generation started waning by the end of 1960s and practically vanished in the 1970s. This was on account of the then nonavailability of materials for fabrication of electrodes and ducts, compatible with the temperature and gas flow speeds,

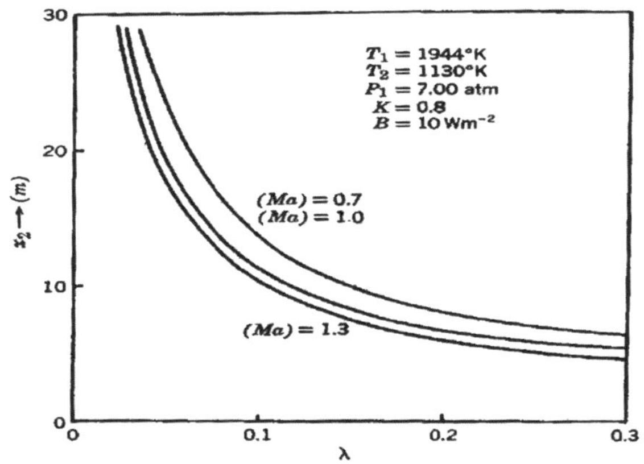

Fig. 12.4 Variation of duct length x_2 with λ and M (after Sodha and Bendor [11], curtsey authors and publishers IOP)

characteristic of the MHD duct designs. The nonavailability and cost of a superconducting system to produce large magnetic field over the duct length was also discouraging. However, since then the tremendous progress in the field of materials, including superconductors, warrants a fresh look at the material aspects.

Dust of BaO and other materials with a particle size of 0.05 μm (or less) has been commercially available for a very long time and the situation has further improved with the advent of nanotechnology. The uniform suspension of fine dust (size 0.05 μm or less) is perfectly feasible on account of negligible slip [10, 11] between the particles and the gas. The tendency of the particles to coalesce is countered on account of the positive charge due to thermionic emission. In cooler parts of the system, ultrasonics should be helpful. The turbulent flow also tends to keep the suspension uniform.

For a long time gas dust suspension loops have been in operation for pneumatic transmission; most of these systems are characterized by larger particle size (1.0 μ), high λ, and low flow velocities. The experimental work on dust suspension reactor (Schluderberg et al. [9]) and dust fuel reactor (Krucoff [6]) leads to the conclusion that closed cycle reactor systems, using dust suspension, are feasible.

Thus the use of dust in the MHD power generation system does not apriori present a road block; however, a lot of work is needed to get a feasible system running.

References

1. W.B. Coe, C.L. Eisen, Elec. Eng. **79**, 997 (1960)
2. T. Honma, K. Fushimi, Jap. J. Appl. Phys. **5**, 238 (1966)
3. T. Honma, O. Nomura, A. Kanai, Bull-Electrotech. Lab. (Japan) **32**, 83 (1968)
4. A.T. Hooper, D. Newby, A.H. Russell, *Electricity from MHD,* vol. 1, (IAEA, Vienna, 1966), p. 631
5. J.H. Huth, in *Energy Conversion for Space Power*, ed. by N.W. Snyder. (Academic Press, New York, 1961)

6. D. Krucoff, Nucleonics, **17**, 100 (1951)
7. F. Mori, K. Fushimi, T. Honma, *Electricity from MHD*, vol. 1 (IAEA, Vienna, 1966), p. 643
8. J. Neuringer, J. Fluid Mech. **7**, 28 (1960)
9. D.C. Schluderberg, R.L. Whitelaw, R.W. Carlson, Nucleonics, **19**, 67 (1961)
10. M.S. Sodha, E. Bendor, *Symposium on MHD Electrical Power Generation*, vol. 2 (ENEA, Paris, 1964a), p. 289
11. M.S. Sodha, E. Bendor, Brit. J. Appl. Phys. **15**, 1031 (1964b)
12. G.W. Sutton, General Electric Report, **R59SD**, 432 (1959)
13. D.T. Swifthook, J.K. Wright, J. Fluid Mech. **15**, 97 (1963)
14. B. Waldie, I. Fells, Phil. Trans. Soc. London, **A261**, 490 (1967)
15. B. Waldie, I. Fells, *International Symposium on MHD Electrical Power Generation*, Warsaw, IAEA Vienna, Paper **SM 107/127** (1968)
16. S. Way, Westinghouse Science Paper, **6-40509-2P1** (1960)
17. E.P. Zimin, Z.G. Mikhnevich, V.A. Popov, *Electricity from MHD*, vol. 3 (IAEA Vienna, 1966), p. 97

Chapter 13
Rocket Exhaust Complex Plasma

13.1 Introduction

Baghdady and Ely [1] have pointed out that observations on signal strength, received from a rocket highlight the fact that during certain phases of the flight severe attenuation of the order of tens of db occurs. For example, the early Polaris firings were characterized by signal blackouts over a period of tens of seconds, resulting in loss of link to the missile; even the range safety command "destruct" and control command for telemetry could not be implemented during this period. These severe electromagnetic propagation effects were attributed to the rocket exhaust tail, comprising of dense plasma (e.g., Brake and Howell [4]).

Various mechanisms to explain the electron density in a rocket exhaust, being far in excess of that predicted by the application of Saha's equation to the gaseous phase were proposed but not analyzed in sufficient detail. Einbinder [6] and Smith [15] have explained the electron density in hydrocarbon flames, which is far in excess of that predicted by Saha's equation, on the basis of thermionic emission from the surface of carbon particles, present in the flame; the kinetics of a dust-electron cloud was worked out taking into account the charge distribution on the particles. Sodha et al. [16] noted the work of Sehgal [13] on the nature of solid particles in the exhaust of solid propellant rockets; Sehgal [13] concluded that.

(i) all particles are nearly spherical.
(ii) the particle surfaces are smooth and continuous without pores or pokes.
(iii) the average diameter D_0 of the particles is related to the rocket chamber pressure by the relation.

$$ln\ P_0 = a + bD_0.$$

Bernhardt et al. [2] have considered the role of ice particles, (formed by condensation of water vapor in the space shuttle exhaust) in the back scatter from the exhaust. For the parameters of the plasma, the approximately 5.6 nm ice particles had little charge and there were few of them; hence no significant effects of the ice dusty plasma were observed.

M. S. Sodha, *Kinetics of Complex Plasmas*, Springer Series on Atomic, Optical, and Plasma Physics 81, DOI: 10.1007/978-81-322-1820-3_13,

Platov and Kosch [11] have considered the effect of sublimation of ice particles in the dynamics of gas cloud formations. The dynamics depends on the size distribution of the particles, their lifetime and initial velocities. The size of the particles is determined by the associated heat and mass transfer. The role of dust in the associated optical phenomena has been discussed. Platov et al. [12] have classified the gas dust formations formed by rocket exhausts in the upper atmosphere/ionosphere.

Large rocket exhausts cause large ionosphere holes, usually attributed to the chemical reaction between the ionosphere gases and the rocket exhaust. Little effort has been made to examine the role of the dust particles in the formation of the holes.

Shukla and Mamun [14] have given typical parameters of complex rocket exhaust and flame plasmas. The electron density (n_e) and temperature (T_e), the number density (n_d) and radius (a) of the dust particles are 10^{13} cm^{-3}, 3000 K, 10^8 cm^{-3}, and 0.1 μm for rocket exhausts and 10^{12} cm^{-3}, 2000 K, 10^8 cm^{-3} and 0.01 μm for the flames, respectively; the neutral gas molecular density is 10^{18} cm^{-3} and 10^{19} cm^{-3} respectively for the two cases. Shukla and Mamun [14] have also given the composition, radius, and density of the dust particles in earth's surroundings; it is seen that the dust from rocket exhausts is many orders of magnitude more abundant than other types of dust. The rocket exhaust dust typically consists of dirty ice particles with radius 5×10^{-3} μm and number density 3×10^4 cm^{-3}.

13.2 Composition of Rocket Exhausts

Sodha et al. [16] have investigated the composition of a system of atoms/molecules, ions, dust particles, and electrons in thermal equilibrium; such an analysis is useful in appreciating the role of low work function dust in the enhancement of the electron density and the reduction of electron density by addition of dust of material with high work function.

Denoting a particle (ion or dust) with charge Ze of the sth species (atom/molecule, ion, or dust) by A_Z^s one has

$$\sum_{Z=-\infty}^{Z=l_s} \left[A_Z^s\right] = \left[A^s\right], \tag{13.1}$$

$$\sum_{s=1}^{s=m} \sum_{Z=-\infty}^{Z=l_s} Z\left[A_Z^s\right] = n_e \tag{13.2}$$

and

$$\frac{n_e\left[A_Z^s\right]}{\left[A_{Z-1}^s\right]} = K_Z^s, \tag{13.3}$$

where

$[A_Z^s]$ is the concentration of A_Z^s

A^s can be gaseous or dust particle

n_e is the electron density

K_Z^s is the equilibrium constant, given by Saha's equation as

$$K_Z^s = \left(\omega_Z^s/\omega_{Z-1}^s\right)2.\left(2\pi m_e k_B T/h^2\right)^{3/2}\exp\left[-\varepsilon_s^Z/k_B T\right], \tag{13.4}$$

ω_Z^s is the statistical weight of the ground state of A_Z^s

m_e is the electronic mass

k_B is Boltzmann's constant

h is Planck's constant

ε_s^Z is the energy needed for change from A_{Z-1}^s to A_Z^s and l_s is the number of electrons in A^s.

For a numerical appreciation of the role of dust particles in a rocket exhaust plasma, Sodha et al. [17] analyzed a simple model of rocket exhaust plasmas characterized by the three following assumptions:

(i) The gaseous medium comprises of a single species of atoms and singly charged ions and electrons.
(ii) There are two species of dust—one having low work function and the other high work function and all dust particles of the same species have uniform radius and charge.
(iii) The plasma is in thermal equilibrium.

Thus the general (13.1) to (13.4) can be simplified to

$$\begin{aligned}
n_0 + n_i &= n, \\
n_e &= n_i + Z_1 n_1 + Z_2 n_2 \\
(n_e n_i/n_0) &= K_0 = 2\left(2\pi m_e k_B T/h^2\right)^{3/2}(\omega_1/\omega_0)\exp(-eV/k_B T), \\
n_e &= K_1 \exp(-Z_1\alpha_1) \\
n_e &= K_2 \exp(-Z_2\alpha_2) \\
K_{1,2} &= 2\left(2\pi m_e k_B T/h^2\right)^{3/2}\exp\left(-e\phi_{1,2}/k_B T\right)
\end{aligned}$$

and

$$K_0 = 2\left(2\pi m_e k_B T/h^2\right)^{3/2}\exp(-eV/k_B T).$$

Table 13.1 Effect of dust particles on electron density in a typical rocket exhaust (after Sodha et al. [16], curtsey authors and publishers IOP)

S. No	N_1 (cm^{-3})	n_e (cm^{-3})	N_2 (cm^{-3})
1	0	10^9	–
2	1.53×10^8	10^{10}	4.30×10^6
3	2.03×10^9	10^{11}	5.70×10^7
4	2.94×10^{10}	10^{12}	8.26×10^8
5	5.35×10^{11}	10^{13}	1.51×10^{11}

The above equations lead to the following equation, whose solutions give n_e and hence other parameters.

$$n_e = [n_e K_0/K_0 n_e] + (n_1/\alpha_1)\ \ln(K_1/n_e) + (n_2/\alpha_2)\ \ln(K_2/n_e), \tag{13.5}$$

where $\alpha_{1,2} = e^2/a_{1,2}k_BT$.

Table 13.1 presents the electron density, as a function of n_1, the number density of particles of dust with a low work function (for $n_2 = 0$) and the values of n_2 the number density of particles of high work function, which restores the electron density to the value in absence of dust. The relevant parameters are:

$$V = 10\ \text{eV}, T = 2318\ \text{K},\ n = 10^{19}\text{cm}^{-3},\ \text{radius}\ a = 480\ \text{Å}, \phi_1 = 3\ \text{eV},\ \phi_2 = 15\ \text{eV}.$$

It can be concluded that the presence of dust can cause a large change in electron density in an exhaust plasma. The simple theory explains electron density in far excess of that predicted by application of Saha's equation to the gaseous component. Further dust of high work function can drastically reduce the electron density. This theory is based on (13.3), which is not valid when ions are taken into account. An improved theory is based on (5.1a, 5.1b, 5.1c and 5.1d), (5.3), (5.4). A satisfactory theory, which also accounts for size distribution of dust has been developed by Sodha et al. [17].

13.3 Impact of Rocket Exhausts on Ionosphere and Upper Atmosphere

13.3.1 Early Work

Booker [3] suggested that powered rocket flights through the F region of ionosphere caused an extensive and drastic reduction of electron density (commonly known as a hole), which may persist for a period of the order of 1 h. This fact was confirmed subsequently in a number of rocket flights; the hole persisted after the flight for long times of the order of 1 h.

One of the most spectacular demonstration of the effect was the launch of Skylab by Saturn V carrier rocket in 1973 (Mendilo et al. [8, 9], which caused large-scale depletion of the ionosphere. Simulations Mendilo and Forbes [10] based on the reactions of H_2 and H_2O gases from the exhaust with the ionospheric constituents explained that the disturbances in the ionosphere could cover an area of the order of one million square kilometer and occur at distances of the order of a thousand kilometer from the flight path. All these observations were made by the radio methods.

Launch of powerful rockets and operation of space craft are characterized by exhaust of complex plasma, consisting of gases (neutral, electron, ions) and dust with a size distribution. This complex plasma causes the formation of cloud-like structures, which have been extensively studied (e.g., Platov et al. [12]).

13.3.2 Optical Observations

The large-scale optical phenomena, associated with the flight of powerful rockets in the upper atmosphere may be classified [5, 12] as follows:

13.3.2.1 Stratospheric Phenomena (Height 40–50 km)

These phenomena occur on account of the discharge of excess fuel components after the separation of the first stage of the rocket and the exhaust of the brake engines of the separated stages. The main features of the exhaust formation are:

(i) Small expansion and hence growth, limited by diffusion. (The total mass of exhausted material in the stratosphere was estimated to be of the order of 500 kg by Chernouss [5]. This amount is indicative of the pollution at these heights.)
(ii) long lifetime and
(iii) high brightness.

13.3.2.2 Turbo Pause Phenomena (Height 100–120 km)

These phenomena, observable in twilight occur on account of the scattering of light by an extended cloud of the exhaust material. The scattered light is very bright and can be observed visually at a distance of 1,000 km. These clouds with cross-sectional size of 100–200 km expand with a velocity of 2 km/s. The location of the phenomenon is determined by the path length of the solid particles in the exhaust.

At altitudes below 100 km, the exhaust trail retains a small cross-sectional size and is hence optically bright. For heights above 120 km. the solid particles expand freely and the trail is less bright.

13.3.2.3 Large-Scale Dynamic Phenomena (Height 150 km or more)

These classes of phenomena are caused by certain modes of operation of the rocket engine. For example, during the shutoff of solid fuel rocket motors, there is a sudden drop of pressure in the combustion chamber, resulting in practically instant ejection of large quantities (hundreds of kilogram) of fuel components and incomplete combustion products in the atmosphere. The clouds thus created can in exceptional cases rise to a height of 700 km and get a cross-sectional size of about 1,500 km and an expansion velocity of 2–3 km/s.

13.3.2.4 Weak Optical Phenomena

Another class of gas dust structures is the conical formation behind working rocket engines at large heights (200–400 km), under dark night conditions. The characteristic size is 200–400 km. The optical effect is due to the scattering of sunlight by the ice particles, formed by the condensation of water vapor in the exhaust. The size of the formation depends on the life time of ice particles, determined by the rate of sublimation for typical conditions the life time is 100–200 s with an expansion velocity of 3 km/s, the characteristic size is 300–600 km.

13.3.3 Nature of Dust

As mentioned before, ice particles are formed in the rocket exhausts on account of condensation of water vapor, which is a product of combustion. Due to sublimation, the ice particles have a limited life. Platov and Kosch [11] have presented a theory of sublimation of such particles.

The combustion in solid propellant rocket motors provides an exhaust, which contains dust of Al or Al_2O_3, having a long life. All longtime optical phenomena are thus associated with the use of solid fuel rockets.

The observed luminosity can be explained by assuming a 5–10 % condensation of water vapor into ice dust with a particle size of 100 Å (Kung et al. [7]).

13.3.4 Chemical Kinetics of Electron/Ion Depletion by Rocket Exhausts

The rocket exhaust molecules, H_2O and H_2, undergo charge–exchange reactions with the O^+ ions of the ionosphere, and the resulting molecular ions react with electrons with O^+ ions; the sequence is (Zinn et al. [18])

$$O^+ + H_2O \rightarrow H_2O^+ + O, \tag{13.6}$$

followed by

$$H_2O^+ + e^- \rightarrow H + OH. \tag{13.7}$$

Reaction (13.7) is about 10^5 times faster than the direct recombination of electrons with O^+ ions, i.e.,

$$O^+ + e^- \rightarrow O + h\nu. \tag{13.8}$$

Reaction (13.6) is much faster than either of the normally occurring F-layer charge-transfer reactions

$$O^+ + O_2 \rightarrow O_2^+ + O \tag{13.9}$$

or

$$O^+ + N_2 \rightarrow NO^+ + N.$$

The OH radical formed in reaction (13.7) can react further with O^+. The reaction is

$$O^+ + OH \rightarrow O_2^+ + H,$$

which is followed by rapid neutralization of the O_2^+ viz

$$O_2^+ + e^- \rightarrow O(^1D) + O.$$

In most cases, however, the OH radical is destroyed by reaction with atomic oxygen, i.e.,

$$OH + O \rightarrow H + O_2 \tag{13.10}$$

before reactions (13.6) and (13.7) proceed to a significant extent.

The result of the two cycles viz (i) Reactions (13.6) and (13.7) and (ii) Reactions (13.8), (13.10), and (13.7) in sequence, is the destruction of between one and two electron–ion pairs by each H_2O molecule.

Similar processes occur with other common rocket-exhaust products, such as H_2 or CO_2; with H_2, the sequence is

$$O^+ + H_2 \rightarrow OH^+ + H$$

followed by

$$OH^+ + e^- \rightarrow \left\{ \begin{matrix} O + H \\ O(^1D) + H \end{matrix} \right\}.$$

The state $O(^1D)$ is metastable; electronically excited oxygen atom decays primarily by emission of 630.0 nm radiation.

13.3.5 Role of Dust

The role of dust in optical phenomena has been extensively investigated; however, the role of dust in the evaluation of electron density in the ionosphere-rocket exhaust interaction has not been investigated to a significant extent. It may be appreciated that the dust will in general reduce the electron density on account of high rate of electron accretion on the dust particles and hence it may play a significant role in the formation of ionospheric holes by rocket exhausts. However, during day time the photo electric emission from dust (particularly dirty ice particles, with low work function) tends to enhance the electron density.

Hence, a careful investigation of the kinetics of upper atmosphere-rocket exhaust system taking into account significant electron production/annihilation and associated processes is called for; dust must be taken as an important constituent in these studies.

References

1. E.J. Baghdady, O.P. Ely, Proc. IEEE **54**, 1134 (1966)
2. P.A. Bernhardt, G. Ganguli, M.C. Kelley, W.E Swartz, *J. Geophys. Res.* **100**, **23811** (1995)
3. H.G. Booker, J. Geophys. Res. **66**, 1073 (1961)
4. W.H. Brake, E.F.S. Howell, Radio frequency propagation to and from ICBMS and IRBMS. In: *Proceedings of IRE National Symposium on Space Electronics and Telemetry* (1959)
5. S.A. Chernauss, A.S. Kirrilov, Yu. V. Platov, *Optical Features of Rocket-Exhaust Products Interaction with the Upper Atmosphere, 17th European Space Agency Symposium on European Rocket and Balloon Programs and Related Research, Sandefford, Norway* (May 10 to June 2, 2005), ESA-SP-590
6. H. Einbinder, J. Chem. Phys. 26, 948 (1957)
7. R.T.V. Kung et al., AIAA J. **13**, 432 (1975)
8. M. Mendilo, G.S. Hawkins, JA Klobuchar, J. Geophys. Res. **80**, 2217 (1975a)
9. M. Mendilo, G.S. Hawkins, J.A. Klobuchar, Science **187**, 343 (1975)
10. M. Mendilo, J.A. Forbes, J. Geophys. Res. **83**, 151 (1978)
11. YuV Platov, M.J. Kosch, J. Geophys. Res. **108**, 1434 (2003)
12. YuV Platov, S.A. Chernouss, M.J. Kosch, J. Spacecr. Rocket. **41**, 667 (2004)
13. R. Sehgal, 9th *Symposium (international) on Combustion: Abstracts* (The Combustion Institute, Pittsburgh, 1962)
14. P.K. Shukla, A.A. Mamun, *Introduction to Dusty Plasma Physics* (Institute of Physics Publishing, Bristol, 2001). p. 17
15. F.T. Smith, J. Chem. Phys. **28**, 746 (1958)
16. M.S. Sodha, C.J. Palumbo, JT Daley, Brit. J Appl. Phys. **14**, 916 (1963)
17. M.S. Sodha, S.K. Mishra, S. Misra, IEEE Trans. Plasma Sci. **39**, 1141 (2011)
18. J. Zinn, C. Dexter Sutherland, S.N. Stone, L.M. Duncan, R. Behnke, *Ionospheric effects of rocket exhausts, Report DOE/ER- 0082, Dist Category UC-34b* (Los Alamos Scientific Laboratory, Los Alamos, 1980)

Chapter 14
Kinetics of Complex Plasmas with Liquid Droplets

14.1 Introduction

The kinetics of complex plasmas with liquid droplets has been discussed in the context of two applications. First, [14] the alkali metal vapor, used as the working gas in a nuclear reactor-based MHD power generation cycle, can partially condense into droplets under certain operating conditions. The wet vapor, so formed can be considered to be a complex plasma with neutral atoms, ions electrons, and charged sodium liquid droplets. The role of droplets is to enhance the electron density by thermionic electron emission from the surface and reduction of electron density due to electron accretion on the surface. A satisfactory analysis of the kinetics of such a process is not available.

The other application [15] is the use of water spray for reduction of electron density in the plasma sheath in front of a hypersonic vehicle in the upper atmosphere. The twin application to rocket exhausts has not been considered.

In this chapter, we briefly consider both the applications.

14.2 Wet Alkali Metal Vapor (After Smith [14])

Smith [14] has analyzed electron conduction in a wet alkali metal vapor in an electric field ($\upsilon\, x\, \boldsymbol{B}$ in an MHD generator). Smith [14] considered the complex plasma to comprise of neutral gas at temperature T_g, electrons at temperature T_e, and ions and droplets at the temperature T_d. The simplified basic equations are

$$n_e n_i / n_0 = 2\left(2\pi m_e k_B T_e / h^2\right)^{3/2} \exp(-eV_i / k_B T_e), \tag{14.1a}$$

where n_e, n_i, n_o are the electron, ion, and neutral atom densities, and V_i is the ionization potential of the alkali metal. Equation (14.1a) implies that even in non-equilibrium conditions Saha's equation is valid when the temperature occurring in the equation refers to the electron temperature [7]; this assumption does not have a theoretical basis but is based on empirical observations [5] in a small range of

M. S. Sodha, *Kinetics of Complex Plasmas*, Springer Series on Atomic, Optical, and Plasma Physics 81, DOI: 10.1007/978-81-322-1820-3_14,

parameters. The charge balance on the droplets, for negative charge, on droplets was given by

$$n_{\mathrm{ec}} = 4\pi a^2 n_{\mathrm{th}}. \tag{14.2a}$$

Equation (14.2a) omits an important term n_{ic}, corresponding to the rate of ion accretion on the droplets.

The temperature of the droplets is given by

$$n_{\mathrm{ec}} k_B (T_e - T_d) = n_o \pi a^2 \left(8\pi k T_g / m_o\right)^{1/2} \left(T_d - T_g\right)\cdot \tag{14.3a}$$

The left-hand side represents heat transfer from the electrons to a droplet while the right-hand side represents the heat lost by the droplet to the gas. As explained in Part I of this book, there should be factors of 2 and 3/2 on the left- and right-hand side. This equation does not take into account the exchange of energy in condensation and evaporation from the droplets and collision of neutral atoms and ions with the droplets.

The energy equation, corresponding to segmented electrode geometry for electrons is

$$K(1-K)\frac{e}{m v_e} \upsilon^2 B^2 = 2 v_o \delta \frac{3}{2}\left(k_B T_e - k_B T_g\right) + n_{\mathrm{ec}} 2 k_B (T_e - T_d). \tag{14.4}$$

where K = E/vB

The charge neutrality may be expressed by

$$n_e = n_i + Z n_1. \tag{14.5a}$$

The number of Potassium atoms in the gas per unit volume is $N_{Na} = p/k_B T$ where p the pressure. If the number of atoms in the liquid state is a fraction f of N_{Na}, the number of droplets having n_1 atoms is $n_d = fN_{na}/n_1$. The above set of equations enables one to evaluate n_e and hence evaluate the transport parameters as indicated in Chap. 8.

Figure 14.1 shows the dependence of n_e/n_{eo} with the electron temperature where n_{e0} is the electron density in absence of droplets. The ionization potential and work function of the alkali metal (potassium) are 4.34 and 2.15 eV, respectively. The vapor corresponds to K vapor at 800 K and 0.1 atmospheres with 75 % moisture.

The inadequacies in [14] theory have been pointed out before in this section.

Kovacs [8] has considered situations, where the electron density gets enhanced by droplets on account of the rate of thermionic emission, exceeding the rate of electron accretion.

Based on the previous chapters in the book, one can formulate a theory of electrical conductivity of Potassium vapor at a temperature T, less than T_0 the temperature at which the vapor is saturated.

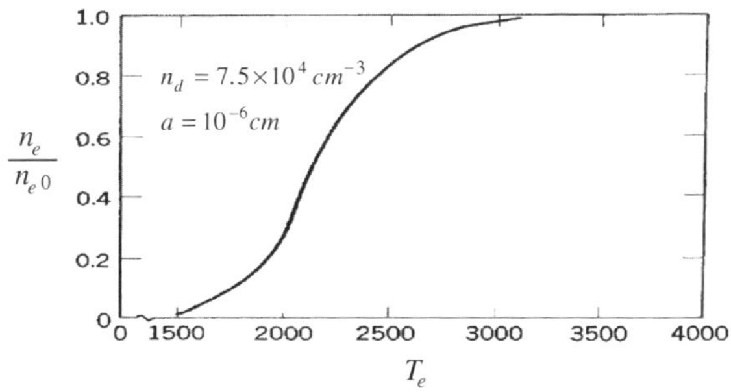

Fig. 14.1 Dependence of electron density on electron temperature in wet Potassium vapor

The density $\rho(T)$ of the saturated vapor at temperature T is given by

$$\rho(T) = p_s(T) \cdot (M/RT), \tag{14.6a}$$

where $p_s(T)$ is the saturation vapor pressure at temperature T (see Lide [10]), M is the molecular weight, and R is gas constant.

Since the vapor is saturated at the temperature T, the mass of the condensate per unit volume is

$$\rho(T_0) - \rho(T) \tag{14.6b}$$

Hence, the number n_d of solid Potassium particles of radius a, per unit volume is given by

$$n_d = [\rho(T_0) - \rho(T)]/\left(4\pi a^3 \rho_s/3\right), \tag{14.6c}$$

where ρ_s is the density of solid Potassium.

The number of Potassium atoms/ions in the vapor per unit volume is

$$N(T) = p_s(T)/k_B T. \tag{14.6d}$$

The kinetics of the complex Potassium plasma is given by

$$N = n_0 + n_i \tag{14.7}$$

$$n_0 n_i/n_e = 2\left(2\pi m_e k_B T/h^2\right)^{3/2} \exp(-eV_i/k_B T) \tag{14.1b}$$

$$n_d Z + n_i = n_e \tag{14.5b}$$

and

$$n_{ic}(Z) + n_{ee}(Z) - n_{ec}(Z) = 0 \tag{14.2b}$$

where the symbols have used meanings.

The simultaneous solution of these four equations can yield, n_e, n_i, n_o and Z.

The rationale behind the four equations is explained in detail in Part I of the book.

14.3 Reduction of Electron Density in a Plasma by Injection of Water Droplets

14.3.1 Early Work

Evans [4, 13] has analyzed a successful hypersonic flight experiment for relief of radio black out by the injection of a water spray in the ionized sheath in front of a re-entering (hypersonic) vehide in the atmosphere [3]. It is well known that the electron density in the plasma sheath is very much larger than that predicted by equilibrium considerations or application of Sahas' equation to the gaseous component. The electrons are removed from the plasma by accretion on the water droplets.

The kinetics of complex plasmas, with dust of solid particles has been discussed in sufficient detail in this book; the analysis is mainly in the context of metallic or metal-like dust. There are two essential differences between the behavior of solid particles and liquid droplets, which are dielectric in nature. Since the nature of the image force on an electron is different for metallic and dielectric spheres, the electron potential energy function $V(r)$ is different in the two cases, which manifests in different expressions (values) for the reduction in the work function due to a negative charge on the droplet/particle. This reduction (due to Schottky effect) is reflected in different expressions for the emission from and accretion of electrons on the surface.

Aisenberg et al. [1] have proposed that the thermal evaporation of ions from the surface of a droplet may be an important mechanism for the loss of charge by a negatively charged liquid droplet. Considering the liquid surface to be a plane one, in an electric field, these researchers evaluated the reduction of latent heat, corresponding to negative ions, taking the image force on an ion (outside the liquid) into account. Further, the ratio of the rate of evaporation of ions to that of molecules was assumed to be $\exp(|\Delta L|/k_BT_d)$ where ΔL is the reduction in latent heat due to charge on the droplets. The effect of the large abundance of neutral molecules on the surface as compared to ions was not considered.

Sodha and Evans [15] evaluated the reduction in work function by a negative charge on the droplets and also the electric field emission of electrons; it was seen that the role of electric field emission is negligible under typical working conditions. Further, the role of evaporation of ions from the surface of a charged water droplet was also revaluated, taking into account the relative abundance of ions and molecules on the surface of the droplet; These results may be used to investigate the kinetics of complex plasma, with a suspension of water droplets.

14.3.2 Reduction of Potential Energy Barrier

As may be seen from Chap. 3, plasma electrons need an energy larger than $-Ze^2/a$ to be able to accrete on the surface of a spherical particle of radius a with a charge $-Ze(Z > 0)$. The potential energy of an electron outside a dielectric sphere of radius a with charge $-Ze$ is given by (Menzel [16])

$$V(r) = \frac{Ze^2}{r} - \frac{e^2}{a} \cdot \frac{K-1}{K+1} \sum_1^{\infty} [1 + 1/n(1+K)]^{-1} (a/r)^{2n+2}, \tag{14.8a}$$

where K is the dielectric constant of the material of the particle viz water.

For water K is large and hence

$1 + 1/n(1 + K) \approx 1$; thus the above equation simplifies to

$$V(\xi) = \left(e^2/a\right)[(K-1)/(K+1)]U(\xi) \tag{14.8b}$$

where

$$U(\xi) = -Z'\xi^{-1} - \xi^{-4}/\left(1 - \xi^{-2}\right),$$

$$Z' = Z(K+1)/(K-1)$$

and

$$\xi = (r/a).$$

The maximum value U_m of $U(\xi)$ is given by

$$\mathrm{dU}/d\xi|_{\xi=\xi_m} = 0$$

and

$$U_m = U(\xi = \xi_m).$$

The reduction in the barrier height is

$$\Delta V = \left(e^2/a\right)(-Z' - U_m)(K-1)/(K+1). \tag{14.8c}$$

The dependence of $-Z' - U_m$ *on* $-Z'$ is tabulated in Table 14.1.

14.3.3 Rate of Accretion of Electrons on Droplets

The rate of accretion of electrons on a negatively charged particle is given by (Chap. 3)

$$n_{ec} = \pi a^2 (8kT_e/m_e\pi)^{1/2} \exp\left(Ze^2/ak_BT_e\right) \text{ when } Z<0. \tag{14.9a}$$

Table 14.1 Dependence of $-Z^{/}-U_m$ ***on*** $-Z^{/}$ (After Sodha and Evans [15], curtsey authors and publishers Springer)

$-Z'$	4498	2220	1248	553.5	198.0	100.0	48.06	20.37	10.75	6.357
$-Z'-U_m$	98.3	64.9	48.3	31.6	18.3	12.5	8.26	4.94	3.20	2.31

As seen in the previous section, an electron needs a minimum energy $V_m(\text{not} - Ze^2/a)$ to accrete on the surface. Hence, the above equation gets modified to

$$n_{ee} = \pi a^2(8kT_e/m_e\pi)^{1/2}\exp(-V_m/k_BT_e), \tag{14.3b}$$

where $V_m = (e^2/a)[(K - 1)/(K + 1)]U_m$ and U_m has been tabulated in Table 14.1.

14.3.4 Emission of Negative Ions from a Charged Droplet Due to Evaporations

14.3.4.1 Saturation Vapor Pressure Over Droplet Surface

Saturation vapor pressure p_s is an important parameter in the evaluation of the rate of evaporation. The saturation vapor pressure over a drop of radius a carrying a charge Ze is given by (Glosios [6]; Scharrer 17)

$$ln(p_s/p_{so}) = (M/\rho RT_d)\left[(2\sigma/a) + \left(Ze^2/8^4\right)\right],$$

where

p_{so} is the saturation vapor pressure over uncharged plane water surface, (Lide [10])

M is the molecular weight of water,

ρ is the density of water, and

σ is the surface tension of water.

For parameters of interest, the right-hand side of the above equation is very small and hence

$$p_s \approx p_{\text{so}}.$$

14.3.4.2 Evaporation of Neutral Molecules from the Droplet

Using the kinetic theory of evaporation [12], the number of neutral molecules evaporating per unit time from the surface of a droplet of radius a is

$$n_n = 4\pi a^2 \alpha_{ac}(n_s/4)(8k_B T_d/m_0\pi)^{1/2},$$

where $n_s = p_s/k_B T_d$ is the number of water molecules per unit volume in saturated water vapor at temperature T_d, m_o is the mass of a water molecule and α_{ac} is the accommodation factor.

Hence, putting $p_s = n_s k_B T_d$ in the above equation.

$$n_n = \pi a^2 p_s \alpha_{ac}(8/m_o \pi k_B T_d)^{1/2}, \quad (14.9b)$$

For water at about 100 °C, $\alpha_{ac} = 0.08$ [2]. However, for water droplets in high temperature gases $\alpha_{ac} = 0.8$ [9]. The parameter p_s is just the atmospheric pressure and the liquid is boiling.

14.3.4.3 Number of Neutral Molecules on the Surface of the Droplet

Assuming that the molecules have a typical dimension d, the number of molecules on the surface is

$$n_n^{/} \approx 4\pi a^z/d^2.$$

where d is a typical dimension of the molecule such that

$$\Lambda d^3 \approx M/\rho,$$

where Λ is Avagadro's number and ρ is the density of water.

Hence

$$n_n^{/} \approx 4\pi a^2 (M/\rho\Lambda)^{-2/3}. \quad (14.10)$$

For other shapes of molecules the RHS of (14.10) gets multiplied by a factor of the order of unity.

14.3.4.4 Lowering of Surface Barrier

The molecules in the droplet have to overcome a potential energy barrier to get evaporated. Following the logic in Sect. 14.3.2, for negative ions, evaporating from the charged drop, the barrier gets lowered by

$$\chi = Ze^2/a - V_m = k_B T_d \alpha_d^{/}\left(-Z^{/} - U_m\right),$$

where

$$\alpha_d^{/} = (K-1)\left(e^2/ak_B T_d\right)/(K+1).$$

14.3.4.5 Ion Evaporation Rate

The rate of ion evaporation from the surface n_{ie}^{-} is given by

$$n_{ie}^{-} / n_n = \left(|Z|/n_n^{\prime}\right) \exp(\chi/k_B T_d),$$

since the ions reside on the surface.

Substituting for n_n from (14.4), for $n_n^{\prime}$ from (14.5a), and for χ from (14.6a), (14.6b), (14.6c), (14.6d) in the above equation one obtains

$$n_{\mathrm{ie}}^{-} = \alpha_{\mathrm{ac}}|Z|p_s(M/\Lambda\rho)^{2/3}(2\pi m_o k_B T_d)^{-1/2}\exp\left[\alpha_d^{\prime}\left(-Z^{\prime} - U_m\right)\right]. \quad (14.11)$$

14.3.5 Kinetics of Complex Plasma with Water Droplets

Sodha and Evans [15] have made a preliminary analysis of the kinetics of a plasma sheath with water droplets. Misra et al [11] have recently developed an updated version of the kinetics of a plasma with an easily ionizable CO_2 like gas, water vapor, and suspended water droplets. The fundamentals of the kinetics viz the charge balance on the droplets and number as well as energy balance of the constituents (neutral atoms, positive and negative ions, and the droplets) have been incorporated in the formulation along with the maintenance of the plasma by the ionization of molecules of the easily ionizable gas and recombination of corresponding ions and electrons. The charge neutrality is a consequence of the number balance of electrons/ions and the charge balance on the droplets and hence has not been considered separately. The expression (14.11) for n_{ie}^{-}, the rate of evaporation of negative ions from the surface of the droplets as derived by Sodha and Evans [15] is an important input in the problem.

In the present analysis [11]; which ignores second-order effects one considers a suspension of water droplets in a CO_2 like plasma, which in the absence of the droplets has an electron density n_{eo} and temperature T_{eo}, singly charged positive ions with a density $n_{\mathrm{io}} = n_{\mathrm{eo}}$ and temperature T_0. The ions and neutral atoms are assumed to be at the same temperature T_0 which is a reasonable assumption on account of efficient energy exchange between the ions and the neutral atoms. In the presence of water droplets (which get negatively charged on account of the accretion of electrons/ions) one also has molecules and negatively charged ions of water in the complex plasma.

If one considers this complex plasma to be moving in the x direction with a velocity $\boldsymbol{i}v$ one can write a system of equations to characterize the kinetics of the system as follows.

Charging of droplets

$$\frac{\mathrm{dZ}}{\mathrm{dt}} = n_{\mathrm{ie}}^{-} + n_{\mathrm{ic}}^{+} - n_{\mathrm{ic}}^{-} - n_{\mathrm{ec}},$$

where

$$\frac{\mathrm{d}}{\mathrm{dt}} = \frac{\partial}{\partial t} + \upsilon \frac{\partial}{\partial x},$$

n_{ic}^{-}, n_{ic}^{+}, and n_{ec} are respectively the accretion rates of negative ions, positive ions, and electrons on a droplet, and n_{ie}^{-} is the rate of negative ion evaporation from the droplet. For steady state and stationary plasma

$$\mathrm{d}/\mathrm{d}t = \mathrm{d}/\mathrm{d}\tau$$

when $\tau = x/\upsilon_x$ for the steady state moving plasma and $\tau = t$ for a stationary plasma.

Hence, one may write (14.9a) and subsequent equations in terms of $\frac{\mathrm{d}}{\mathrm{d}\tau}$; thus

$$(\mathrm{dZ}/\mathrm{d}\tau) = \mathrm{n}_{\mathrm{ie}}^{-} + \mathrm{n}_{\mathrm{ic}}^{+} - \mathrm{n}_{\mathrm{ic}}^{-} - \mathrm{n}_{\mathrm{ec}} \tag{14.12}$$

Number balance of electrons

$$(\mathrm{dn_e}/\mathrm{d}\tau) = \left(\beta n_o - \alpha n_e n_i^{+}\right) + \left(\alpha_d n_i^{-} - \alpha_a N_o n_e\right) - n_d n_{\mathrm{ec}}\,. \tag{14.13}$$

where

βn_0 is the rate of production of electron-positive ion pairs per unit volume,
$\alpha_{\mathrm{ei}} n_e n_i^{+}$ is the rate of recombination of positive ions with electrons per unit volume,
$\alpha_d n_i^{-}$ is the rate of electron detachment from negative water ions per unit volume,
$\alpha_a N_0 n_e$ is the rate of electron attachment to water molecules per unit volume,
N_0 is the number density of water molecules,
n_{ec} is the rate of electron accretion on a droplet, and
n_d is the number of droplets per unit volume.

Number balance of positive ions (***of ionizing gas***)

$$(\mathrm{dn_i^{+}}/\mathrm{d}\tau) = (\beta \mathrm{n_o} - \alpha_{\mathrm{ei}} \mathrm{n_e} \mathrm{n_i^{+}}) - \mathrm{n_d} \mathrm{n_{ic}^{+}} \tag{14.14}$$

Number balance of negative ions of water

$$(\mathrm{dn_i^{-}}/\mathrm{d}\tau) = \mathrm{n_d} \mathrm{n_{ie}^{-}} + (\alpha_\mathrm{a} \mathrm{N_o} \mathrm{n_e} - \alpha_\mathrm{d} \mathrm{n_i^{-}}) - \mathrm{n_d} \mathrm{n_{ic}^{-}} \tag{14.15}$$

Conservation of sum of positive ion density and neutral molecular density (***of ionizing gas***)

$$n = n_0 + n_i^{+} \tag{14.16}$$

Number balance of neutral water molecules

$$\frac{\mathrm{dN_o}}{\mathrm{d}\tau} = \alpha_d n_i^- - \alpha_a N_o n_e - n_d N_{\mathrm{oc}} + n_d n_n. \tag{14.17}$$

Energy balance of electrons

$$\frac{\mathrm{d}}{\mathrm{d}\tau}\left(\frac{3}{2}k_B n_e T_e\right) = \left[\beta n_o \varepsilon_e - \alpha n_e n_i^+ (3k_B T_e/2)\right] + \left[\alpha_d n_i^- \varepsilon_{di} - \alpha_a N_o n_e (3k_B T_e/2)\right] - n_d n_{\mathrm{ec}} \varepsilon_{\mathrm{ec}} \tag{14.18}$$

where $\varepsilon_e = (3/2)k_B T_{\mathrm{eo}}$ is the energy a the electrons ion produced by ionization, in recombination and attachment of on electron an energy $(3/2)k_B T_e$ is lost, ε_{di} is the energy of the electron produced by detachment from negative water ions and $\varepsilon_{\mathrm{ec}}$ is the mean energy (at large distance from the drop) of accreting electrons on the droplet.

Energy balance of neutral molecule and ions

We have so far considered monoatomic gases and their ions and hence the mean energy $(3/2)k_B T$ was justified. However, water vapor corresponds to a mean energy of $(7/2)k_B T$ per molecule as per data [10] on specific heat at constant pressure. Similarly, for the other CO_2 like gas, corresponding to combustion products we use a value $6k_B T$, corresponding to CO_2. Keeping this in mind the energy balance of neutral molecules and ions may be expressed as

$$\begin{aligned}\frac{\mathrm{d}}{\mathrm{d}\tau}&\left[(7k_B T_o/2)\left(N_o + n_i^-\right) + 6k_B T_o\left(n_o + n_i^+\right)\right] = \beta n_o\left[\varepsilon_i^+ - (6k_B T_o)\right]\\ &+ \alpha n_e n_i^+\left[(3k_B/2)T_e + I_p\right] - n_d n_{\mathrm{ic}}^+\left[\varepsilon_{\mathrm{ic}}^+ - (6k_B T_d/2)\right]\\ &+ n_d n_{\mathrm{ie}}^-\left\{\left[(7k_B T_d/2) - Ze^2/a\right]\right\} - n_d n_{\mathrm{ic}}^- \varepsilon_{\mathrm{ic}}^- - \alpha_d n_i^- \varepsilon_{\mathrm{di}}\\ &- N_{\mathrm{oc}}(7k_B/2)T_0 - n_n n_d \cdot 7k_{\mathbf{B}}(T_0 - T_d)/2\end{aligned} \tag{14.19}$$

where (i) $\varepsilon_i^+ = 6k_B T_{io}$ is the energy of ions produced by ionization of neutral molecules with mean energy $6k_B T_o$ (ii) an electron of energy $3/2k_B T_e$ and an ion of energy $6k_B T_o$ produce a neutral molecule of energy $[3/2k_B T_e + 6k_B T_o + I_p]$, and hence the net gain in energy of ions plus neutral molecules is $[3/2k_B T_e + 6k_B T_o + I_p] - 6k_B T_o = (3/2)k_B T_o + I_p$ (I_p is the ionization potential), (iii) in accretion a positive ion with mean energy ε_{ic}^+ is lost and a neutral molecule with energy $6k_B T_o$ gets generated (iv) a negative water ion evaporates and acquires an energy $(7k_B T_d/2)$ at the surface of the droplet, gets pushed by the negative charge on the droplet and thus has an energy $(7k_B T_d/2) - Ze^2/a$ at large distance from the droplet, (v) ε_{ic}^- is the mean energy of accreting negative water ions, (vi) an electron gets detached from a negative ion with an energy $\varepsilon_{\mathrm{di}}$, and (vii) N_{oc} is the rate of accretion of water molecules on the surface of the droplets and $(7k_B T_o/2)$ is the corresponding energy per molecule, and (viii) n_n water molecules get evaporated at temperature T_d per unit time per droplet.

In the evaluation of ε_{ic}^{-} and ε_{ic}^{+} we have to add $(7k_BT/2) - (3k_BT/2)$ and $(6k_BT) - (3k_BT/2)$ to the usual expressions (Chap. 3) for ions of water vapor and ionizable gas respectively because the non-translational energy is not affected by accretion. To obtain the energy of the ions on the surface $Z\alpha$ has to be added to the expressions for $\varepsilon_{\rm ic}^{+}$ and $\varepsilon_{\rm ic}^{-}$.

Energy balance of droplets

$$-L\rho\frac{\mathrm{d}}{\mathrm{d}\tau}\left(\frac{4}{3}\pi a^3\right) = n_{\rm ec}\varepsilon_{\rm ec}^{s} + n_{\rm ic}^{+}\varepsilon_{\rm ic}^{s+} + n_{\rm ic}^{-}\left(\varepsilon_{\rm ic}^{s-} + L_h\right) + 6k_Bn_{\rm oc}(T_o - T_d) - \varepsilon\sigma\left(T_d^4 - T_o^4\right)4\pi a^2 + N_{\rm oc}[(7/2)k_BT_0 + ML/\Lambda] \quad (14.20)$$

where

M	is molecular weight of water,
Λ	is Avagadro's number,
$L_h = ML/\Lambda$	is the latent heat of vaporization/condensation per molecule/ion, and
L	is the latent heat of vaporization per unit mass.

Simultaneous solution of (14.12)–(14.18) can be obtained, using the Mathematica software.

In steady state, the value of parameters Z, n_e, n_i^{+}, n_{w}^{-}, n_o, n_w, T_e, and T, it can be determined by simultaneous solution of (14.9a, 14.9b)–(14.18) as $t \to \infty$; the steady state values are, in general, independent of initial conditions. The initial conditions are at $\tau = 0$ are $Z = 0$, $n_e = n_{\rm eo}$, $n_i^{+} = n_{\rm io}^{+}$, $n_{w}^{-} = 0$, $n_o = n_{\rm oo}$, $n_w = 0$, $T_e = T_{\rm eo}$, $a = a_0$ and $T = T_{\rm oo}$, while $T_d = 100$ °C.

For ease of computation we have put $\chi = 0$

For a numerical appreciation of the results, the following standard parameters have been chosen.

$$a_o = 10^{-4}\,\text{cm}, n_{\rm eo} = 10^{14}\,\text{cm}^{-3}, n_{\rm io}^{+} = 10^{14}\,\text{cm}^{-3}, n_{\rm oo} = 10^{3}n_{\rm eo}, T_{\rm eo} = 5000\,\text{K},$$
$$T_{\rm io} = T_o = 2000\,\text{K}, T_d = 373\,\text{K}, T_w = 273\,\text{K}, L = 450\,\text{C/g}, I_p = 13.77\,\text{eV},$$
$$\Lambda = 6.06\times10^{23}, M = 18, \alpha_{\rm ac}, \rho_s = 1\,\text{atmosphere}, \rho = 1\,\text{gm/cc}$$

Figure 14.2 illustrates the transient evolution of the dimensionless radius of the water droplet and its dependence on the number density of the water droplets.

Figure 14.3 represents the time dependence of the electron density on the number density of droplets. The electron density of the plasma decreases with increasing density of the water droplets, due to larger number of water droplets, available for accretion of electrons. When the droplets get fully evaporated, the electron density attains a steady value.

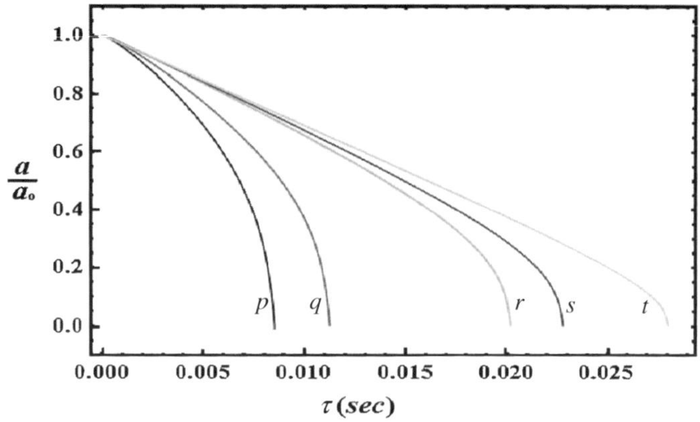

Fig. 14.2 Dependence of radius of droplet on τ; p, q, r, s, t refer to n_d (5×10^4, 10^5, 5×10^5, 10^6, and 5×10^6) cm^{-3}, respectively (after Misra et al. [11])

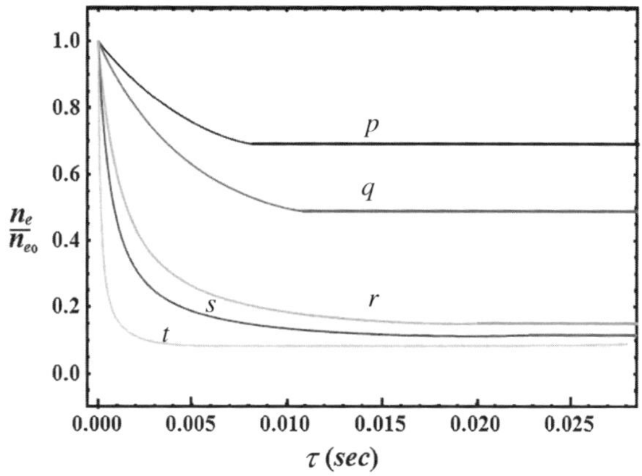

Fig. 14.3 Dependence of electron density on τ; p, q, r, s, t refer to $n_d = (5 \times 10^4, 10^5, 5 \times 10^5, 10^6$, and $5 \times 10^6)$ cm^{-3}, respectively (after Misra et al. [11])

References

1. S. Aisenberg, R.W. Chang, P.N. Hu, *Modificetion of Plama by Rapidly Evaporating Liquid Additives, Semiannual Report of Contract No. DAHCO4-68C-0031*, (Space Sciences Inc., Waltham, Mass., USA, 1969), p. 02154
2. T. Alty, Philos. Mag. **15**, 82 (1933)
3. W.F. Cuddihy, I.E. Beckwith, L.C. Schroeder, *NASA TM X-**1092*** (1963) declassified (Dec. 30, 1970)
4. J.S. Evans, *NASA TM X* 1186 (1965) declassified (Dec. 30, 1970.)
5. J.S. Evans, *Third Symposium on Plasma Sheath-Plasma Electromagnetics of Hypersonic Light,* vol III, (Air force Cambridge Research Lab., Bedford, Mass USA, 1967)
6. T. Glosios, Ann. Phys. **34**, 446 (1939)

7. J. Kerrebrock, *Second Symposium on Engineering Aspects of MHD* (Columbia University Press, New York, 1962), p. 327
8. C. Kovacs, P. Halasz, J.W. Hansen, Acta. Tech. Hungr. **56**, 383 (1966)
9. S.C. Kurzius, F.H. Raab, *Vaporization and Decomposition of Condidate Reenty Blackout Suppresants in Low pressure Plasmas, NASA CR-1330* (Aerochem Research Lab Inc, Princeton N.J., 1969)
10. D Lide (Editor), *CRC Handbook of Physics and Chemistry,* (CRC Press, Boca Raton, 2004)
11. S. Misra, S.K. Mishara, M.S. Sodha, Phys. Plasmas, **20**, 123701 (2013)
12. J.R Partington, *An Advanced Tractise on Physical Chemistry*, vol. 2, (Wiley, New York, 1962), p. 291
13. J.S. Evans in report AFCRL-67-0280 (Vol. III). *Special Report No 64 (III)*, ed. by W. Rotman, Moore, R. Papa, J. Lennon, pp. 343–361
14. J.M. Smith, AIAA J. **3**, 648 (1965)
15. M.S. Sodha, J.S. Evans, Appl. Sci. Res. **29**, 380 (1974)
16. D.H. Monzel, Fundamental Formulas of Physics, Vol. 1, 214 (1960)
17. L. Scharrer, Ann. Phys. **35**, 619 (1935)

Chapter 15
Growth of Particles in a Plasma

15.1 Introduction

The presence of particles in a plasma reactor is of serious concerns to microelectronic industry, material science, and other areas. Dust in plasma reactors causes irrecoverable defects and line shorts in large-scale integrated circuits. However, the emphasis has recently changed from considering dust as a process killer to recognizing it as a desired feature with numerous applications like (i) plasma-assisted assembly of carbon-based nanostructures, (ii) plasma-enhanced CVD of nanostructure silicon-based films, (iii) high rate deposition of clusters and particles on nanostructure films, etc. A discussion of the deleterious and beneficial aspects of complex plasma in micro electronic industry has been given in the classic books by Vladimirov et al. [4] and Ostrikov [6]. For a rational design of such applications, it is necessary to have an appreciation of the physics and chemistry of the formation and growth of the dust particles in a plasma.

Perrin and Hollenstein [2] have authored an excellent chapter on sources and growth of particles; these authors have covered the time sequence from (i) the formation of primary cluster of atoms up to a critical size, followed by nucleation, (ii) growth of small particles by condensation ($a < 5$ nm), (iii) coagulation of small particles into macroscopic particles by condensation ($a < 50$ nm), and (iv) independent growth of the macroscopic particles ($a > 50$ nm) by condensation of neutral atoms and ions. With the background of Part I of this book, we will confine our attention to a simple model of the independent growth of the embryonic particles ($a > 5$ nm) by condensation of neutral atoms and ions.

Haaland et al. [1] analyzed a simple model of the complex plasma and growth of particles. These authors considered a plasma consisting of (i) neutral atoms/molecules of A and B, (ii) embryonic particles (A) with charge $-e$ (to enable efficient accretion of positive ions), (iii) ions of A and B, and (iv) electrons. The atoms and ions of A condense on the particles while atoms/ions of B can exchange charge and energy in encounters with dust particles. The initial charge $-e$ on the embryonic particles initiates the acquisition of ions of A.

M. S. Sodha, *Kinetics of Complex Plasmas*, Springer Series on Atomic, Optical, and Plasma Physics 81, DOI: 10.1007/978-81-322-1820-3_15,

However, Haland et al. [1] assumed that the plasma parameters are independent of time; this assumption violates the principles of number and energy balance. In this context, Sodha et al. [3] analyzed this model, on the basis of number and energy balance of the constituents and the charge balance on the particles; the charge neutrality follows from the number balance of the constituents and has therefore been not separately considered.

15.2 Basic Equations

In the context of a complex plasma flowing along the x direction with a velocity υ_x, the operator

$$\frac{\mathrm{d}}{\mathrm{d}t} = \frac{\partial}{\partial t} + \upsilon_x \cdot \frac{\partial}{\partial x}. \tag{15.1a}$$

For the two specific cases considered herein viz. steady state $\left(\frac{\partial}{\partial t} \to 0\right)$and stationary ($\upsilon_x = 0$) plasma (15.1a) reduces to

$$\frac{\mathrm{d}}{\mathrm{d}t} = \frac{\mathrm{d}}{\mathrm{d}\tau}, \tag{15.1b}$$

where

$$\tau = t \text{ for stationary plasma and } \tau = x/\upsilon_x \text{ for the steady state.} \tag{15.1c}$$

The initial ($\tau = 0$) radius of the dust particles a_0 is the minimum radius to hold a charge $-e$ and may be obtained by equating the rates of accretion of electrons and ions on the particle.

$$n_{\mathrm{ec}} = n_{\mathrm{icA}} + n_{\mathrm{icB}}.$$

Substituting for the accretion rates from Chap. 3 in the above equation one obtains

$$\gamma_e n_{e0}(T_{e0}/m_e)^{1/2} \exp\left(-e^2/a_0 k_B T_{e0}\right) = \left(1 + e^2/a_0 k_B T_{i0}\right) \left[n_{iA0}(T_{i0}/m_{iA})^{1/2} + n_{iB0}(T_{i0}/m_{iB})^{1/2}\right], \tag{15.2}$$

where

$n_{ec,}$ n_{icA} and n_{icB} are the rates of accretion of electrons and ions A and B on the particle, γ_e is the sticking coefficient of electrons to the surface of the particles, $\tau = 0$.
T_{e0}/T_{i0} are the initial ($\tau = 0$) electron/ion temperature, m_e, m_{iA} and m_{iB} are the masses of an electron ion A and ion B.

n_{e0}, n_{iA0} and n_{iB0} are the initial ($\tau = 0$) electron and ion densities.
The initial radius a_0 may be obtained from a numerical solution of (15.2).

15.2.1 Charging of Dust Grains

The charge on the particle is governed by the equation

$$(\mathrm{d}Z/\mathrm{d}\tau) = n_{iAc} + n_{iBc} - \gamma_e n_{ec}, \tag{15.3}$$

where the expressions for n_{ec} and n_{ic} (*A*, *B*) have been given in Chap. 3.

15.2.2 Number Balance of Electrons

$$(\mathrm{d}n_e/\mathrm{d}\tau) = (\beta_A n_A + \beta_B n_B) - (\alpha_A n_e n_{iA} + \alpha_B n_e n_{iB}) - \gamma_e n_d n_{ec}, \tag{15.4}$$

where

β_j is the coefficient of ionization of the constituent neutral atoms due to an external agency (e.g. R.F. discharge) and
$\alpha_j(T_e) = \alpha_{j0}(300/T_e)^{\kappa}$ cm^3/s is the coefficient of recombination of electrons and ions (Gurevich [5]).

15.2.3 Number Balance of Ions

$$(\mathrm{d}n_{iA}/\mathrm{d}\tau) = \beta_A n_A - \alpha_A n_e n_{iA} - n_d n_{iAc} \tag{15.5a}$$

and

$$(\mathrm{d}n_{iB}/\mathrm{d}\tau) = \beta_B n_B - \alpha_B n_e n_{iB} - n_d n_{iBc}. \tag{15.5b}$$

It may be remembered that all accreting ions either get deposited on the particle or get converted to neutral atoms by giving charge to the particle.

15.2.4 Number Balance of Neutral Atoms

$$(\mathrm{d}n_A/\mathrm{d}\tau) = \alpha_A n_e n_{\mathrm{iA}} - \beta_A n_A + n_d(1 - \gamma_{\mathrm{iA}}) n_{\mathrm{iAc}} - n_d \gamma_A n_{Ac}, \tag{15.6a}$$

and

$$(\mathrm{d}n_B/\mathrm{d}\tau) = \alpha_B n_e n_{iB} - \beta_B n_B + n_d n_{iBc}, \tag{15.6b}$$

where

$n_{jc} = \pi a^2 (8k_B T_n/m_j \pi)^{1/2} n_j$ is the neutral collection current to a dust particle; and
γ_A/γ_{iA} are the sticking coefficient for ions and atoms of A.

Equations (15.6a) and (15.6b) refer to the growth of the number density of neutral atoms (A and B). The first term of the right-hand side shows the gain in neutral particle density per unit time on account of electron–ion recombination in plasma while the next term corresponds to the decrease in the neutral density per unit time due to ionization. The third term corresponds to the gain in the neutral density per unit time on account of neutralization of the ions collected on the surface of the dust grain. The last term in (15.6a) refers to the accretion of the neutral atoms of species A on the dust particles. It may be remembered that the third term in (15.6a) corresponds to incident neutral atoms on the particle, not depositing on the particles but just giving their charge to the particle and getting converted to neutral atoms.

15.2.5 Radius of the Particles

$$\frac{\mathrm{d}}{\mathrm{d}\tau}\left(\frac{4}{3}\pi a^3 \rho\right) = (m_A \gamma_A n_{AC} + m_{iA} \gamma_{iA} n_{iAc}), \tag{15.7}$$

where

ρ is the density of the particle material; and
n_{AC}/n_{iAC} denote the rates of accretion of neutral atoms and positive ions of A.

15.2.6 Energy Balance of Electrons

$$\frac{\mathrm{d}}{\mathrm{d}\tau}\left(\frac{3}{2} n_e k_B T_e\right) = (\beta_A n_A \varepsilon_A + \beta_B n_B \varepsilon_B) - (3k_B/2)(\alpha_A n_e n_{Ai} + \alpha_B n_e n_{Bi}) T_e - n_{ec} \varepsilon'_{ec}(Z) \tag{15.8}$$

where

T_n is the temperature of the neutral atomic species,
T_d is the temperature of the dust particles,
ε_j is the mean energy of electrons produced by ionization of neutral atoms,

$$\varepsilon^l_{ec}(Z) = \varepsilon^s_{ec}(Z) - Ze^2/a$$

is the mean energy of the electrons, (at large distance from the grain surface) collected by the dust particles,

$$\varepsilon^s_{ec}(Z) = 2k_BT_e$$

is the mean energy of electrons at the surface, collected by the dust particles.

The first term on the right-hand side in (15.8) represents the power gained per unit volume by the electrons through ionization of the neutral atoms/molecules. The second term represents the energy loss per unit volume per unit time due to recombination with ions in the plasma. The last term is the power loss per unit volume by accreting electrons.

We have ignored the negligible loss of electron energy on account of elastic collisions with ions, neutral atoms, and dust particles on account of the large mass ratio.

15.2.7 Energy Balance for Ions

$$\frac{\mathrm{d}}{\mathrm{d}\tau}\left(\frac{3}{2}(n_{Ai}+n_{Bi})k_BT_i\right) = (\beta_A n_A \varepsilon_{iA} + \beta_B n_B \varepsilon_{iB}) - (3k_B/2)(\alpha_A n_e n_{iA} + \alpha_B n_e n_{iB})T_i - n_d(n_{iAc}\varepsilon^l_{iAc} + n_{iBc}\varepsilon^l_{iBc}) \tag{15.9}$$

where

$$\varepsilon^l_{ijc}(Z) = \left(\frac{2 - Z\alpha_{\mathrm{ji}}}{1 - Z\alpha_{\mathrm{ji}}}\right)k_BT_i,$$

is the mean energy of ions (at large distance from the grain surface), collected by the dust particles (Chap. 3).

The next two terms refer to the energy loss per unit volume per unit time due to the electron-ion recombination and accretion of ions on the surface of the dust particles.

15.2.8 Energy Balance for Neutral Species

$$\frac{\mathrm{d}}{\mathrm{d}\tau}\left(\frac{3}{2}(n_A+n_B)k_BT_n\right) = [(3k_B/2)(\alpha_A n_e n_{iA} + \alpha_B n_e n_{\mathrm{iB}})(T_e+T_i) + (\alpha_A n_e n_{\mathrm{iA}} I_{\mathrm{pA}} + \alpha_B n_e n_{\mathrm{iB}} I_{\mathrm{pB}})] + (3k_B/2)n_d[(1-\gamma_{\mathrm{iA}})n_{\mathrm{iAc}} + n_{\mathrm{iBc}}]T_d - (3k_B/2)n_d n_{\mathrm{Ac}}\gamma_A T_n - (3k_B/2)(\beta_A n_A + \beta_B n_B)T_n - E_{\mathrm{diss}}, \tag{15.10}$$

where

I_{pj} is the ionization energy of the constituent atomic species,

$$E_{\text{diss}} = (E_{A,\text{diss}} + E_{B,\text{diss}})$$

$E_{j,\text{diss}}$ is the energy dissipated per unit volume per unit time by neutral atoms to the surrounding atmosphere, and
T_a is the ambient temperature.

The dissipation energy may be reasonably assumed to be proportional to the difference between the temperature of the neutral atomic species and the ambient temperature. Thus one obtains

$$E_{j,\text{diss}} = E_{j,\text{diss0}}\left[\frac{(T_j - T_a)}{(T_{j0} - T_a)}\right] \tag{15.11}$$

The constant E_{diss0} may easily be obtained by imposing the ambient conditions of the complex system in (15.11) for both the constituent neutral species.

The first two terms on the right-hand side of (15.10) refer to the power gained per unit volume by the neutral species due to the recombination of electrons and ions. The third term is the power gain by neutral species per unit volume due to formation of neutrals at the surface of the dust grain on account of electron and ion collection currents. The next term corresponds to the power loss per unit volume due to the sticking accretion on the dust grains. The last but one term refers to the thermal energy lost per unit volume per unit time by neutral atoms due to ionization. It is assumed that the total energy thus gained by the neutral atoms gets dissipated to the surroundings; the last term refers to the power dissipation rate per unit volume by neutral atoms to the surroundings.

15.2.8.1 Energy Balance for the Dust Particles

$$\begin{aligned}\frac{d}{d\tau}\left(\frac{4}{3}\pi a^3 \rho C_p T_d\right) &= n_{ec}\gamma_e \varepsilon_{ec}^s + (3k_B/2)n_{Ac}[\gamma_A T_n + \left[n_{iAc}(\varepsilon_{iAc}^s + I_{pA}) + n_{iBc}(\varepsilon_{iBc}^s + I_{pB})\right] \\ &\quad - (3k_B/2)[(1-\gamma_{Ai})n_{iAc} + n_{iBc}]T_d - 4\pi a^2[\in \sigma(T_d^4 - T_a^4) \\ &\quad + \left[n_A(8k_B T_n/\pi m_A)^{1/2} + n_B(8k_B T_n/\pi m_B)^{1/2}\right]k_B(T_d - T_n)]\end{aligned} \tag{15.12}$$

where

$$\varepsilon_{ijc}^s(Z) = \left(\frac{2 - Z\alpha_{ji}}{1 - Z\alpha_{ji}} - Z\alpha_{ji}\right)k_B T_i$$

is the mean energy of collected ions at the surface of the dust particle,

C_p is the specific heat of material of the dust particle at constant pressure,
$\in$ is the emissivity of the material of the dust grains; and
σ is the Stefan–Boltzmann constant.

The first two terms on the right-hand side of (15.12) are the rate of energy transferred to the dust particle due to the sticking electron and neutral atom accretion. The third term is the power carried away by the neutral species per unit (generated by recombination of accreted ions and electrons) from the dust grains. The last term describes the rate of energy dissipation of the dust particles per unit volume through radiation and energy gained conduction to the host gas.

15.2.9 Numerical Results and Discussion

Let us consider the growth of graphite particles in an acetylene flame dusty plasma with R.F. discharge; the thermionic emission from the particles has been neglected. The computations have been made to investigate the dependence of the size and charge on the dust grains and other relevant parameters viz. n_e, $n_{\rm iA}$, $n_{\rm iB}$, n_A, n_B, T_e, T_i, T_n on the parameter τ for different values of n_d and γ_A by simultaneous solution of (15.4), (15.5a, b), (15.6a, b), (15.7), (15.8), (15.9), (15.10), (15.11), and (15.12) with appropriate boundary conditions viz. at $\tau = 0$ viz. $n_d = 10^6\ \text{cm}^{-3}$, $n_{\rm iA0} = 0.6 n_{e0}$, $n_{\rm iB0} = 0.4 n_{e0}$, $n_{A0} = n_{B0} = 5 \times 10^{10}\ \text{cm}^{-3}$, $n_{e0} = 10^9\ \text{cm}^{-3}$, $T_{e0} = 0.5$ eV, $T_{i0} = 2500$ K, and $T_{n0} = T_d = 2000$ K.

Further the other relevant parameters used in this investigation are as follows:

$$m_{ia} \approx m_a = 12\ \text{amu},\ m_{ib} \approx m_b = 20\ \text{amu},\ \alpha_{A0} = \alpha_{B0} \approx 10^{-7}\ \text{cm}^3/s,\ \in = 0.6,$$
$$\gamma_{iA} = \gamma_A = 1,\ C_p = 7 \times 10^6\ \text{ergs/gK},\ I_{pA} = 11.26\ \text{eV},\ I_{pB} = 10\ \text{eV},\ \varepsilon_A = 6.2\ \text{eV},$$
$$\varepsilon_B = 10.7\ \text{eV},\ \varepsilon_{iA} = 7.3\ \text{eV},\ \varepsilon_{iB} = 12.2\ \text{eV},\ \varepsilon_{A,\text{diss0}} = 42.9\ \text{eV},\ \varepsilon_{B,\text{diss0}} = 19.6\ \text{eV},$$
$$\kappa = -1.2,\ a_0 = 8.0 \times 10^{-8}\ \text{cm and } \rho = 2.5\ \text{g/cm}^3.$$

Figure 15.1 illustrates the dependence of the radius a of the dust particle on the parameter τ for different values of number density of the embryonic dust particle n_d. The figure indicates that the size of the dust grain increases with τ and attains a saturation value corresponding to depletion of the atoms and ions of A. For larger values of n_d the number of neutral atoms and ions of A, available for accretion per particle is small and hence the radius of the particle saturates at a lower value of a and sooner.

Figure 15.2 displays the dependence of the charge on the particle $-Ze$ with n_d.

Figure 15.3 indicates the dependence of saturation radius a and the saturation charge $-Ze$ on n_d.

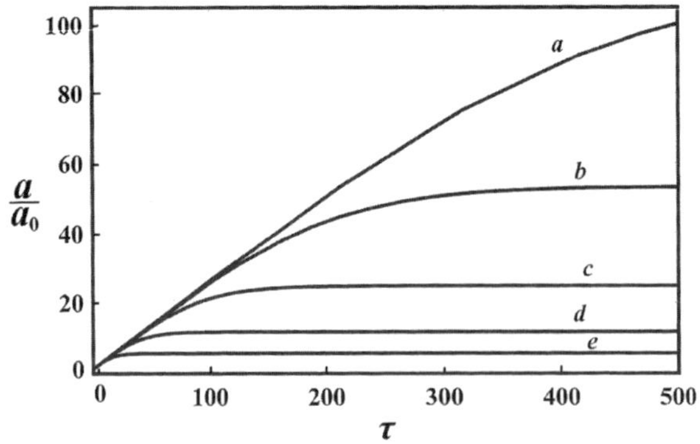

Fig. 15.1 Dependence of the radius of the particles (a/a_0) on the parameter τ: $n_{iA} = 0.6n_{e0}$, $n_{iB} = 0.4n_{e0}$, $n_{A0} = n_{B0} = 5 \times 10^{10}\ \text{cm}^{-3}$, $n_{e0} = 10^9\ \text{cm}^{-3}$, $T_{e0} = 0.5$ eV, $T_{i0} = 2500$ K, $T_{n0} = T_{d0} = 2000$ K at $\tau = 0$ and $\gamma_e = \gamma_{iA} = \gamma_A = 1$, $\gamma_B = \gamma_{iB} = 0$. The *curves a, b, c, d,* and *e* correspond to $n_d = 10^2$, 10^3, 10^4, 10^5, and $10^6\ \text{cm}^{-3}$, respectively (after Sodha et al. [4] curtsey authors and publishers AIP)

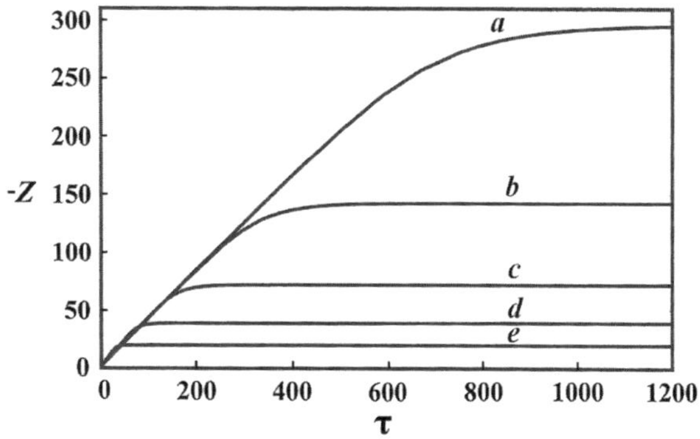

Fig. 15.2 Charge on particles as a function of τ for parameters, corresponding to Fig. 15.1. The *curves a, b, c, d* and *e* correspond to $n_d = 10^2$, 10^3, 10^4, 10^5, and $10^6\ \text{cm}^{-3}$, respectively (after Sodha et al. [4] curtsey authors and publishers AIP)

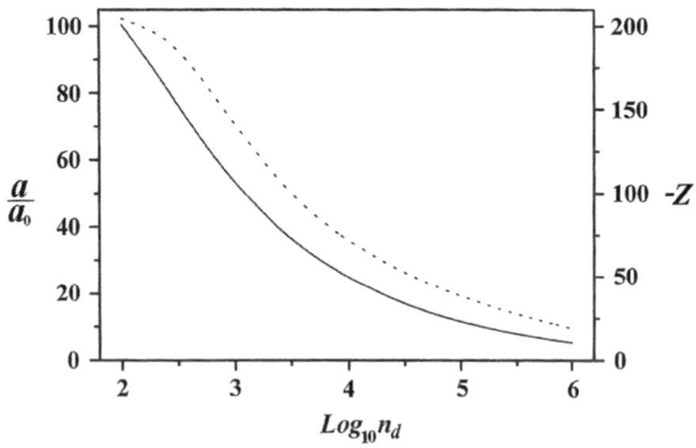

Fig. 15.3 Dependence of steady state of radius (a/a_0) and charge ($-Ze$) on the particles on the density of dust particles n_d; the parameters are the same as in Fig. 15.1 (after Sodha et al. [4] curtsey authors and publishers AIP)

References

1. P. Haaland, A. Garscadden, B. Ganguly, Appl. Phys. Lett. **69**, 904 (1996)
2. M.J. Perrin, C. Hollenstein, in *Dusty Plasmas: Physics, Chemistry and Technological Impacts in Plasma Processing,* ed. by A. Boucherville *Sources and Growth of Particles* (Wiley, New York, 1999)
3. M.S. Sodha, S. Misra, S.K. Mishra, S. Srivastava, J. Appl. Phys. **107**, 103307 (2010)
4. S.V. Vladimirov, K. Ostrikov, A.A. Samarian, *Physics and Applications of Complex Plasmas* (Imperial College Press, London, 2005)
5. A.V. Gurevich, Some nonlinear phenomena in the ionosphere (Springer, New York, 1978)
6. K. Ostrikov, Plasma Nanoscience, Wiley VCH (2008)

Chapter 16
Electrostatic Precipitation

16.1 Introduction

Suspended particles in air may be caused by industrial processes and use of wood or coal as a fuel. These particles are a menace from the health and aesthetics point of view. This has caused a widespread concern about the removal of such particles and enforcement of increasingly severe regulations.

There are a number of methods for removal of the suspended particles, but only the electrostatic precipitation is within the scope of this book on complex plasmas viz. particle charging, particle collection, and removal of the collected dust. We may begin this study by first considering corona discharge, which is a unipolar corona, comprising of neutral atoms/molecules and positive or negative ions.

16.2 Corona Discharge (After White [4], Oglesby and Nichols [2])

The unipolar corona, employed for electrostatic precipitation is a gas discharge, characterized by stability and self-maintenance. The discharge is maintained between a fine wire and a cylinder or a plate. The electrode spacing is a few centimeters and the gas pressure is around one atmosphere. The gas is ionized by the impact of high energy electrons, produced by acceleration due to the high electric field around the wire. An accelerated electron causes a number of electron–ion pairs, to be produced. In case of a negative corona, the positive ions are drawn to the wire and released as molecules. The electrons attach to molecules to form negative ions; thus a negative corona gets produced.

It is of interest to derive a simple current–voltage relationship in case of a negative corona discharge between a wire and a coaxial cylinder. The starting point is the Poisson equation viz.

$$\nabla^2 V = -4\pi e\, n_i^-,$$

M. S. Sodha, *Kinetics of Complex Plasmas*, Springer Series on Atomic, Optical, and Plasma Physics 81, DOI: 10.1007/978-81-322-1820-3_16,

which for a cylindrical symmetry reduces to

$$\frac{\mathrm{d}^2 V}{\mathrm{d}r^2} + \frac{1}{r}\frac{\mathrm{d}V}{\mathrm{d}r} + 4\pi e\, n_i^- = 0. \tag{16.1}$$

The current I corresponding to unit length of wire is given by

$$I = 2\pi r n_i^- e \mu_i^- E, \tag{16.2}$$

where

$E = -(\mathrm{d}V/\mathrm{d}r)$	is the electric field
μ_i^-	is the ionic mobility, and
n_i^-	is the negative ion density.

From (16.1)–(16.3) one obtains

$$rE\frac{\mathrm{d}E}{\mathrm{d}r} + E^2 - 2I/\mu_i^- = 0,$$

which on integration yields.

$$E = -(\mathrm{d}V/\mathrm{d}r) = \left[\left(2I/\mu_i^-\right) + C^2/r^2\right]^{1/2}, \tag{16.3}$$

where C is a constant.

At $r = r_0$ (outer radius of visible glow region), $E = E_0$ and hence $C = r_0\left(E_0^2 - 2I/\mu_i^-\right)^{1/2}$.

Using the above value of C and integrating (16.3), one obtains

$$V = r_0 E_0 \Big\{ \ln(a/b) + 1 - \left[1 + \left(2I/\mu_i^-\right)\left(b^2/E_0^2 r^2\right)^{1/2}\right] + \ln\left[(1/2)\left(1 + \sqrt{1 + \left(2I/\mu_i^-\right)\left(b^2/E_0^2 r_0^2\right)}\right)\right]\Big\}. \tag{16.4}$$

16.3 Particle Charging (After White [4])

16.3.1 Two Distinct Processes

There are two distinct mechanisms for charging of dust particles in a corona. The first phenomenon is the ion attachment in the electric field. This process in known as field charging and is the dominant one for particles with radius, larger than 0.5 μ. A second process, known as ion diffusion is the accretion of ions by the charged particles, as discussed in Chap. 3 of this book; this process is dominant for particles with radii less than 0.2 μ. In the range of radius of the particles 0.2 μ to 0.5 μ both the processes should be taken into account.

16.3.2 Field Charging

When a conducting sphere (the particle) of radius a is in an initially uniform electric field E_0, the electric field at any point on the sphere can be shown to be given by

$$E_1 = 3E_0 \cos\ \theta;$$

the coulomb field due to the charge $-Ze$ on the particle is

$$E_2 = -Ze/a^2.$$

Hence, the total electric field E is given by

$$E = E_1 + E_2 = 3E_0 \cos\ \theta - Ze/a^2 \tag{16.5}$$

The total electric flux entering the surface is given by

$$\psi = \int E.2\pi a^2 \sin\ \theta\ \mathrm{d}\theta. \tag{16.6}$$

The above integral is to be evaluated over the regions of θ where E is positive. Thus,

$$\psi = 3\pi a^2 E_0 \left(1 - Ze/3E_0 a^2\right)^2 \pi e Z_s (1 - Z/Z_s). \tag{16.7}$$

The saturation charge on the particle $-Z_s e$ corresponds to $\psi = 0$, where

$$Z_s = 3E_0 a^2/e. \tag{16.8a}$$

For dielectric material dust, (16.8a) should be replaced by

$$Z_s = \left(3E_0 a^2/e\right)[1 + 2(K - 1)/(K + 2)], \tag{16.8b}$$

where K is the dielectric constant of the dust material.

The charging current to the sphere is

$$I = en_i^- \mu_i^- \psi = \frac{\mathrm{d}}{\mathrm{d}t}(Ze).$$

Substituting for ψ from (16.7) in the above equation simplifies to

$$\frac{\mathrm{d}(Z/Z_s)}{(1 - Z/Z_s)^2} = \mathrm{d}(t/t_0),$$

where $t_0 = 1/\pi n_i^- \mu_i^-$.

Integrating the above equation with the initial boundary condition $Z = 0$ at $t = 0$ one obtains

$$Z/Z_s = (t/t_0)/(1 + t/t_0). \tag{16.9}$$

For $(t/t_0) = 1,\ Z/Z_s = 1/2$; thus, t_0 can be considered as characteristic time of charging.

16.3.3 Ion Diffusion (Ion Accretion)

The charge $-Ze$ on a particle due to negative ion accretion (Chap. 3) is given by

$$\frac{\mathrm{d}Z}{\mathrm{d}t} = n_i^- \upsilon_i^- \exp\left(-Ze^2/akT_i\right) = n_{i,c}^-$$

where $\upsilon_i^- = (8k_BT_i/m_i\pi)^{1/2}$ is the mean speed of the negative ions.

Integrating the above equation with $Z = 0$ at $t = 0$ one gets

$$Z = \left(ak_BT_i/e^2\right)\ln(1 + t/T_0), \tag{16.10}$$

where $T_0 = k_BT_i/\pi a\upsilon_i^- n_i^- e^2$.

Equation (16.10) indicates a monotonic increase of Z with increasing t which is not physically tenable; this is caused by considering n_i^- as constant, which is only true when the number of particles is very small.

16.3.4 Magnitude of Charge, Acquired by a Particle Due to Electric Field and Ion Diffusion (Accretion)

Table 16.1 displays the time dependence of the charge Z on the particles acquired by the two processes for the following typical set of parameters, $n_i^- = 5 \times 10^7\ \mathrm{cm}^{-3}, E_0 = 2\,\mathrm{kV/cm}$ and $K = 3$ which are typical of electrostatic precipitation.

Fair agreement between theory and experiment has been found by White [4] and Liu et al. [1]. As stated earlier, it may be noticed from Table 16.1 that the field charging is the dominant process of charging of particles for large radius while for smaller particles ion accretion is the dominant process.

16.4 Particle Collection (After White [4])

16.4.1 Limitation of Theory

Collection of charged dust particles on a suitable surface is the second essential step in electrostatic precipitators. This may be achieved by continuation of the corona electric field configuration or an applied electrostatic potential difference between non-discharging electrodes. In this section, a simple theory of the

Table 16.1 Charging of particle in negative corona: dependence of Z on time

Particles radius μ	Field charging				Accretion			
	Period of exposure (s)				Period of exposure (s)			
	0.01	0.1	1	∞[a]	0.01	0.1	1	10
0.1	0.7	2	2.4	2.5	3	7	11	15
1.0	72	200	240	250	70	110	150	190
10	7,200	20,000	24,400	25,000	1,100	1,500	1,900	2,300

The charge on the particle is $-Ze$ (after Sodha and Guha [3], curtsey authors)
[a] Limiting charge

collection of charged particles on the positive electrode has been given; complications like particle entrainment and disturbances of the corona are neglected.

16.4.2 Drift of Particles

The charged particles will acquire a velocity w along the electric field E_p (perpendicular to the collecting surface) given by

$$\frac{m\mathrm{d}w}{\mathrm{d}t} = -ZeE_p - 6\pi\eta a.w/[1 + A(\lambda/a)],$$

where

m is the mass of the particles
η is the viscosity of the gas
λ is the molecular mean free path and
A is a constant, equal to 0.86 for air at NTP, corresponding to $\lambda = 0.1\mu$ (Cunningham correction).

The solution of this equation with the boundary condition $w = 0$ at $t = 0$ is given by

$$w/w_s = 1 - \exp[-6\pi\eta at/m\ A(1 + \lambda/a)]$$

For durations even much less than the retention time in the collection stage, the exponential term tends to zero and hence for our purpose.

$$w = w_s, \tag{16.11a}$$

where

$$w_s = Z_s eE_p/6\pi\eta a[1 + A\ (\lambda/a)] = E_0 E_p a/2\pi\eta[1 + A_0(\lambda/a)]. \tag{16.11b}$$

When the value of Z_s from (16.8a) is substituted; the expression gets multiplied [(16.8b)] by $\{1 + 2(K - 1)/K + 2\}$ for dielectric particles.

16.4.3 Collection Efficiency

Consider an element of precipitator of thickness dx along the direction of the flow. The loss of dust particles per unit time in the corresponding volume A_c dx is $p\mathrm{d}x\, wn_d$ and hence the loss of particles per unit time per unit volume is $(pw/A_c)n_d$.

The continuity equation for the dust particles may be expressed as:

$$\frac{\mathrm{d}}{\mathrm{d}t}n_d = -(pw/A_c)n_d,$$

where

$$\frac{\mathrm{d}}{\mathrm{d}t} = \frac{\partial}{\partial t} + \upsilon\,\frac{\partial n_d}{\partial x}.$$

In the steady state $(\partial/\partial t) = 0$ and

$$\frac{\upsilon.\,\mathrm{d}n_d}{\mathrm{d}x} = -(pw/A_c)n_{d.}$$

For the initial condition $n_d = n_{d_0}$ at $x = 0$ the solution of the above equation is

$$n_d = n_{d0}\exp\,(-x/x_0) \tag{16.12a}$$

where

$$x_0 = \upsilon A_c/pw.$$

Putting $x = \upsilon t$, the above expression may be put in the form

$$n_d = n_{d0}\exp(-t/t_0), \tag{16.12b}$$

where

$$t_0 = A_c/pw.$$

The collection efficiency η may be expressed as:

$$\eta = \frac{n_{d0} - n_d}{n_{d0}} = 1 - \exp(-x/x_0). \tag{16.13}$$

The mechanical removal of dust from the collector plate is not in the scope of the book.

References

1. B.Y. Liu, K.T. Whitby, H.S. Yu, J. App. Phys. **38**, 1592 (1967)
2. S. Oglesby Jr, G.B. Nichols, *Electrostatic Precipitation* (Marc el Dekker, New York, 1978)
3. M.S. Sodha, S. Guha in *Physics of Colloidal Plasmas,* ed. by A. Simon, W.B. Thompson. Advances in Plasma Physics, Vol. 4, (Inter Science Publishers, New York, 1971), p. 219
4. H.J. White, *Industrial Electrostatic Precipitation* (International Society for Electrostatic Precipitation, Birmingham, 1963)

About the Author

The author is presently Visiting Professor at the Centre of Energy Studies IIT Delhi and has worked in universities and industries in India, USA, and Canada. He has served three universities as Vice-Chancellor and in IIT Delhi as Professor, Head of Department, Dean, Deputy Director, etc. Prof. Sodha has been Ph.D thesis advisor to over 75 candidates who successfully completed the program. He has published over 600 papers in international journals of repute in plasma physics, optics, semiconductors, energy, combustion and ballistics, etc., in addition to authoring/editing 11 books.

The author is a Fellow of Indian National Science Academy and National Academy of Sciences, India and has received the S.S. Bhatnagar prize for physical science in 1974.

The honor of Padmashri was conferred on Prof. Sodha by the President of India in 2003.

M. S. Sodha, *Kinetics of Complex Plasmas*, Springer Series on Atomic, Optical, and Plasma Physics 81, DOI: 10.1007/978-81-322-1820-3,

Index

M. S. Sodha, *Kinetics of Complex Plasmas*, Springer Series on Atomic, Optical, and Plasma Physics 81, DOI: 10.1007/978-81-322-1820-3,

Printed by Printforce, the Netherlands